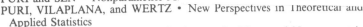

Probability and Mathematical Statistics (C
  PRESS • Bayesian Statistics
  PURI and SEN • Nonparametric M
  PURI and SEN • Nonparametric M
  PURI, VILAPLANA, and WERTZ • New Perspectives in Theoretical and
    Applied Statistics
  RANDLES and WOLFE • Introduction to the Theory of Nonparametric
    Statistics
  RAO • Linear Statistical Inference and Its Applications, *Second Edition*
  RAO • Real and Stochastic Analysis
  RAO and SEDRANSK • W.G. Cochran's Impact on Statistics
  RAO • Asymptotic Theory of Statistical Inference
  ROBERTSON, WRIGHT and DYKSTRA • Order Restricted Statistical
    Inference
  ROGERS and WILLIAMS • Diffusions, Markov Processes, and
    Martingales, Volume II: Îto Calculus
  ROHATGI • An Introduction to Probability Theory and Mathematical
    Statistics
  ROHATGI • Statistical Inference
  ROSS • Stochastic Processes
  RUBINSTEIN • Simulation and The Monte Carlo Method
  RUZSA and SZEKELY • Algebraic Probability Theory
  SCHEFFE • The Analysis of Variance
  SEBER • Linear Regression Analysis
  SEBER • Multivariate Observations
  SEBER and WILD • Nonlinear Regression
  SEN • Sequential Nonparametrics: Invariance Principles and Statistical
    Inference
  SERFLING • Approximation Theorems of Mathematical Statistics
  SHORACK and WELLNER • Empirical Processes with Applications to
    Statistics
  STOYANOV • Counterexamples in Probability

Applied Probability and Statistics
  ABRAHAM and LEDOLTER • Statistical Methods for Forecasting
  AGRESTI • Analysis of Ordinal Categorical Data
  AICKIN • Linear Statistical Analysis of Discrete Data
  ANDERSON and LOYNES • The Teaching of Practical Statistics
  ANDERSON, AUQUIER, HAUCK, OAKES, VANDAELE, and
    WEISBERG • Statistical Methods for Comparative Studies
  ARTHANARI and DODGE • Mathematical Programming in Statistics
  ASMUSSEN • Applied Probability and Queues
  BAILEY • The Elements of Stochastic Processes with Applications to the
    Natural Sciences
  BARNETT • Interpreting Multivariate Data
  BARNETT and LEWIS • Outliers in Statistical Data, *Second Edition*
  BARTHOLOMEW • Stochastic Models for Social Processes, *Third Edition*
  BARTHOLOMEW and FORBES • Statistical Techniques for Manpower
    Planning
  BATES and WATTS • Nonlinear Regression Analysis and Its Applications
  BECK and ARNOLD • Parameter Estimation in Engineering and Science
  BELSLEY, KUH, and WELSCH • Regression Diagnostics: Identifying
    Influential Data and Sources of Collinearity
  BHAT • Elements of Applied Stochastic Processes, *Second Edition*
  BLOOMFIELD • Fourier Analysis of Time Series: An Introduction
  BOX • R. A. Fisher, The Life of a Scientist
  BOX and DRAPER • Empirical Model-Building and Response Surfaces
  BOX and DRAPER • Evolutionary Operation: A Statistical Method for
    Process Improvement
  BOX, HUNTER, and HUNTER • Statistics for Experimenters: An
    Introduction to Design, Data Analysis, and Model Building

(*continued on back*)

Introduction to
the Theory of
Coverage Processes

Introduction to the Theory of Coverage Processes

PETER HALL
The Australian National University
Canberra, Australia

WILEY

JOHN WILEY & SONS

New York   Chichester   Brisbane   Toronto   Singapore

Library of Congress Cataloging in Publication Data:

Hall, Peter (Peter G.)
    Introduction to the theory of coverage processes/Peter Hall.
        p. cm.—(Wiley series in probability and mathematical
statistics. Probability and mathematical statistics)
        ISBN 0-471-85702-5
        1. Point processes. I. Title. II. Title: Coverage processes.
III. Series.
QA274.42.H35 1988
519.2'3—dc19   87-31876
                                                                         CIP

Printed in the United States of America

10 9 8 7 6 5 4 3 2 1

Preface

For our purposes, a coverage process is a countable sequence of sets in Euclidean space, for example, on the real line or in the Cartesian plane. It should have a stochastic aspect, perhaps through the fact that the "centers" of the sets (e.g., their centroids) comprise a stochastic point process, or because the sets themselves, were they to be centered at the origin, would be random sets in the sense of Matheron (1975).

A formal, mathematical definition of a coverage process may be given as follows. Let $\mathcal{P} \equiv \{\xi_1, \xi_2, \ldots\}$ be a countable collection of points in k-dimensional Euclidean space, and $\{S_1, S_2, \ldots\}$ be a countable collection of nonempty sets. Define $\xi_i + S_i$ to be the set $\{\xi_i + \mathbf{x} : \mathbf{x} \in S_i\}$. Then $\mathcal{C} \equiv \{\xi_i + S_i : i = 1, 2, \ldots\}$ is a *coverage process*. The union of all sets in $\mathcal{C}$ is a *germ-grain model*. The sequence $\mathcal{P}$ may be a stochastic point process (Karr 1986), and the S_i's may be random sets. Should $\mathcal{P}$ be a stationary Poisson point process, and the S_i's be independent and identically distributed random sets independent also of $\mathcal{P}$, then $\mathcal{C}$ is a *Boolean model*. There is also a discrete, or lattice-based, version of a Boolean model.

We define ξ_i to be the "center" of the set $\xi_i + S_i$, even though it might not be a real center in any practical sense. If the sets S_i have countable numbers of elements then $\mathcal{C}$ is hardly more than a point process. More interesting examples are those where the sets S_i have positive measure, or at least have an uncountable infinity of elements. The *vacancy* $V = V(\mathcal{R})$ within a subset $\mathcal{R}$ of k-dimensional space, equals the amount of $\mathcal{R}$ that is not covered by any of the sets in $\mathcal{C}$:

$$V = \int_{\mathcal{R}} I\left\{ \mathbf{x} \notin \bigcup_i (\boldsymbol{\xi}_i + S_i) \right\} d\mathbf{x},$$

where I denotes indicator function. Vacancy is an important descriptive characteristic of a coverage process, and has the added advantage of being mathematically tractable.

The means by which coverage processes are generated are as diverse as their potential applications. Examples include queueing theory, where one-dimensional sets S_i denote service times; ballistics, where two-dimensional sets represent overlapping shell craters; and physical chemistry, where three-dimensional sets represent nonoverlapping molecules. The notion of a coverage process is broader even than that of a distribution in classical statistical theory, and a wide-ranging treatment would require a book of encyclopedic proportions.

In this monograph we direct attention at processes in the continuum, and concentrate on the case of two or more dimensions. By working in the continuum we are granted access to powerful tools from stochastic geometry, and avoid the tedium of treating individual lattice types. By giving relatively little attention to one-dimensional processes we get away from those results that are essentially geometric interpretations of classical statistical theory about spacings of order statistics, and into a theory that has a genuine spatial flavor. For example, percolation occurs only in two or more dimensions.

This decision on direction involves sacrifices. It means that we skirt several of the well-known and often very pretty problems about the distribution of segments on the real line or of arcs on the circle. For example, one of the oldest problems in the theory of coverage, that of determining the chance that an interval or circle is completely covered by a given number of randomly placed segments or arcs, is relegated to an exercise at the end of Chapter 1. Our concession to the great amount of literature on the special case of one dimension is Chapter 2 on processes of segments, queues, and counters. We hope that by discussing one-dimensional coverage problems there, in the context of classical stochastic processes, we add a depth and breadth to the subject that is lacking in some earlier accounts.

Our aim is to produce an introduction to coverage processes that emphasizes distributional properties, rather than simply moments. We could have provided a more mathematically elegant but less statistically satisfying account by confining attention to means and variances, discussing, for example, the expected number of sets that intersect a given fixed set. The reader will find some formulas of this type, but they are generally a prelude to distributional results, often limit theorems, which give qualitative descriptions of quantities such as the *distribution* of the number of intersections or of the number of clumps. Our account tries to flesh out the "first moment" skeleton which is provided by, for example, the theory of integral geometry.

An attractive and tantalizing feature of coverage process problems is that they draw on methodology from many, many parts of probability theory and mathematics—examples include queueing theory, renewal processes, point processes, shot noise processes, branching processes, percolation, random set theory, integral geometry, differential geometry, and all types of stochastic limit theory. Actual applications run the gamut from image processing to industrial safety. In our treatment we emphasize this stimulating diversity.

Chapters 1 and 2 serve as introductions to concepts and tools that play major roles in later chapters. Chapter 1 investigates coverage problems that sometimes lie outside the scope of the rest of the monograph, for example, lattice processes (Sections 1.2 and 1.3) and parking and packing problems (Section 1.10). Processes of Poisson-centered line segments are studied in Chapter 2. Chapter 3 employs the concept of vacancy to unravel some of the mysteries of coverage processes in general Euclidean space. The notion of a random set is discussed there, and Boolean models are defined. Chapter 4 examines problems of counting and clumping in Euclidean space, and treats applications of these concepts in the theory of estimation. Elements of inference are discussed in Chapter 5, mainly from a theoretical viewpoint but with several analyses of data. Exercises at the end of each chapter complement and expand on topics treated there. Six Appendixes introduce theory and results needed throughout the monograph.

It was originally intended to present all technical arguments at similar levels of rigor and detail. However, there turned out to be a small number of places where certain results could profitably be included in the text, but had tedious and complex proofs that would hardly benefit the average reader. It was decided to break with a (roughly) uniform level of rigor and omit detailed arguments in these few places, principally in parts of Sections 2.6 and 2.7 (Chapter 2) and Sections 3.5 and 3.6 (Chapter 3).

The general level of technical expertise required during theoretical arguments is that of a graduate course in probability theory. Key results such as the central limit theorem, ergodic theorem, and laws of large numbers are reviewed in the Appendixes. Important theoretical tools, such as the theory of marked point processes, are introduced in the text as the need arises (see particularly Chapter 4, Section 4.1). On occasion a little topology and metric space theory is unavoidable, for example, when introducing the notion of a random set and discussing its properties (Chapter 3, Section 3.1).

References in the text are kept to a minimum, and the reader is referred to notes at the end of each chapter for most bibliographic details. These aim to briefly trace the history of coverage problems, and to give sources for recent work.

Several references stand out as major sources of bibliographic detail on the theory of coverage processes and related topics. They include monographs by Kendall and Moran (1963), Santaló (1976), Solomon (1978), and Stoyan, Kendall, and Mecke (1987) on geometric probability and stochastic geometry, the *Applied Probability* series of reviews of those topics (Moran 1966a, 1969; Little 1974; Baddeley 1977), and Stoyan's (1979b) review of applied stochastic geometry. Naus (1979) provides a bibliography of articles on clusters, clumps, and coincidences, into which categories some of the work on coverage processes falls. Baddeley (1982) gives an engagingly lucid, stimulating introduction to stochastic geometry, together with a selected reading list. Monographs on the statistical analysis of spatial patterns include Bartlett (1975) (see also Bartlett 1974), Ripley (1981), and Diggle (1983). Bartlett (1974, 1975) and Diggle (1983) are concerned primarily with analysis of point patterns. Ripley (1981) provides a very useful account of the general theory and application of spatial statistics, with an extensive bibliography. See also Ripley (1984, 1986). Roach (1968) gives the first monograph-length treatment of stochastic coverage processes, centered around problems of clumping and clustering.

PETER HALL

Canberra, Australia
May, 1988

Acknowledgments

As a Vacation Scholar at the Australian National University in 1973, I had the good fortune to be introduced to geometric probability by two masters, Roger Miles and Pat Moran. With their inspiration and guidance I spent two months learning about random sets, integral geometry, geometric models for plant competition, and other random geometric phenomena of which the average statistics student is sadly unaware. Pat's influence is strong in this monograph, but his contribution goes well beyond mere influence. He critically read drafts of all chapters and the appendixes, pointing out errors ranging from grammatical slips to technical blunders. He clarified many obscurities, and freely gave me the benefit of his insight. I am deeply grateful.

Other colleagues contributed in various ways. They include A. J. Baddeley, T. C. Brown, D. J. Daley, L. Devroye, P. J. Diggle, E. J. Hannan, L. Holst, E. B. Jensen, P. N. Kokic, G. M. Laslett, J. Møller, J. H. T. Morgan, B. D. Ripley, R. L. Smith, H. Solomon, D. Stoyan, and M. Westcott. I am very grateful to them all, and to my wife for her encouragement.

Ann Milligan typed the manuscript immaculately, with unfailing good cheer. She is a godsend to any author, particularly to one whose spelling is as bad as mine. Val Lyon drew the diagrams beautifully from my scruffy sketches.

P.H.

Contents

Notation

1. ALPHABETICAL NOTATION

α: Mean shape content—e.g., mean shape length in one dimension (Chapter 2), mean shape area in two dimensions. When shapes are fixed (e.g., line segments of fixed length in Chapter 2), notation a is often used instead of α.

β: Mean shape perimeter—e.g., $\beta = 2$ for a line segment if coverage process is in one dimension, β denotes mean length of boundary if process is in two dimensions.

$\mathcal{C}$: Coverage process—that is, countable collection of random subsets of $\mathbb{R}^k$.

$I = I(\mathcal{S})$: Indicator function of $\mathcal{S} \subseteq \mathbb{R}^k$, defined by $I(\mathcal{S})(\mathbf{x}) = 1$ if $\mathbf{x} \in \mathcal{S}$, $I(\mathcal{S}) = 0$ otherwise. In a slight abuse of notation we sometimes write $I(\mathcal{E})$ for the indicator of a random event $\mathcal{E}$. For example,

$$I(\text{tossed coin comes down heads}) = \begin{cases} 1 & \text{if coin is heads} \\ 0 & \text{otherwise.} \end{cases}$$

$\mathcal{K}$:   Class of all compact subsets of $\mathbb{R}^k$. $\mathcal{K}'$ denotes class of all nonempty compact subsets of $\mathbb{R}^k$.

λ: Intensity of point process driving a coverage process.

$N(\mu,\sigma^2)$:   Normal distribution with mean μ and variance σ^2.

$\mathcal{P}$:   Point process. Usually the elements of $\mathcal{P}$ are denoted by ξ_i, $i \geq 1$.

$\mathbb{R}, \mathbb{R}^k$:   Real line, k-dimensional Euclidean space, respectively.

$\mathcal{R}$:   fixed subset of $\mathbb{R}^k$, usually a region within which observations are made.

ρ: Hausdorff metric, defined by

$$\rho(\mathcal{S}_1,\mathcal{S}_2) \equiv \inf\left\{t > 0 : \mathcal{S}_1 \subseteq \mathcal{S}_2^t \;\; \text{and} \;\; \mathcal{S}_2 \subseteq \mathcal{S}_1^t\right\}$$

where $\mathcal{S}_1, \mathcal{S}_2 \subseteq \mathbb{R}^k$. See nonalphabetical notation below for definition of $\mathcal{S}^t$.

s_k: Surface content of k-dimensional unit-radius sphere; $s_k \equiv 2\pi^{k/2}/\Gamma(k/2)$.

S: Random subset of $\mathbb{R}^k$, usually generic form of one of the random shapes (grains) generating a coverage process.

$\mathcal{S}$:   Nonrandom subset of $\mathbb{R}^k$.

$\mathcal{T}(r)$:   Closed sphere of radius r centered at the origin.

v_k: Content ("volume") of k-dimensional unit-radius sphere; $v_k \equiv \pi^{k/2}/\Gamma(1 + \frac{1}{2}k)$.

V: Vacancy, or uncovered content, within a given region. Specifically, $V(\mathcal{R})$ denotes vacancy within region $\mathcal{R}$.

$\bar{v}(\mathcal{S}), \underline{v}(\mathcal{S})$:   Contents of smallest sphere containing $\mathcal{S} \subseteq \mathbb{R}^k$, and largest sphere contained in $\mathcal{S}$, respectively.

$$\bar{v}(\mathcal{S}) = \inf\left\{\|\mathcal{T}(r)\| : r \geq 0 \text{ and } \mathcal{S} \subseteq \mathbf{x} + \mathcal{T}(r) \text{ for some } \mathbf{x} \in \mathbb{R}^k\right\},$$

$$\underline{v}(\mathcal{S}) = \sup\left\{\|\mathcal{T}(r)\| : r \geq 0 \text{ and } \mathbf{x} + \mathcal{T}(r) \subseteq \mathcal{S} \text{ for some } \mathbf{x} \in \mathbb{R}^k\right\}.$$

χ: Indicator function for uncovered parts of a coverage process. If $\mathcal{C} = \{\xi_i + S_i, i \geq 1\}$ is a coverage process, then $\chi(\mathbf{x}) = 1$ if for all i, $\mathbf{x} \notin \xi_i + S_i$, and $\chi(\mathbf{x}) = 0$ otherwise.

$\mathbb{Z}, \mathbb{Z}^k$: Set of all integers, set of all k-tuples of integers, respectively.

Ω: Set of all k-dimensional unit vectors. Denoted by Ω_k if dimension k could be ambiguous.

2. NONALPHABETICAL NOTATION

Notation involving sets
Let $S \subseteq \mathbb{R}^k$, $c \in \mathbb{R}$ and $\mathbf{x} \in \mathbb{R}^k$.

Translation: $\mathbf{x} + S \equiv \{\mathbf{x} + \mathbf{y} : \mathbf{y} \in S\}$; $\mathbf{x} + S$ is "centered at" $\mathbf{x}$.

Rescaling: $cS \equiv \{c\mathbf{y} : \mathbf{y} \in S\}$.

Complementation: $\tilde{S} = S^\sim \equiv \{\mathbf{x} \in \mathbb{R}^k : \mathbf{x} \notin S\}$.

Closure: $\bar{S} = S^- \equiv \{\mathbf{x} \in \mathbb{R}^k : \mathbf{x}$ is an accumulation point (limit point) of $S\}$.

Interior: S° = union of all open sets contained in S.

Boundary: $\partial S = \bar{S} \cap (S^\circ)^\sim$.

Surface content: $\|\partial S\|_{k-1} \equiv \lim_{\epsilon \to 0} (2\epsilon)^{-1} \|(\partial S)^\epsilon\|$, provided the limit exists. (See below for definition of S^t, $t > 0$.)

Superscript: For integers l, S^l denotes l-fold product of S, a subset of $\mathbb{R}^{kl}$. For arbitrary positive t,

$$S^t \equiv \left\{\mathbf{x} \in \mathbb{R}^k : |\mathbf{x} - \mathbf{y}| < t \text{ for some } \mathbf{y} \in S\right\}.$$

Radius:

$$\text{rad}(S) \equiv \text{radius of smallest sphere containing } S$$
$$= \{v_k^{-1} \bar{v}(S)\}^{1/k}.$$

See above for definitions of v_k, $\bar{v}(S)$.

Other notation
Euclidean metric: For $\mathbf{x} = (x_1, \ldots, x_k) \in \mathbb{R}^k$, define $|\mathbf{x}| \equiv (x_1^2 + \cdots + x_k^2)^{1/2}$.

Shapes and figures: A "plane" in k-dimensional Euclidean space
                     is a $(k-1)$-dimensional hyperplane, or
                     $(k-1)$-flat, unless the dimension of the
                     hyperplane is specifically qualified. Likewise,
                     "sphere" means "hypersphere" and "cube"
                     means "hypercube." The names "disk,"
                     "sphere," and "cube" refer to the *solid*
                     figures, not the surfaces (or boundaries) of
                     those figures.

Introduction to
the Theory of
Coverage Processes

CHAPTER ONE
Examples, Concepts, and Tools

1.1. INTRODUCTION

In principle, a stochastic coverage process might be thought of as any random mechanism governing the positioning and configuration of random sets on the line, or in the plane, or in space, or indeed in any metric space. The study of coverage processes includes that of queueing theory, where the random sets represent service times of individuals and may overlap if there is more than one server; of sampling techniques for spatial patterns or lattices, where a sampling transect or sampling hoop is cast at random upon the pattern under investigation; and of the molecular structure of liquids, where nonoverlapping particles are arranged randomly in three-dimensional space. Coverage processes have military applications, in which processes of overlapping circles represent craters formed by a salvo of weapons, medical applications, where overlapping circles represent antibodies attached to a virus, and uses in many other areas.

The purpose of modeling such phenomena is to obtain insight into the physical processes which generated them, often with the aim of corroborating a plausible hypothesis or predicting behavior in altered circumstances. For example, we might wish to explain the observed packing density of grains in sand, or compute the probability that a target will be completely destroyed by a salvo of shells, or discover whether antibodies attach randomly and uniformly to

1

the surface of a virus. In addition, mathematical models for spatial phenomena promise a richness and variety rarely equaled in other areas of probability theory and mathematical statistics. They have intrinsic mathematical interest.

The scope of any introduction to the theory is necessarily restricted. To a large extent, the work in this monograph is confined to studies of random overlapping sets uniformly distributed in one or more dimensions. We usually confine attention to the continuum. However, in the present chapter we discuss a wider range of processes, on the lattice and in the continuum, and indicate their applications. These examples serve to motivate concepts and to introduce tools for the more specific tasks that lie ahead. Concepts include vacancy, defined in Section 1.5, and tools include Poisson processes, discussed in Section 1.7. Many additional tools will be introduced in later chapters, as needs arise.

Our plan is to develop first the case of random patterns and coverage processes defined on lattices, usually the square lattice. From there we graduate to coverage processes in the continuum, for a nonrandom number of sets. Finally we examine continuum processes with random numbers of sets.

We begin (Sections 1.2 and 1.3) by studying properties of coverage patterns on lattices. Section 1.2 introduces a problem connected variously to digitizing an image, to sampling of plants grown in an array, and to an unsolved problem in number theory. Section 1.3 discusses simple but sometimes paradoxical properties of clumping. Patterns formed from random functions defined on a lattice are described in Section 1.4. Sections 1.5 and 1.6 examine patterns created by centering given numbers of overlapping random sets in the continuum. The effect of making the number of sets random is also discussed. Emphasis in Section 1.5 is on uniformly distributed centers and in Section 1.6 on nonuniform centers, particularly ballistic problems. We discuss difficulties caused by discontinuities at the edge of the pattern, and introduce the torus convention. This work leads naturally to coverage problems in infinite space, where the spatial pattern extends indefinitely in all directions. To model such patterns we introduce elements of the theory of Poisson processes (Section 1.7) and of Boolean models (Section 1.8). Sometimes Boolean models are sampled by sectioning them. Stereology (Section 1.9) is the science of reconstructing properties of the original pattern from lower dimensional sections. Section 1.10 examines packing and parking problems, concerned with random patterns formed from nonoverlapping sets. Finally, Section 1.11 investigates the probability of covering a sphere by circular caps, introducing techniques that will be employed later in Chapters 2 and 3.

Topics are arranged so that new sections are motivated as much as possible by material in preceding sections. It should be mentioned that fundamental concepts, such as the definition of a Boolean model, will be treated in more depth and more detail in later chapters. To this extent, the present chapter is a sampler of the theory of coverage processes, surveying problems and outlining techniques for solving them.

We close this section with a brief list of commonly used notation and a formal definition of a coverage process. As a rule, the reader will be reminded of the more unusual notation each time it is pressed into use. An index of notation at the end of the monograph summarizes the main notation.

The notations $\mathbb{R}$ and $\mathbb{Z}$ denote the set of all real numbers and the set of all integers, respectively. Thus, $\mathbb{R}^k$ and $\mathbb{Z}^k$ represent Euclidean space of k dimensions and the k-dimensional integer lattice, respectively; $\mathbb{Z}^2$ is the common square lattice with integer vertices. If $S \subseteq \mathbb{R}^k$ then

$$\mathbf{x} + S \equiv \{\mathbf{x} + \mathbf{y} : \mathbf{y} \in S\}$$

denotes the translate of S by $\mathbf{x}$, often called the *Minkowski sum* of $\{\mathbf{x}\}$ and S, sometimes written by other authors as $S + \mathbf{x}$, $\mathbf{x} \oplus S$, or $\{\mathbf{x}\} + S$. The *complement* of S in $\mathbb{R}^k$ is denoted by $\tilde{S}$ or $S^{\sim}$, and the *closure* (or set of limit points) of S by $\overline{S}$ or S^-. The *indicator* or *indicator function* of S is the function $I(S)$ defined by

$$I(S)(\mathbf{x}) \equiv \begin{cases} 1 & \text{if } \mathbf{x} \in S \\ 0 & \text{otherwise.} \end{cases}$$

In a slight abuse of notation we write $I(\mathcal{E})$ for the indicator of a random event $\mathcal{E}$. For example,

$$I\,(\text{tossed coin comes down heads}) = \begin{cases} 1 & \text{if coin is heads} \\ 0 & \text{otherwise.} \end{cases}$$

If $S \subseteq \mathbb{R}^k$ and l is restricted to be a positive integer then S^l denotes the l-fold product of S, a subset of $\mathbb{R}^{kl}$. The content, or Lebesgue measure, of S is written as $\|S\| = \|S\|_k$:

$$\|S\| = \int_S d\mathbf{x},$$

assuming S is a Borel subset of $\mathbb{R}^k$. Throughout, "measurable" means "Borel measurable" unless otherwise qualified. We say that

S is *Riemann measurable* if the integral defining $\|S\|$ exists as an ordinary Riemann integral. This entails boundedness of S. The ordinary Euclidean norm of $\mathbf{x} = (x_1, \ldots, x_k) \in \mathbb{R}^k$ is defined by

$$|\mathbf{x}| = (x_1^2 + \cdots + x_k^2)^{1/2}.$$

A "plane" in k-dimensional Euclidean space is a $(k-1)$-dimensional hyperplane, or $(k-1)$-flat, unless the dimension of the hyperplane is specifically qualified. Likewise, "sphere" means "hypersphere" and "cube" means "hypercube." The terms "disk," "sphere," and "cube" refer to the *solid* figures. For example, the k-dimensional (closed) sphere of radius r centered at the origin is

$$\mathcal{T}(r) \equiv \left\{ \mathbf{x} \in \mathbb{R}^k : |\mathbf{x}| \leq r \right\},$$

not $\{\mathbf{x} \in \mathbb{R}^k : |\mathbf{x}| = r\}$. The latter is referred to as the *surface* or *boundary* of $\mathcal{T}(r)$, and denoted by $\partial \mathcal{T}(r)$. A circle is the boundary of a disk. The k-dimensional content of $\mathcal{T}(1)$ is denoted by $v_k = \pi^{k/2}/\Gamma(1 + \frac{1}{2}k)$, and the $(k-1)$-dimensional content of $\partial \mathcal{T}(1)$ is denoted by $s_k = kv_k$.

The normal distribution with mean μ and variance σ^2 is written as $N(\mu, \sigma^2)$.

We may formally define a coverage process as follows. Let $\mathcal{P} = \{\xi_1, \xi_2, \ldots\}$ be a countable collection of points in k-dimensional Euclidean space, and $\{S_1, S_2, \ldots\}$ be a countable collection of nonempty sets. Then $\mathcal{C} = \{\xi_i + S_i : i = 1, 2, \ldots\}$ is a coverage process. The sequence $\mathcal{P}$ may be a stochastic point process on a lattice or in the continuum, and the S_i's may be random sets. Should $\mathcal{P}$ be a stationary Poisson point process in the continuum, and the S_i's be independent and identically distributed random sets independent also of $\mathcal{P}$, then $\mathcal{C}$ is a *Boolean model* (Section 1.8). Boolean models have lattice-based analogues (Section 1.4), and have generalizations in which "random sets" are replaced by "random functions." The union of all the sets in a coverage process is a *germ-grain model*; the points ξ_i are the "germs" and the sets S_i are the "grains."

1.2. NUMBER OF LATTICE POINTS COVERED BY A RANDOM DISK

Imagine casting a disk of radius r randomly onto the plane. Let $N(r)$ equal the number of points of the form (j, k), with j and k both

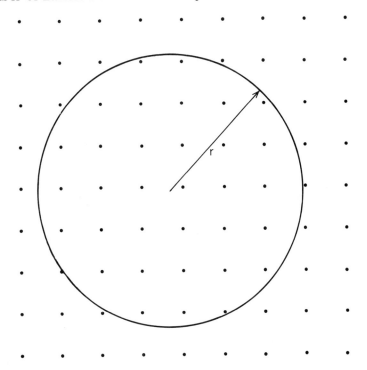

Figure 1.1. Number of lattice points covered by a random disk. In this example, $N(r) = 33$.

integers, which are covered by the disk. What is the distribution of $N(r)$? In particular, what are its mean and variance?

   These questions are related to a classic, unsolved problem in number theory, as we shall shortly relate. For the moment we tackle the problem as posed, following the development in Kendall (1948). The problem is of intrinsic interest in that form, particularly if we consider the perimeter of the disk to be a "sampling hoop" cast at random upon a group of plants arrayed in a square lattice. It also has application in graphical integration, where one may compute the area of a disk by counting the number of squares with covered centers in a fine grid laid over the disk. Analysis of random sets in the continuum often proceeds by digitizing the sets using just such a grid.

   Put

$$I_r(u,v) \equiv \begin{cases} 1 & \text{if } u^2 + v^2 \leq r^2 \\ 0 & \text{otherwise.} \end{cases}$$

Then the number of integer lattice points (j,k) covered by a (closed) disk of radius r centered at (u,v), equals

$$N(r; u,v) \equiv \sum_j \sum_k I_r(j-u, k-v),\qquad (1.1)$$

where the sum is over all pairs of integers (j,k). The random variable $N(r)$ has the distribution of $N(r; U,V)$, where U and V are independently and uniformly distributed on the interval $(0,1)$. Henceforth we take $N(r) \equiv N(r; U,V)$. This variable has mean

$$E\{N(r)\} = \int_0^1 \int_0^1 \left\{ \sum_j \sum_k I_r(j-u, k-v) \right\} du\, dv$$
$$= \sum_j \sum_k \int_{j-1}^{j} \int_{k-1}^{k} I_r(u,v)\, du\, dv$$
$$= \int_{-\infty}^{\infty} \int_{-\infty}^{\infty} I_r(u,v)\, du\, dv$$
$$= \pi r^2.\qquad (1.2)$$

We calculate the mean square of $N(r)$ via the Fourier series of the bounded function $N(r; u,v)$. The (m,n)th Fourier coefficient equals

$$a_{mn} \equiv \int_0^1 \int_0^1 N(r; u,v) \exp\{2\pi i(mu + nv)\}\, du\, dv$$
$$= \sum_j \sum_k \int_{j-1}^{j} \int_{k-1}^{k} I_r(u,v) \cos\{2\pi(mu+nv)\}\, du\, dv$$
$$= \int_{-\infty}^{\infty} \int_{-\infty}^{\infty} I_r(u,v) \cos\{2\pi(mu+nv)\}\, du\, dv$$
$$= \iint_{u^2+v^2 \le r^2} \cos\{2\pi(mu+nv)\}\, du\, dv,$$

using symmetry and periodicity properties of sine and cosine functions. To evaluate the last-written integral, make the change of variable

$$x = (mu+nv)(m^2+n^2)^{-1/2}, \qquad y = (-nu+mv)(m^2+n^2)^{-1/2}.$$

Then $u^2 + v^2 = x^2 + y^2$, and

$$a_{mn} \doteq \int\int_{x^2+y^2 \leq r^2} \cos\left\{2\pi(m^2+n^2)^{1/2}x\right\} dx\, dy$$

$$= 4\int_0^r \cos\left\{2\pi(m^2+n^2)^{1/2}x\right\} dx \int_0^{(r^2-x^2)^{1/2}} dy$$

$$= 4\int_0^r (r^2-x^2)^{1/2} \cos\left\{2\pi(m^2+n^2)^{1/2}x\right\} dx$$

$$= 4r^2\int_0^{\pi/2} \sin^2\theta \cos\left\{2\pi r(m^2+n^2)^{1/2}\cos\theta\right\} d\theta,$$

on changing variable from x to $\theta = \cos^{-1}(x/r)$. The last-written integral is an expression for the Bessel function J_1, and leads to the identity

$$a_{mn} = r(m^2+n^2)^{-1/2} J_1\left\{2\pi r(m^2+n^2)^{1/2}\right\} \qquad (1.3)$$

provided $(m,n) \neq (0,0)$; see Abramowitz and Stegun (1965, p. 360) or Whittaker and Watson (1927, p. 366).

Parseval's theorem (Indritz 1963, pp. 238–239) implies that

$$E\left\{N(r)^2\right\} = \int_0^1\int_0^1 N^2(r; u,v)\, du\, dv$$

$$= \sum_m\sum_n a_{mn}^2,$$

where the sum is over all pairs of integers (m,n). We know from (1.2) that

$$a_{00} = E\left\{N(r)\right\} = \pi r^2,$$

and so

$$\mathrm{var}\left\{N(r)\right\} = \sum_m\sum_n{}' a_{mn}^2, \qquad (1.4)$$

where $\sum_m\sum'_n$ denotes summation over pairs (m,n) with $(m,n) \neq (0,0)$.

A problem of some interest is the asymptotic behavior of $\mathrm{var}\{N(r)\}$ as r increases. The Bessel function J_1 is bounded, and for large positive x satisfies

$$J_1(x) = (2/\pi x)^{1/2} \cos\left\{x - (3\pi/4)\right\} + O(x^{-3/2});$$

see Abramowitz and Stegun (1965, p. 364) or Whittaker and Watson (1927, p. 368). Therefore by (1.3) and (1.4),

$$\text{var}\{N(r)\} = r\pi^{-2} \sum_m \sum_n{}' (m^2 + n^2)^{-3/2}$$

$$\times \cos^2\left\{2\pi r(m^2 + n^2)^{1/2} - (3\pi/4)\right\} + O(1)$$

$$= r(2\pi^2)^{-1} \sum_m \sum_n{}' (m^2 + n^2)^{-3/2}$$

$$\times \left[1 - \sin\left\{4\pi r(m^2 + n^2)^{1/2}\right\}\right] + O(1).$$

It is clear from this formula that $\text{var}\{N(r)\}$ is of order r as r increases, but that $r^{-1}\text{var}\{N(r)\}$ does not have a limit. Indeed, $r^{-1}\text{var}\{N(r)\}$ oscillates between a lower bound of zero and an asymptotic upper bound of

$$\pi^{-2} \sum_m \sum_n{}' (m^2 + n^2)^{-3/2} \simeq 0.9153.$$

In any event,

$$E|N(r) - \pi r^2| = O(r^{1/2}) \qquad (1.5)$$

as r increases.

The quantity $n(r) \equiv N(r; 0, 0)$ equals the number of integer pairs (j, k) satisfying $j^2 + k^2 \le r^2$. A little algebra shows that $n(r)/\pi r^2 \to 1$ as r increases. It is known that

$$n(r) - \pi r^2 = O(r^{(13/20)+\epsilon})$$

as $r \to \infty$, for each $\epsilon > 0$ (Hua 1942). Hardy (1916) conjectured that

$$n(r) - \pi r^2 = O(r^{(1/2)+\epsilon})$$

for each $\epsilon > 0$, although the veracity of this statement remains unconfirmed. We may easily show from (1.5) that if r_m, $m \ge 1$, diverges to infinity sufficiently quickly then for almost all real numbers $(u, v) \in \mathbb{R}^2$,

$$|N(r_m; u, v) - \pi r_m^2| = O\left\{r_m^{1/2}(\log r_m)^\epsilon\right\}$$

for each $\epsilon > 0$, as $m \to \infty$. To prove this result it suffices to show that

$$|N(r_m) - \pi r_m^2| = O\left\{r_m^{1/2}(\log r_m)^\epsilon\right\} \qquad (1.6)$$

almost surely, where $N(r_m) = N(r_m; U, V)$. Taking $r_m \equiv \exp(e^m)$ we find that

$$\sum_{m=1}^{\infty} P\left\{|N(r_m) - \pi r_m^2| > r_m^{1/2}(\log r_m)^{\epsilon}\right\}$$

$$\leq \sum_{m=1}^{\infty} \text{var}\{N(r_m)\} r_m^{-1}(\log r_m)^{-2\epsilon}$$

$$\leq \text{const.} \sum_{m=1}^{\infty} \exp(-2\epsilon m)$$

$$< \infty.$$

Formula (1.6) follows from these inequalities and the Borel-Cantelli lemma (Billingsley 1979, p. 46; Chung 1974, p. 73).

It is of interest to give an intuitive explanation of why $\text{var}\{N(r)\} = O(r)$. Notice from (1.1) that

$$N(r) - E\{N(r)\} = \sum_j \sum_k \{I_r(j - U, k - V) - \mu(j, k)\}, \quad (1.7)$$

where

$$\mu(j, k) \equiv E\{I_r(j - U, k - V)\}.$$

Most of the summands in the double series at (1.7) are identically zero. Indeed, only pairs (j, k) which lie within the annulus

$$\left\{(j, k) : (r - 2^{1/2})^2 \leq j^2 + k^2 \leq (r + 2^{1/2})^2\right\}$$

can make a nonzero contribution to the series. The number of such pairs is of order r as r increases. If the nonzero summands were uncorrelated then since they number only $O(r)$, the right-hand side of (1.7) would have variance $O(r)$. The fact that $\text{var}\{N(r)\}$ is indeed $O(r)$ reflects an asymptotic lack of correlation among the summands.

1.3. CLUMPS OF MARKED VERTICES IN A SQUARE LATTICE

Here we study the formation of clumps of marked vertices in a square lattice, where each vertex is marked independently of all others and with common probability. This process is the simplest discrete analogue of the continuum Boolean model. See Section 1.8

and Chapter 3, Section 3.1, for definitions of the latter, and Section 1.4 for more general discrete Boolean models. The present process has applications in many settings, most recently in image processing and related information sciences. Discrete Boolean models are used extensively to model the formation of aggregates, such as polymers, in random media. They are relatively easy to analyze, in that properties of clumps can be derived by enumeration of possible configurations. Therefore they form a convenient introduction to continuum processes. The basic features of clumping and clustering discussed here will be encountered again in Chapter 4 when we discuss clumping in a continuum Boolean model.

Let $\mathcal{L}$ be the square lattice of pairs of integers (j, k). The elements of $\mathcal{L}$ are called *vertices* or *sites*. Neighboring vertices are joined by *edges* or *bonds*, as in Figure 1.2. Suppose each vertex is marked with probability p, independently of all other vertices. Each edge linking two marked vertices is retained, but all other edges are deleted. The result is a collection of clumps of vertices. Henceforth when we refer to an edge, we mean one of the edges that has not been deleted. If p is sufficiently large (in fact, if $p > p_0 \simeq 0.59$) then the probability that any given vertex is part of an infinite clump of vertices is strictly positive. This is the problem of *site percolation*, and is discussed in Appendix VI. Many other questions connected with clumping in the plane may be formulated, and some are of considerable practical interest. For example, we might consider the problem of determining the expected number of clumps per unit area of the plane.

We define a *clump of order n* to be a set of n vertices connected by edges, with no edges connecting them to any other vertex (see Figure 1.2). Note that clump order refers to the number of vertices in the clump, not the number of edges. An infinite clump is of course an infinite connected set of vertices. So that we might talk about the position of a clump, it is convenient to assign each finite clump a "center." We take the center to be the "lowest right-hand vertex": that vertex furthest to the right (if this is unique) or the lowest of the far right vertices (if there is more than one). There are problems in defining the "center" of an infinite clump, for with probability 1 such a clump will not have a well-defined lowest right-hand vertex. Thus, in working with lowest right-hand vertices we are effectively ignoring infinite clumps. Unless we bear this in mind we shall be afflicted by awkward paradoxes, as we shall show at the end of this section.

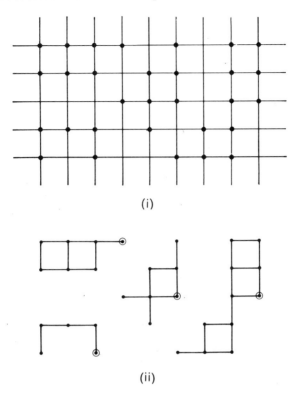

(i)

(ii)

Figure 1.2. Clumps of vertices and edges in a square lattice. Part (i) shows relative positions of marked vertices, indicated by dots. To obtain part (ii), each edge not joining two marked vertices was deleted. Lowest right-hand vertex of each clump is ringed. There are four clumps, of orders 5, 7, 7, and 11. (Clump order refers to number of *vertices*, not number of edges.)

We should point out that some authors use definitions of clumps and clump orders different from our own. For example, our clumps would be called "isolated clumps" by Roach (1968). These differences have no bearing on the problem of infinite clumps.

The various configurations of clumps of a given order may be enumerated, and divided into groups of topologically identical configurations. Two clumps are topologically identical if one can be obtained from the other by a combination of translation, rotation, and reflection. For example, there are 19 different clumps of order four having a given lowest right-hand vertex. These may be divided into five classes, within each of which all clumps are topologically identical. The classes contain 1, 2, 4, 4, and 8 elements, respectively.

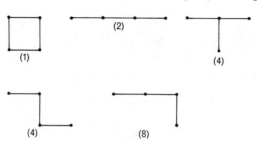

Figure 1.3. Topologically distinct clumps of order four. In parentheses under each clump is written the number of different but topologically identical clumps having the same lowest right-hand center as that clump. All the topologically distinct varieties are illustrated. The total number of different possible clumps of order four having a given lowest right-hand center, is therefore $1 + 2 + 4 + 4 + 8 = 19$.

TABLE 1.1   Numbers of Clumps of Various Orders

Clump Order, n	n_t	m_1	m_2	m_4	m_8	Total Number of Order n Clumps, $\sum_i i m_i$
1	1	1	0	0	0	1
2	1	0	1	0	0	2
3	2	0	1	1	0	6
4	5	1	1	2	1	19
5	12	1	1	5	5	63
6	35	0	2	13	20	216
7	108	0	4	20	84	760

Here n_t equals the number of topologically distinct clumps of order n, and m_i equals the number of topologically distinct clumps for which there are precisely i topologically identical but otherwise different clumps with the same lowest right-hand vertex. Thus, $n_t = \sum_i m_i$, and $\sum_i i m_i$ equals the total number of different clumps with the same lowest right-hand vertex. (Note that the number of topologically identical but otherwise different clumps with the same right-hand vertex must equal 1, 2, 4, or 8.) [After Roach (1968, p. 79).]

Topologically distinct members of each class are represented in Figure 1.3. Note that for any clump order, any class of topologically identical but distinct clumps with the same lowest right-hand center must contain precisely 2^m elements, where $m = 0, 1, 2,$ or 3.

Table 1.1 lists the numbers of topologically distinct clumps of various orders, together with the number of distinct clumps with given lowest right-hand vertex in each class of topologically identical clumps. By studying each possible configuration the probability $\pi(i)$ that a given vertex is the right-hand center of a clump of order i may be calculated. Taking $q \equiv 1 - p$, these probabilities are listed below for $1 \leq i \leq 7$:

$$\pi(1) = pq^4$$
$$\pi(2) = p^2 q^5 2q$$
$$\pi(3) = p^3 q^6 (4q + 2q^2)$$
$$\pi(4) = p^4 q^7 (9q + 8q^2 + 2q^3) \tag{1.8}$$
$$\pi(5) = p^5 q^8 (1 + 20q + 28q^2 + 12q^3 + 2q^4)$$
$$\pi(6) = p^6 q^9 (4 + 54q + 80q^2 + 60q^3 + 16q^4 + 2q^5)$$
$$\pi(7) = p^7 q^{10} (22 + 136q + 252q^2 + 228q^3 + 100q^4 + 20q^5 + 2q^6).$$

For the time being, neglect the possibility that infinite clumps can form. Then $\sum_{i \geq 1} \pi(i)$ equals the probability p^* that a given vertex is the lowest right-hand center of some clump. We shall construct upper and lower bounds to p^*. Notice that

$$\sum_{i=1}^{\infty} i\pi(i) = \text{probability that a given vertex is marked} = p.$$

Therefore

$$p - \sum_{i=1}^{7} i\pi(i) = \sum_{i=8}^{\infty} i\pi(i)$$
$$> 8 \sum_{i=8}^{\infty} \pi(i), \tag{1.9}$$

whence

$$\sum_{i=1}^{7} \pi(i) < p^* < \sum_{i=1}^{7} \pi(i) + \frac{1}{8}\left\{ p - \sum_{i=1}^{7} i\pi(i) \right\}. \tag{1.10}$$

We may establish an alternative lower bound to p^* by arguing as follows. A vertex P will be called a *local* lowest right-hand vertex (local l.r.h.v., for short) if it is marked, if the vertex immediately below P is unmarked, and if the vertex immediately to the right of P is unmarked. A vertex Q will be called a left-hand corner vertex if it, the vertex immediately to its right, and the vertex immediately below it are all marked. A vertex R will be called a square corner vertex if it is in the top left-hand corner of a 1×1 or 2×2 square,

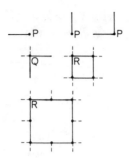

Figure 1.4. Types of vertex. The three vertices P are local lowest right-hand vertices, vertex Q is a left-hand corner vertex, and the two vertices R are square corner vertices. Edges indicated by solid lines are not deleted; edges indicated by dashed lines may be deleted or not.

as depicted in Figure 1.4. Note that a vertex can be both a left-hand corner vertex and a square corner vertex. A little consideration shows that for any clump containing a finite number of vertices,

(number of local l.r.h.v.'s)+(number of square corner vertices)

$\leq$ (number of left-hand corner vertices) + 1   (1.11)

(see Figure 1.5). The probabilities that a given vertex is a local l.r.h.v., or a square corner vertex, or a left-hand corner vertex, or a l.r.h.v. are, respectively, pq^2, $p^4 + p^8 q$, p^3, and p^*, the first three probabilities following by simple enumeration. The expected number of vertices of any one of these types occurring in a set $\mathcal{V}$ of vertices equals the number of elements of $\mathcal{V}$ multiplied by the respective probability. Arguing thus we see that formula (1.11) reduces to

$$pq^2 + (p^4 + p^8 q) < p^3 + p^*,$$

or

$$p(q^2 + p^3 + p^7 q - p^2) < p^*.   (1.12)$$

(Strict inequality results from the fact that strict inequality occurs in (1.11) with positive probability.) Combining (1.10) and (1.12) we obtain the bounds,

$$r_1(p) \equiv \max\left\{ \sum_{i=1}^{7} \pi(i),\ p(q^2 + p^3 + p^7 q - p^2) \right\}$$

$$< p^*$$

$$< \sum_{i=1}^{7} \pi(i) + \frac{1}{8}\left\{ p - \sum_{i=1}^{7} i\pi(i) \right\} \equiv r_2(p),   (1.13)$$

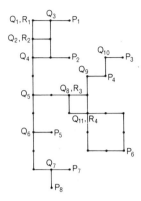

Figure 1.5. A clump with vertex types indicated. There are 8 local lowest right-hand vertices (numbered P_1 to P_8), 11 left-hand corner vertices (Q_1 to Q_{11}) and four square corner vertices (R_1 to R_4). Note particularly that $8 + 4 \leq 11 + 1$.

where the probabilities $\pi(i)$ are defined at (1.8).

Figure 1.6 depicts the functions $r_1(p)$ and $r_2(p)$. The lower bound $r_1(p)$ takes the value

$$p(q^2 + p^3 + p^7 q - p^2)$$

for $p < 0.57$, and the value

$$\sum_{i=1}^{7} \pi(i)$$

for $p > 0.57$. Now, $p = p_c \simeq 0.59$ is the critical probability beyond which the chance that a given vertex is part of an infinite clump is strictly positive; see Appendix VI. If we define p^* to be the chance that a given vertex is the lowest right-hand center of a clump of *finite* order then inequalities (1.13) remain true for $p > p_c$, as we now show. Clearly

$$\sum_{i=1}^{7} \pi(i) < p^*$$

for all p, in particular for $p > p_c$. On the other hand, when $p > p_c$ inequalities (1.9) should be changed to

$$p - \sum_{i=1}^{7} i\pi(i) > \sum_{i=8}^{\infty} i\pi(i) > 8 \sum_{i=8}^{\infty} \pi(i),$$

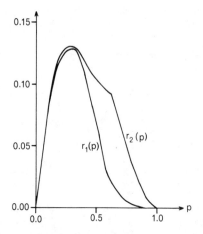

Figure 1.6. Upper and lower bounds to p^*. Curves represent lower bound $r_1(p)$ and upper bound $r_2(p)$ to the expected number $p^*(p)$ of finite-order clumps per unit area of the plane, when the probability that a given vertex is marked equals p. [After Roach (1968).]

but it is still true that

$$p^* = \sum_{i=1}^{\infty} \pi(i) < \sum_{i=1}^{7} \pi(i) + \frac{1}{8}\left\{ p - \sum_{i=1}^{7} i\pi(i) \right\}.$$

We may interpret p^* as the expected number of clumps of *finite* order per unit area in the plane.

Let $\mathcal{R}$ be a large rectangle in the plane. The expected order of an arbitrary clump may be defined by an ergodic argument as the almost sure limit

$$\nu^* \equiv \lim_{\mathcal{R} \uparrow \mathbb{R}^2} \frac{\text{number of vertices within } \mathcal{R}}{\text{number of clumps with l.r.h.v. in } \mathcal{R}},$$

it being assumed that $\mathcal{R}$ grows as a scale multiple of a fixed rectangle. (The ergodic theorem is discussed in Appendix II.) Writing $\|\mathcal{R}\|$ for the area of $\mathcal{R}$, and recalling that vertices are points of the integer lattice, we see that with probability 1,

$$\|\mathcal{R}\|^{-1}(\text{number of marked vertices within } \mathcal{R}) \to p$$

and

$$\|\mathcal{R}\|^{-1}(\text{number of clumps with l.r.h.v. within } \mathcal{R}) \to p^*$$

as $\mathcal{R}$ increases. Therefore

$$\nu^* = p/p^*.$$

Paradoxically, ν^* is finite for all $0 < p < 1$, even though for $p > p_0$ there is strictly positive probability that a clump containing a given vertex is of infinite order. Remember that ν^* equals the expected order of an "arbitrary" clump. There is no completely satisfactory definition of an arbitrary clump when clumps can have infinite order, and our specification of ν^* implicitly ignores *all* infinite clumps. Similar difficulties will arise in Chapter 4 (especially Section 4.6) when we discuss clump order in a continuum Boolean model.

1.4. RANDOM SETS AND RANDOM FUNCTIONS DEFINED ON LATTICES

A Boolean model, on a lattice or in the continuum, is obtained by centering independent and identically distributed random sets at random places in space. See Figure 1.7 for an example of the lattice case, and Section 1.8 for discussion of the continuous case. This process of overlapping sets results in each point of space being classified as either "uncovered" or "covered," giving rise to a simple 0-1 dichotomy. For example, in the special case encountered in the previous section the random sets were all singletons, and a vertex was covered if and only if it was marked. Random function models generalize such processes by assigning to each part of each random "set" a general intensity, and conferring upon each point in space an intensity equal to the sum of the intensities of all the random "sets" that "cover" it. For example, if each set assigns points the value 1 on the set and 0 off it, then the total intensity at any given point in space equals the total number of sets covering that point. These random function models are special cases of shot noise models, to which references are given in the bibliographic notes at the end of this chapter.

Let $\mathcal{S}_0$ be a large set, such as the integer lattice $\mathbb{Z}^2$. A subset $\mathcal{S}$ of $\mathcal{S}_0$ can be thought of as a function $f : \mathcal{S}_0 \to \{0,1\}$ for which

$$f(x) = \begin{cases} 1 & \text{if } x \in \mathcal{S} \\ 0 & \text{otherwise.} \end{cases}$$

A *fuzzy set* results from "blurring" this indicator function. Strictly speaking, the blur should result in a new function f satisfying $0 \le f(x) \le 1$ for all x, and often $f(x)$ is close to 0 well away from $\mathcal{S}$, close to 1 in the middle of $\mathcal{S}$, and midway between 0 and 1 around the edges of $\mathcal{S}$. But there is no theoretical reason for insisting that f be between 0 and 1 everywhere, so we may replace the

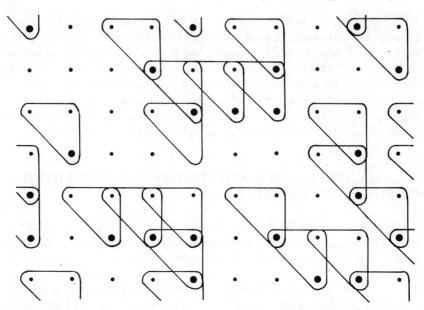

Figure 1.7. Representation of Boolean model on square lattice. The "random sets" are sets of three vertices. The "centers" of these triples (the lowest right-hand vertices, say) are randomly marked points on the lattice, marked independently of one another and with constant probability p, as discussed in Section 1.3. Marked points are indicated by large dots. Vertices contained in one or another of the random sets are said to be covered; all other vertices are uncovered.

set $\mathcal{S}$ by a general function $f : \mathcal{S}_0 \to \mathbb{R}$. Indeed there exist practical circumstances where it is desirable to replace $\mathbb{R}$ by something else, such as $\mathbb{R}^3$ if we are modeling a color photographic image with three primary colors (see the discussion of image models three paragraphs below). However, the simple case where f maps $\mathcal{S}_0$ to $\mathbb{R}$ provides ample illustration of the important features of random function models.

We begin by describing a random function model on the integer lattice $\mathbb{Z}^2$. Let F (a random version of f) be a random, measurable function from $\mathbb{Z}^2$ to $\mathbb{R}$, satisfying

$$E\left\{ \sum_{\xi \in \mathbb{Z}^2} |F(\xi)| \right\} < \infty. \tag{1.14}$$

Let $\{\xi_j, j \geq 1\}$ be the sequence of marked vertices obtained by marking each vertex of $\mathbb{Z}^2$ with probability p independently of all other points, exactly as in Section 1.3. Write $F_1, F_2, \ldots$ for a sequence of independent copies of F, independent also of $\{\xi_j, j \geq 1\}$. We may regard the total intensity function

$$t(\xi) \equiv \sum_j F_j(\xi - \xi_j), \qquad \xi \in \mathbb{Z}^2, \tag{1.15}$$

as the result of superimposing the sequence of intensities $\{F_j(-\xi_j), j \geq 1\}$. Condition (1.14) ensures that the series in (1.15) is absolutely convergent for all $\xi \in \mathbb{Z}^2$, with probability 1.

As an extreme example we might take S to be a random subset of the elements of $\mathbb{Z}^2$, and F to be the "indicator function"

$$F(\xi) \equiv \begin{cases} +\infty & \text{if } \xi \in S \\ 0 & \text{otherwise.} \end{cases}$$

Then $F_1, F_2, \ldots$ may be regarded as "indicators" of independent copies $S_1, S_2, \ldots$ of S, and $t(\xi)$ equals infinity or 0 depending on whether or not ξ is contained in at least one of the random sets

$$\xi_j + S_j \equiv \left\{ \xi_j + \xi : \xi \in S_j \right\}.$$

The sequence of sets $\mathcal{C} \equiv \{\xi_j + S_j, j \geq 1\}$ comprises a Boolean model in $\mathbb{Z}^2$ (see the opening paragraph of this section), and $t(\xi)$ is the "indicator" of the pattern of covered and uncovered regions resulting from $\mathcal{C}$. [Admittedly condition (1.14) fails for this example, but since $F \geq 0$ then the question of convergence of series such as (1.15) does not arise.]

We may regard the points $\xi \in \mathbb{Z}^2$ as the pixels (picture elements) in a digitized image, and $t(\xi)$ [or a function of $t(\xi)$ such as $\exp\{t(\xi)\}$] as the intensity of the image at pixel ξ. Advantages of the random function image model over the simpler model discussed in Section 1.3 are that it admits a general gradation of intensity and it allows for correlation between image intensity at neighboring pixels. This dependence is important. Pixel correlation gives an image its most striking features, such as regularly shaped boundaries, which image processing techniques endeavor to enhance or restore. If the image is in color then the random function F might map $\mathbb{Z}^2$ into $\mathbb{R}^3$, where the components of $\mathbb{R}^3$ correspond to the three primary colors. Images recorded by satellite scanners often contain many more than three bands, or colors.

One valuable feature of random function models is that their covariance functions have particularly simple Fourier transforms. This is important to image processing theory, which is still dominated by techniques based on squared-error loss and Fourier inversion. To make this point more clearly, strengthen condition (1.14) to

$$E\left\{\sum_{\xi\in\mathbb{Z}^2}|F(\xi)|\right\}^2 < \infty. \tag{1.16}$$

Then the stochastic process described by (1.15) is second-order stationary, in that $E\{t^2(\mathbf{O})\} < \infty$ and the functions

$$\mu_1 \equiv E\{t(\eta)\} \quad \text{and} \quad \mu_2(\xi) \equiv E\{t(\eta)t(\xi+\eta)\}$$

do not depend on $\eta \in \mathbb{Z}^2$. Since the definition of $\mu_2(\xi)$ remains unchanged if we replace η by $\eta - \xi$, then $\mu_2(\xi) = \mu_2(-\xi)$ for all $\xi \in \mathbb{Z}^2$. For our particular process t we have

$$\begin{aligned}
\mu_1 &= E\{t(\mathbf{O})\} \\
&= E\left\{\sum_{\xi\in\mathbb{Z}^2} F(\xi)I(\xi \text{ marked})\right\} \\
&= p\sum_{\xi\in\mathbb{Z}^2} E\{F(\xi)\},
\end{aligned} \tag{1.17}$$

and similarly the covariance function

$$a(\xi) \equiv \mu_2(\xi) - \mu_1^2$$

satisfies

$$a(\xi) = p\sum_{\eta\in\mathbb{Z}^2} E\{F(\eta)F(\xi+\eta)\}. \tag{1.18}$$

Since

$$\begin{aligned}
\sum_{\xi\in\mathbb{Z}^2}|a(\xi)| &\le p\sum_{\xi}\sum_{\eta} E\left\{|F(\eta)F(\xi+\eta)|\right\} \\
&= pE\left[\left\{\sum_{\xi}|F(\xi)|\right\}^2\right] < \infty,
\end{aligned}$$

then the covariance function has a well-defined Fourier transform:

$$\begin{aligned}
\alpha(\boldsymbol{\theta}) = \alpha(\theta_1,\theta_2) &\equiv \sum_{(\xi_1,\xi_2)\in\mathbb{Z}^2} \exp\{i(\theta_1\xi_1+\theta_2\xi_2)\}a(\xi_1,\xi_2) \\
&= pE\left\{|\phi(\boldsymbol{\theta})|^2\right\},
\end{aligned} \tag{1.19}$$

where

$$\phi(\boldsymbol{\theta}) \equiv \sum_{(\xi_1, \xi_2) \in \mathbb{Z}^2} \exp\{i(\theta_1 \xi_1 + \theta_2 \xi_2)\} F(\xi_1, \xi_2)$$

denotes the Fourier transform of F. Exercise 1.5 outlines some of the implications of these formulas in the theory of image processing.

1.5. INTERPENETRATING SPHERES DISTRIBUTED RANDOMLY IN SPACE

Let n points $\xi_1, \ldots, \xi_n$ be distributed independently and uniformly within a large region $\mathcal{R}$, a subset of $\mathbb{R}^k$. For convenience we take $\mathcal{R}$ to be a cube. At each point ξ_i center a k-dimensional sphere of unit radius. Allow the n spheres to overlap (interpenetrate) at will. We write

$$\mathcal{T} \equiv \left\{ \mathbf{x} \in \mathbb{R}^k : |\mathbf{x}| \leq 1 \right\}$$

for the (closed) unit sphere centered at the origin, and

$$\mathbf{x} + \mathcal{T} \equiv \{ \mathbf{x} + \mathbf{y} : \mathbf{y} \in \mathcal{T} \}$$

for the (closed) unit sphere centered at $\mathbf{x}$. A point $\mathbf{x} \in \mathcal{R}$ is said to be *covered* by the sequence of random sets $\{\xi_i + \mathcal{T}, 1 \leq i \leq n\}$ if it is contained in at least one of the spheres $\xi_i + \mathcal{T}$. The *vacancy* within $\mathcal{R}$, $V = V(\mathcal{R})$, equals the amount of uncovered content of $\mathcal{R}$.

In this section we explore properties of V. One of our aims is to elucidate the differences between the case where n is fixed and that where $n = N$ is a random variable. The latter case is important if the region $\mathcal{R}$ represents a small portion of a much larger set, within which a given number of points is distributed independently and uniformly. We work with general k, and so our results apply equally to problems of line segments distributed on the line, circles in the plane, and three-dimensional spheres in $\mathbb{R}^3$.

Let χ be the indicator function of uncovered points:

$$\chi(\mathbf{x}) \equiv \begin{cases} 1 & \text{if } \mathbf{x} \text{ is not covered} \\ 0 & \text{if } \mathbf{x} \text{ is covered.} \end{cases}$$

Then vacancy V is given by

$$V \equiv \int_{\mathcal{R}} \chi(\mathbf{x}) \, d\mathbf{x},$$

and has mean

$$E(V) = \int_{\mathcal{R}} E\{\chi(\mathbf{x})\}\, d\mathbf{x} = \int_{\mathcal{R}} P(\mathbf{x} \text{ not covered})\, d\mathbf{x} \qquad (1.20)$$

and mean square

$$E(V^2) = \int\int_{\mathcal{R}^2} P(\mathbf{x}, \mathbf{y} \text{ both not covered})\, d\mathbf{x}\, d\mathbf{y}. \qquad (1.21)$$

Write

$$\|\mathcal{S}\| = \int_{\mathcal{S}} d\mathbf{x}$$

for the k-dimensional content of a set $\mathcal{S} \subseteq \mathbb{R}^k$. The chance that $\mathbf{x} \in \mathcal{R}$ is uncovered equals the probability that none of the n points $\xi_1, \ldots, \xi_n$ lies within the set $(\mathbf{x} + \mathcal{T}) \cap \mathcal{R}$, and so equals

$$\{1 - \|(\mathbf{x} + \mathcal{T}) \cap \mathcal{R}\| / \|\mathcal{R}\|\}^n.$$

This probability depends to at least some extent on $\mathbf{x}$. It is larger if $\mathbf{x}$ is toward the edge of $\mathcal{R}$ than if $\mathbf{x}$ is toward the center. This asymmetry is a nuisance, technically speaking, since it introduces cumbersome edge-effect terms into our formulas. We may avoid it by using the so-called *torus convention*. This means that we imagine that each sphere that protrudes out one side of the cube $\mathcal{R}$ enters the cube again from the opposite side. Figure 1.8 illustrates the convention in the case of $k = 2$ dimensions. When $k = 1$ the convention is tantamount to considering arcs distributed on the circumference of a circle, rather than line segments distributed on an interval. Provided we assume that the sides of $\mathcal{R}$ are at least two units long, it is impossible for a sphere to intersect itself under the torus convention. In that case the region of $\mathcal{R}$ within which the point ξ_i must lie in order for the sphere $\xi_i + \mathcal{T}$ to cover $\mathbf{x}$ has content precisely $\|\mathcal{T}\|$, from which it follows that

$$P(\mathbf{x} \text{ not covered}) = (1 - \|\mathcal{T}\| / \|\mathcal{R}\|)^n,$$

not depending on $\mathbf{x}$. Therefore by (1.20),

$$E(V) = (1 - \|\mathcal{T}\| / \|\mathcal{R}\|)^n \|\mathcal{R}\|. \qquad (1.22)$$

In calculating the mean square and variance of V we assume that the sides of $\mathcal{R}$ are at least four units long. The probability that neither $\mathbf{x}$ nor $\mathbf{y}$ is covered equals the chance that none of the points $\xi_1, \ldots, \xi_n$ lies within a region that has the content $e(|\mathbf{x} - \mathbf{y}|)$ of the union of two unit spheres with centers distant $|\mathbf{x} - \mathbf{y}|$ apart.

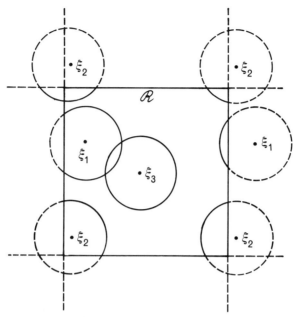

Figure 1.8. The torus convention in the case of $k = 2$ dimensions. The torus convention may be interpreted by thinking of $\mathcal{R}$ as just one member of a lattice of cubes (squares in the present case), and supposing that all points ξ_i are repeated in precisely the same relative positions in all cubes. If spheres (disks in our case) are centered at all points ξ_i in all cubes, then the pattern of spheres protruding into $\mathcal{R}$ is that decreed by the torus convention.

Now, the content of the lens of intersection of two k-dimensional unit spheres centered $2x$ apart equals

$$B(x) \equiv \begin{cases} 2\pi^{(k-1)/2}\{\Gamma(\tfrac{1}{2} + \tfrac{1}{2}k)\}^{-1} \int_x^1 (1 - y^2)^{(k-1)/2}\, dy & \text{if } 0 \le x \le 1 \\ 0 & \text{if } x > 1, \end{cases}$$

$$(1.23)$$

and

$$v_k \equiv B(0) = \|T\| = \pi^{k/2}/\Gamma(1 + \tfrac{1}{2}k)$$

is the content of a unit k-dimensional sphere. In this notation,

$$e(|\mathbf{x} - \mathbf{y}|) = 2v_k - B(|\mathbf{x} - \mathbf{y}|/2),$$

and so

$$P(\mathbf{x}, \mathbf{y} \text{ both not covered}) = [1 - \{2v_k - B(|\mathbf{x} - \mathbf{y}|/2)\}/\|\mathcal{R}\|]^n.$$

Therefore by (1.21),

$$E(V^2) = \int \int_{\mathcal{R}^2} [1 - \{2v_k - B(|\mathbf{x} - \mathbf{y}|/2)\}/\|\mathcal{R}\|]^n \, d\mathbf{x}\, d\mathbf{y}. \qquad (1.24)$$

Variance of vacancy may be computed from (1.22) and (1.24).

Some simplification of formulas for $E(V)$ and var(V) is possible if we take $\mathcal{R}$ to be a large cube. Assume that the density

$$\rho \equiv n/\|\mathcal{R}\|$$

of points per unit content of $\mathcal{R}$ converges to λ as $\mathcal{R}$ increases, and that $0 < \lambda < \infty$. Then by (1.22),

$$E(V) = (1 - n^{-1}\rho v_k)^n \|\mathcal{R}\| \sim \|\mathcal{R}\| \exp(-\lambda v_k),$$

and by (1.22) and (1.24),

$$\text{var}(V) = \int \int_{\mathcal{R}^2} J(\mathbf{x}, \mathbf{y}) \, d\mathbf{x}\, d\mathbf{y} \qquad (1.25)$$

where

$$J(\mathbf{x}, \mathbf{y}) \equiv [1 - \{2v_k - B(|\mathbf{x} - \mathbf{y}|/2)\}/\|\mathcal{R}\|]^n - (1 - v_k/\|\mathcal{R}\|)^{2n}.$$

Write the double integral in the formula for var(V) as

$$\int_{\mathcal{R}} d\mathbf{x} \left(\int_{\{\mathbf{y} \in \mathcal{R}: |\mathbf{x} - \mathbf{y}| \le 2\}} + \int_{\{\mathbf{y} \in \mathcal{R}: |\mathbf{x} - \mathbf{y}| > 2\}} \right) J(\mathbf{x}, \mathbf{y}) \, d\mathbf{y}.$$

This decomposition gives two terms, of which the first is

$$\int_{\mathcal{R}} d\mathbf{x} \int_{\{\mathbf{y} \in \mathcal{R}: |\mathbf{x} - \mathbf{y}| \le 2\}} (\exp[-\rho\{2v_k - B(|\mathbf{x} - \mathbf{y}|/2)\}]$$
$$- \exp(-2\rho v_k) + O(n^{-1})) \, d\mathbf{y}$$

$$= \exp(-2\rho v_k) \int_{\mathcal{R}} d\mathbf{x}$$

$$\times \int_{\{\mathbf{y} \in \mathcal{R}: |\mathbf{x} - \mathbf{y}| \le 2\}} [\exp\{\rho B(|\mathbf{x} - \mathbf{y}|/2)\} - 1] \, d\mathbf{y} + O(1)$$

$$\sim \|\mathcal{R}\| \exp(-2\lambda v_k) 2^k s_k \int_0^1 [\exp\{\lambda B(x)\} - 1] x^{k-1} \, dx,$$

on changing to polar coordinates in the inner integral and writing $s_k \equiv 2\pi^{k/2}/\Gamma(k/2)$ for the $(k-1)$-dimensional surface content of a

k-dimensional unit sphere. The second of the terms is

$$\int_{\mathcal{R}} d\mathbf{x} \int_{\{\mathbf{y}\in\mathcal{R}:|\mathbf{x}-\mathbf{y}|>2\}} \left\{ (1 - 2v_k/\|\mathcal{R}\|)^n - (1 - v_k/\|\mathcal{R}\|)^{2n} \right\} d\mathbf{y}$$

$$= \left\{ \exp(-2\rho v_k - 2\rho^2 v_k^2 n^{-1}) - \exp(-2\rho v_k - \rho^2 v_k^2 n^{-1}) + O(n^{-2}) \right\}$$

$$\times \int_{\mathcal{R}} d\mathbf{x} \int_{\{\mathbf{y}\in\mathcal{R}:|\mathbf{x}-\mathbf{y}|>2\}} d\mathbf{y}$$

$$\sim -\exp(-2\rho v_k - 2\rho^2 v_k^2 n^{-1}) \left\{ \exp(\rho^2 v_k^2 n^{-1}) - 1 \right\} \|\mathcal{R}\|^2$$

$$\sim -\exp(-2\rho v_k)\rho^2 v_k^2 n^{-1} \|\mathcal{R}\|^2$$

$$\sim -\exp(-2\lambda v_k)\lambda v_k^2 \|\mathcal{R}\|.$$

Combining the results from (1.25) down, and noting that $s_k = kv_k$, we conclude that as $\mathcal{R}$ increases,

$$\mathrm{var}(V) \sim \|\mathcal{R}\| v_k \exp(-2\lambda v_k)$$

$$\times \left(2^k k \int_0^1 [\exp\{\lambda B(x)\} - 1] x^{k-1}\, dx - \lambda v_k \right).$$

We pause to combine these large-region formulas for $E(V)$ and $\mathrm{var}(V)$ into a theorem.

Theorem 1.1. *Assume that as the cube $\mathcal{R}$ increases, $n/\|\mathcal{R}\| \to \lambda$, where n equals the number of spheres centered within $\mathcal{R}$ and $0 < \lambda < \infty$. If vacancy is interpreted using the torus convention then*

$$E(V)/\|\mathcal{R}\| \to \exp(-\lambda v_k)$$

and

$$\mathrm{var}(V)/\|\mathcal{R}\| \to v_k \exp(-2\lambda v_k)$$

$$\times \left(2^k k \int_0^1 [\exp\{\lambda B(x)\} - 1] x^{k-1}\, dx - \lambda v_k \right)$$

as $\mathcal{R}$ increases.

A little additional algebra shows that the theorem remains valid without interpreting vacancy according to the torus convention; see Exercise 1.7.

Suppose now that n' points $\xi_1, \ldots, \xi_{n'}$ are distributed independently and uniformly throughout a region $\mathcal{R}'$ containing $\mathcal{R}$, and that the density $n'/\|\mathcal{R}'\|$ of points per unit content of $\mathcal{R}'$ equals λ. At each point center a unit-radius sphere. If $\mathcal{R}'$ is much larger than

$\mathcal{R}$ then for all intents and purposes the spheres that intersect $\mathcal{R}$ have their centers distributed "uniformly in space," with density λ; such a collection of points is called a Poisson process of intensity λ in $\mathbb{R}^k$. (See Section 1.7 for a discussion of Poisson processes.) Let N denote the number of points ξ_i that lie within $\mathcal{R}$, and let V be the vacancy within $\mathcal{R}$. Theorem 1.1 (which holds with or without the torus convention) supplies a large-region formula for the mean and variance of V *conditional on* $N = n$, as n increases with $\mathcal{R}$ in such a manner that $n/\|\mathcal{R}\| \to \lambda$. Of course, the ratio $N/\|\mathcal{R}\|$ converges to λ as $\mathcal{R}$ increases. Now, the unconditional mean and variance of vacancy are given by

$$E(V) = E\{E(V \mid N)\}$$

and

$$\mathrm{var}(V) = E\{\mathrm{var}(V \mid N)\} + \mathrm{var}\{E(V \mid N)\}, \qquad (1.26)$$

and so it is to be expected that $E(V)$ will be asymptotic to $\|\mathcal{R}\| \exp(-\lambda v_k)$ (its value in Theorem 1.1) but $\mathrm{var}(V)$ will be somewhat larger than the value stated in Theorem 1.1, due to the extra term $\mathrm{var}\{E(V \mid N)\}$ in formula (1.26). Indeed, we shall show in Chapter 3 that in the case of Poisson-distributed unit-radius spheres,

$$E(V) = \|\mathcal{R}\| \exp(-\lambda v_k),$$

and for large regions $\mathcal{R}$,

$$\mathrm{var}(V) \sim \|\mathcal{R}\| v_k \exp(-2\lambda v_k) 2^k k \int_0^1 [\exp\{\lambda B(x)\} - 1] x^{k-1} \, dx;$$

see the discussion succeeding Theorem 3.5 in Section 3.4.

In the case of Poisson-distributed spheres, asymptotic normality of vacancy is relatively easy to prove (see Section 3.4 of Chapter 3). The reason for this simplicity is that in the Poisson case the numbers of spheres centered within any sequence of disjoint subsets of $\mathbb{R}^k$ are independent random variables. From that it follows that vacancy within a region $\mathcal{R}$ may be readily approximated by a sum of independent random variables. However, if a fixed number of sphere centers are distributed throughout $\mathcal{R}$ then the numbers of centers within disjoint regions have a joint multinomial distribution, and then it is not a trivial matter to approximate vacancy by a sum of independent variables. Nevertheless, a suitable approximation may be obtained via a conditioning argument and an estimate of the rate of convergence in the central limit theorem, as we now show.

Let V denote vacancy within the cube $\mathcal{R}$ after the independent, uniform distribution of n unit-radius spheres. Write $N(\mu,\sigma^2)$ for a random variable having the normal distribution with mean μ and variance σ^2.

Theorem 1.2.   *Under the conditions of Theorem 1.1,*

$$\{V - E(V)\} / \{\mathrm{var}(V)\}^{1/2} \to N(0,1)$$

in distribution as $\mathcal{R}$ increases.

Once again the theorem remains true without the torus convention. Indeed, our proof is valid with or without that convention.

Proof.   We take $\mathcal{R}$ to be the cube $[0,l]^k$. Divide $\mathcal{R}$ into a lattice of subcubes $\mathcal{R}_i$, each of side length $d+2$. We keep d fixed as l and n increase. Of course, there is likely to be a little of $\mathcal{R}$ left uncovered by the cubes $\mathcal{R}_i$, due to the fact that $d+2$ need not divide l. Inside $\mathcal{R}_i$ construct a concentric subcube $\mathcal{S}_i$ of side length d. The vacancy within $\mathcal{S}_i$ equals

$$V_i \equiv \int_{\mathcal{S}_i} \chi(\mathbf{x})\, d\mathbf{x}.$$

We prove a central limit theorem for

$$V^* \equiv \sum_i V_i,$$

and use that result to deduce a central limit theorem for V. The number $m(n)$ of terms in this series is asymptotic to $\{l/(d+2)\}^k$ as l increases.

Let N_i denote the number of points within $\mathcal{R}_i$. The arguments leading to (1.22) and (1.24) give us

$$E(V_i \mid N_i) = (1 - v_k/\|\mathcal{R}_i\|)^{N_i}\|\mathcal{S}_i\| \tag{1.27}$$

and

$$E(V_i^2 \mid N_i) = \int\int_{\mathcal{S}_i{}^2}[1 - \{2v_k - B(|\mathbf{x}-\mathbf{y}|/2)\}/\|\mathcal{R}_i\|]^{N_i}\, d\mathbf{x}\, d\mathbf{y}. \tag{1.28}$$

[We defined the function B at (1.23).] The variables N_i have a multinomial distribution, with

$$E(\alpha_i^{N_i}) = \{1 + (\alpha_i - 1)p\}^n \tag{1.29}$$

for arbitrary $\alpha_i > 0$, where $p \equiv \|\mathcal{R}_i\|/\|\mathcal{R}\|$. Also,

$$E(\alpha_i^{N_i}\alpha_j^{N_j}) = \{1+(\alpha_i-1)p+(\alpha_j-1)p\}^n = \{1+(\alpha_i+\alpha_j-2)p\}^n \tag{1.30}$$

if $i \neq j$. We may now deduce from (1.27) and (1.28) that

$$E\left\{\sum_i \mathrm{var}(V_i \mid N_i)\right\} = m(n)E[E(V_1^2 \mid N_1)-\{E(V_1 \mid N_1)\}^2]$$

$$= m(n)\int\int_{S_1^2}\Big([1-\{2v_k-B(|\mathbf{x}-\mathbf{y}|/2)\}/\|\mathcal{R}\|]^n$$

$$-\big\{1-(2v_k/\|\mathcal{R}\|)+v_k^2(\|\mathcal{R}\|\|\mathcal{R}_1\|)^{-1}\big\}^n\Big)\,d\mathbf{x}\,d\mathbf{y}. \tag{1.31}$$

A similar argument gives a formula for $E[\{\sum_i \mathrm{var}(V_i \mid N_i)\}^2]$. Subtracting from that the square of (1.31) we deduce that

$$\sigma_n^2 \equiv \mathrm{var}\left\{\sum_i \mathrm{var}(V_i \mid N_i)\right\}$$

$$= m\int\int\int\int_{S_1^4} J_1(\mathbf{x}_1,\mathbf{y}_1,\mathbf{x}_2,\mathbf{y}_2)\,d\mathbf{x}_1\,d\mathbf{y}_1\,d\mathbf{x}_2\,d\mathbf{y}_2$$

$$+ m(m-1)\int\int\int\int_{S_1^2 S_2^2} J_2(\mathbf{x}_1,\mathbf{y}_1,\mathbf{x}_2,\mathbf{y}_2)\,d\mathbf{x}_1\,d\mathbf{y}_1\,d\mathbf{x}_2\,d\mathbf{y}_2,$$

where $|J_1|$ is uniformly bounded and $\sup|J_2| = O(\|\mathcal{R}\|^{-1})$. Since $m = O(\|\mathcal{R}\|)$ then $\sigma_n^2 = O(\|\mathcal{R}\|)$ as $\mathcal{R}$ increases. We shall prove shortly that $E\{\sum_i \mathrm{var}(V_i \mid N_i)\}$ is asymptotic to a constant multiple of $\|\mathcal{R}\|$, and then it follows that

$$\mathrm{var}\left\{\sum_i \mathrm{var}(V_i \mid N_i)\right\}\Big/\left[E\left\{\sum_i \mathrm{var}(V_i \mid N_i)\right\}\right]^2 \to 0$$

as $\mathcal{R}$ increases. Therefore

$$\sum_i \mathrm{var}(V_i \mid N_i)\Big/E\left\{\sum_i \mathrm{var}(V_i \mid N_i)\right\} \to 1 \tag{1.32}$$

in probability.

Remembering that $m(n) \sim \|\mathcal{R}\|\|\mathcal{R}_1\|^{-1}$ as $\mathcal{R}$ increases, we see from (1.31) that

$$E\Big\{\sum_i \mathrm{var}(V_i \mid N_i)\Big\}$$

$$\sim \|\mathcal{R}\|\|\mathcal{R}_1\|^{-1} \exp(-2\lambda v_k) \int\int_{\mathcal{S}_1{}^2} [\exp\{\lambda B(|\mathbf{x}-\mathbf{y}|/2)\}$$

$$- \exp(\lambda v_k^2/\|\mathcal{R}_1\|)]\,d\mathbf{x}\,d\mathbf{y} \tag{1.33}$$

as $\mathcal{R}$ increases. The right-hand side equals a constant multiple of $\|\mathcal{R}\|$. Since $O \le V_i \le \|\mathcal{S}_i\| = \|\mathcal{S}_1\|$ then

$$\sum_i E\Big\{|V_i - E(V_i \mid N_i)|^3 \,\Big|\, N_i\Big\} \le 2\|\mathcal{S}_1\| \sum_i \mathrm{var}(V_i \mid N_i) = O_p(\|\mathcal{R}\|),$$

using (1.32) and (1.33). From this it follows that

$$R_n \equiv \Big[\sum_i E\Big\{|V_i - E(V_i \mid N_i)|^3 \,\Big|\, N_i\Big\}\Big]\Big/\Big\{\sum_i \mathrm{var}(V_i \mid N_i)\Big\}^{3/2}$$

$$= O_p(\|\mathcal{R}\|.\|\mathcal{R}\|^{-3/2}) \to 0 \tag{1.34}$$

in probability.

The Berry-Esseen theorem (Appendix II) asserts that for any sequence of independent random variables $X_1, X_2, \ldots$ with finite third moments, and any $m \ge 1$,

$$\sup_{-\infty < x < \infty} \Big| P\Big\{\sum_{i=1}^m (X_i - EX_i) \le x\Big(\sum_{i=1}^m \mathrm{var}\,X_i\Big)^{1/2}\Big\} - \Phi(x)\Big|$$

$$\le C_0 \Big\{\sum_{i=1}^m E|X_i - E(X_i)|^3\Big\}\Big/\Big\{\sum_{i=1}^m \mathrm{var}(X_i)\Big\}^{3/2}, \tag{1.35}$$

where C_0 is an absolute constant and Φ denotes the standard normal distribution function. In fact, $C_0 = 1$ will do. Let $\mathcal{F}_n$ denote the σ-field generated by $N_1, \ldots, N_m$. Conditional on $\mathcal{F}_n$, the variables $V_1, \ldots, V_m$ are independent, with conditional mean and variance

$$E(V_i \mid \mathcal{F}_n) = E(V_i \mid N_i) \quad \text{and} \quad \mathrm{var}(V_i \mid \mathcal{F}_n) = \mathrm{var}(V_i \mid N_i).$$

With R_n defined as in (1.34), we see from (1.35) that

$$
\sup_{-\infty < x < \infty} \left| P\left[\sum_i V_i \le \sum_i E(V_i \mid N_i) \right.\right.
$$
$$
\left.\left. + x \left\{ \sum_i \text{var}(V_i \mid N_i) \right\}^{1/2} \;\middle|\; \mathcal{F}_n \right] - \Phi(x) \right| \le R_n .
$$

The left-hand side obviously does not exceed unity, and so is actually dominated by $\min(R_n, 1)$. In view of (1.34) and the dominated convergence theorem,

$$
E\{\min(R_n, 1)\} \to 0
$$

as $\mathcal{R}$ increases. It follows that for any constant $c > 0$ and sequence of constants $\{b_n\}$,

$$
\sup_{-\infty < x < \infty} \left| P\left(\sum_i V_i \le cn^{1/2} x + b_n \right) - \Phi(x) \right|
$$
$$
\le \sup_{-\infty < x < \infty} \left| E\left(\Phi\left[\left\{ cn^{1/2} x + b_n - \sum_i E(V_i \mid N_i) \right\} \right.\right.\right.
$$
$$
\left.\left.\left. \times \left\{ \sum_i \text{var}(V_i \mid N_i) \right\}^{-1/2} \right] \right) - \Phi(x) \right| + o(1). \quad (1.36)
$$

Suppose we can prove that for a constant $\alpha > 0$,

$$
n^{-1/2} \left\{ \sum_i E(V_i \mid N_i) - b_n \right\} \to N(0, \alpha^2) \quad \text{in distribution} \quad (1.37)
$$

as $\mathcal{R}$ increases. Remembering from (1.32) and (1.33) that for a constant $\beta > 0$,

$$
n^{-1} \sum_i \text{var}(V_i \mid N_i) \to \beta^2 \quad \text{in probability}
$$

as $\mathcal{R}$ increases, we see that if $c^2 = \alpha^2 + \beta^2$,

$$
E\left(\Phi\left[\left\{ cn^{1/2} x + b_n - \sum_i E(V_i \mid N_i) \right\} \left\{ \sum_i \text{var}(V_i \mid N_i) \right\}^{-1/2} \right] \right) \to \Phi(x)
$$

for each x as $n \to \infty$. Then it follows from (1.36) that

$$
n^{-1/2} \left(\sum_i V_i - b_n \right) \to N(0, c^2) \quad \text{in distribution.} \quad (1.38)
$$

The next step in the proof is to verify (1.37).

In view of (1.27) we may write

$$\sum_i E(V_i \mid N_i) = \|\mathcal{S}_1\| \sum_i r^{N_i},$$

where $r \equiv 1 - v_k/\|\mathcal{R}_1\|$ is fixed. Central limit theorems for elementary functions of multinomials such as this are well-known; see, for example, Holst (1972, Theorem 1). Thus, $\sum_i E(V_i \mid N_i)$ is asymptotically normally distributed with mean

$$b_n \equiv E\left\{\sum_i E(V_i \mid N_i)\right\} = E\left(\sum_i V_i\right)$$

and variance

$$a_n^2 \equiv \mathrm{var}\left\{\sum_i E(V_i \mid N_i)\right\}$$

$$= \|\mathcal{S}_1\|^2 \left[m\left\{1 + (r^2 - 1)\|\mathcal{R}_1\|/\|\mathcal{R}\|\right\}^n \right.$$
$$+ m(m-1)\left\{1 + 2(r-1)\|\mathcal{R}_1\|/\|\mathcal{R}\|\right\}^n$$
$$\left. - m^2\left\{1 + (r-1)\|\mathcal{R}_1\|/\|\mathcal{R}\|\right\}^{2n} \right],$$

using (1.29) and (1.30). We may prove after a little algebra that

$$a_n^2 \sim \|\mathcal{S}_1\|^2 m \left[\exp\left\{\lambda(r^2 - 1)\|\mathcal{R}_1\|\right\} - \left\{1 + \lambda(r-1)^2\|\mathcal{R}_1\|\right\} \right.$$
$$\left. \times \exp\left\{\lambda 2(r-1)\|\mathcal{R}_1\|\right\} \right],$$

which in turn is asymptotic to a constant multiple of n. This proves (1.37).

We may now deduce from (1.38) that $\sum_i V_i$ is asymptotically normally distributed with mean $E(\sum_i V_i)$ and variance

$$n(\alpha^2 + \beta^2) \sim \mathrm{var}\left\{\sum_i E(V_i \mid N_i)\right\} + E\left\{\sum_i \mathrm{var}(V_i \mid N_i)\right\}$$

$$= \mathrm{var}\left\{E\left(\sum_i V_i \mid \mathcal{F}_n\right)\right\} + E\left\{\mathrm{var}\left(\sum_i V_i \mid \mathcal{F}_n\right)\right\}$$

$$= \mathrm{var}\left(\sum_i V_i\right).$$

An argument simpler than that above may be used to prove that

$$E\left\{\left(V - \sum_i V_i\right) - E\left(V - \sum_i V_i\right)\right\}^2$$

is asymptotic to a constant multiple of $\|\mathcal{R}\|$ as $\mathcal{R}$ increases, and that the constant converges to 0 as d (the side length of S_i) diverges to infinity. Therefore V is asymptotically normally distributed with mean $E(V)$ and variance $\text{var}(V)$, as had to be shown. $\square$

1.6. BALLISTICS AND NONUNIFORMLY DISTRIBUTED SETS

The work in the previous section, and indeed throughout most of this monograph, is concerned with sets distributed *uniformly* in some sense. However, there are many circumstances where uniformity is a restrictive assumption. For example, if overlapping sets represent foliage covers of shrubs then their centers may tend to congregate more densely in regions where nutrients are more plentiful. This behavior may be modeled by a coverage process in which shrub centers are represented by points of a nonstationary point process in the plane. A related topic which has generated a very large literature is that of modeling the effect of salvos of weapons fired at a target. Assume that a target occupies a region $\mathcal{R}$ in the plane $\mathbb{R}^2$, and that all of the target is of equal importance. A weapon is aimed in the direction of the target, and impacts at a point $\mathbf{x} \in \mathbb{R}^2$ with likelihood (i.e., density) $f(\mathbf{x})$. The destruction caused by the weapon is represented by a *damage function*, d. Given that the weapon strikes at $\mathbf{x}$, the proportion of that part of the target at $\mathbf{y} \in \mathcal{R}$ which is destroyed equals $d(\mathbf{y} - \mathbf{x})$. If the target $\mathcal{R}$ is of uniform value, which we take to be unity, then the expected total amount of the target destroyed by a single weapon equals

$$e_1 \equiv \int_{\mathcal{R}} d\mathbf{y} \int_{\mathbb{R}^2} d(\mathbf{y} - \mathbf{x}) f(\mathbf{x}) \, d\mathbf{x}.$$

Suppose next that a salvo of n weapons (called a salvo of n *rounds*) is fired. Given that the impact points are $\mathbf{x}_1, \ldots, \mathbf{x}_n$, the destroyed proportion of the target at $\mathbf{y} \in \mathcal{R}$ equals

$$1 - \prod_{i=1}^{n} \{1 - d(\mathbf{y} - \mathbf{x}_i)\}.$$

If the impact points are independently and identically distributed with density f, and if the target is of uniform, unit value, then the expected amount of the target destroyed by the salvo equals

$$e_n \equiv \int_{\mathcal{R}} d\mathbf{y} \int \ldots \int_{(\mathbb{R}^2)^n} \left[1 - \prod_{i=1}^{n} \{1 - d(\mathbf{y} - \mathbf{x}_i)\} \right]$$
$$\times f(\mathbf{x}_1) \ldots f(\mathbf{x}_n) \, d\mathbf{x}_1 \ldots d\mathbf{x}_n.$$

There are many more sophisticated versions of this model of target destruction. For example, the target might be moving, or for some other reason have uncertain location or shape. Or we might suppose that the target region $\mathcal{R}$ is not all of equal value, but is of more importance toward its center than around its edges. The latter change requires only the insertion of a target-value function $t(\mathbf{y})$ into the integrals defining e_1 and e_n. Typically t is taken to be a normal probability density function in the plane, giving rise to a so-called *Gaussian target*. However, we shall progress in the opposite direction, simplifying the model rather than making it more complex. Assume that the damage function d equals 1 inside a disk of radius r and 0 outside that radius. This is the commonly used *cookie-cutter damage function*, modeling a weapon which destroys everything within crater-radius r from the impact point and leaves all else unscathed. Then

$$e_n = \int_{\mathcal{R}} d\mathbf{y} \int \ldots \int_{(\mathbb{R}^2)^n} I(|\mathbf{y} - \mathbf{x}_i| \le r, \text{ some } 1 \le i \le n)$$
$$\times f(\mathbf{x}_1) \ldots f(\mathbf{x}_n) d\mathbf{x}_1 \ldots d\mathbf{x}_n. \tag{1.39}$$

Figure 1.9 illustrates the geometry of this problem.

To connect the formula for e_n with work in Section 1.5, observe that

$$\|\mathcal{R}\| - e_n = \int_{\mathcal{R}} \left\{ 1 - \int_{|y-x| \le r} f(\mathbf{x}) \, d\mathbf{x} \right\}^n d\mathbf{y} \tag{1.40}$$

equals *expected vacancy* within the region $\mathcal{R}$ resulting from placing n disks of fixed radius r into the plane, such that disk centers are independently distributed with density f. Of course, vacancy itself equals

$$V = \int_{\mathcal{R}} I(|\mathbf{y} - \boldsymbol{\xi}_i| > r, \text{ all } 1 \le i \le n) \, d\mathbf{y},$$

where $\boldsymbol{\xi}_1, \ldots, \boldsymbol{\xi}_n$ are independent and identically distributed random variables with density f. A quantity of some practical interest is the *kill probability*—the chance that $V = 0$, or that the target is

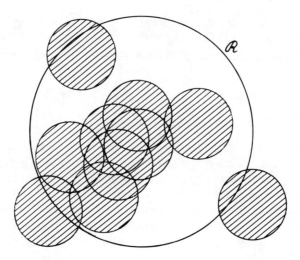

Figure 1.9. Damage to an area target caused by 10 weapons with cookie-cutter damage functions. The area target $\mathcal{R}$ is represented by the large, unshaded disk. Smaller shaded disks represent damage areas caused by individual weapons.

completely destroyed. An upper bound for kill probability is available from inequality (3.7) in Chapter 3, a formula which is valid for general coverage processes:

$$P(V = 0) \leq \mathrm{var}(V)/E(V^2).$$

Applications of the cookie-cutter model frequently involve only moderate salvo sizes (e.g., $n = 10$), and it is common to compute numerical values of quantities such as expected destroyed target fraction (e_n) and kill probability. Notes at the end of this chapter tell where sources may be found.

The notion of kill probability is of most importance when the target $\mathcal{R}$ is a point. For definiteness, take $\mathcal{R}$ to be the origin: $\mathcal{R} = \{\mathbf{O}\}$. Continue to use the cookie-cutter damage function introduced above. Then the probability that the origin is destroyed by a salvo of n rounds—or equivalently, that it is covered by at least one of the overlapping cookie-cutter disks—equals

$$p_n \equiv 1 - \left\{ \int_{|\mathbf{x}|>r} f(\mathbf{x})\,d\mathbf{x} \right\}^n. \tag{1.41}$$

It is commonly assumed that the impact-point density f is bivariate normal, with mean located at the target (here, the origin). Since the integral in (1.41) is over a circular region, and since any bivariate normal distribution may be written in the form of two independent components by rotating the axes suitably, then we may suppose without loss of generality that f has the form

$$f(x_1, x_2) = (2\pi\sigma_1\sigma_2)^{-1} \exp\left\{-(2\sigma_1^2)^{-1}x_1^2 - (2\sigma_2^2)^{-1}x_2^2\right\}, \qquad (1.42)$$

where $\sigma_1 > 0$ and $\sigma_2 > 0$. If $\sigma_1^2 = \sigma_2^2 = \sigma^2$ then

$$\int_{|x|>r} f(\mathbf{x})d\mathbf{x} = (2\pi\sigma^2)^{-1} \int_{x_1^2+x_2^2>r^2} \exp\left\{-(2\sigma^2)^{-1}(x_1^2 + x_2^2)\right\} dx_1 dx_2$$

$$= \sigma^{-2} \int_r^\infty \exp\left\{-(2\sigma^2)^{-1}x^2\right\} x\, dx$$

$$= \exp\left\{-r^2/(2\sigma^2)\right\}.$$

Therefore the probability of destroying a point target by a salvo of n rounds with cookie-cutter damage functions and aiming errors symmetrically and normally distributed about the target, equals

$$p_n = 1 - \exp\left\{-(nr^2)/(2\sigma^2)\right\},$$

where r is the damage radius and σ^2 the aiming error variance. The case where $\sigma_1 \neq \sigma_2$ does not admit such a simple solution.

Let us return to the problem of damaging an area target $\mathcal{R}$ by a salvo of n rounds with cookie-cutter damage functions. It is clear that some error in the aiming of weapons is desirable, even necessary, if we are to destroy a large part of the target. For if the aiming error variance is 0 then all rounds will strike the same part of the target. On the other hand, a large error variance may mean that the majority of rounds will miss the target completely. Thus for a given target $\mathcal{R}$, a given salvo size n, and a given aiming density f admitting a variable scale factor, there is an optimal value of the scale factor which maximizes the expected amount of the target destroyed. Suppose for the sake of argument that $\mathcal{R}$ is a disk of unit

radius centered at the origin, and that n cookie-cutter rounds of
radius r are fired at the origin with accuracy determined by density
f, given at (1.42), with $\sigma_1 = \sigma_2 = \sigma$. From formula (1.39) or (1.40)
we may determine the expected amount of the target destroyed by
a salvo of size n:

$$e_n = \pi - \int_{|y| \le 1} \left[1 - (2\pi\sigma^2)^{-1} \int_{|x-y| \le r} \right.$$

$$\left. \times \exp\left\{ -(2\sigma^2)^{-1}(x_1^2 + x_2^2) \right\} dx_1\, dx_2 \right]^n dy_1\, dy_2.$$

Jarnagin (1965, 1966) has tabulated values of σ which maximize e_n
for given values of r and n. The usefulness of introducing a variation
in aiming points was employed in eight-gun Spitfires during World
War II.

We might model the effect of a large salvo on a large target by
letting n increase to infinity and r decrease to 0 in such a manner
that $n\pi r^2 \to \lambda$, where $0 < \lambda < \infty$. Then

$$\int_{|x-y| \le r} f(\mathbf{x})\, d\mathbf{x} = \pi r^2 f(\mathbf{y}) + o(r^2)$$

$$= n^{-1}\lambda f(\mathbf{y}) + o(r^2)$$

uniformly in $|\mathbf{y}| \le 1$, assuming f is uniformly continuous. Therefore
by (1.39) or (1.40),

$$e_n \to e_\infty(f) \equiv \pi - \int_{|y| \le 1} \exp\left\{ -\lambda f(\mathbf{y}) \right\} d\mathbf{y} \qquad (1.43)$$

as $n \to \infty$, assuming the target $\mathcal{R}$ to be a disk of unit radius centered at the origin. If f is circularly symmetric then we may write
$f(\mathbf{y}) = |\mathbf{y}|^{-1}(2\pi)^{-1}g(|\mathbf{y}|)$, where g is a probability density on $(0, \infty)$.
Suppose $g(x) = \sigma^{-1}g_0(x/\sigma)$, where g_0 is a fixed density and σ is a
variable scale parameter. Then

$$e_\infty = \pi - 2\pi \int_0^1 x \exp\left\{ -\lambda(2\pi\sigma x)^{-1}g_0(x/\sigma) \right\} dx.$$

As expected, $e_\infty = 0$ if either $\sigma = 0$ or $\sigma = \infty$. For a given circularly
symmetric f (e.g., the normal), the value of σ which maximizes e_∞
may be calculated numerically as a function of λ.

A simple heuristic argument suggests that the density f which
maximizes $e_\infty(f)$ in formula (1.43), corresponds to the uniform dis-

tribution on the disk $|\mathbf{y}| \le 1$. This may be confirmed formally, using an argument based on the calculus of variations. Thus,

$$\sup_{f \in \mathcal{F}} e_\infty(f) = \pi(1 - e^{-\lambda/\pi}),$$

where $\mathcal{F}$ is the class of all densities f on $\mathbb{R}^2$. More generally, the same argument may be used to show that the asymptotic expected proportion of a fixed region $\mathcal{R}$ damaged by a salvo of n rounds with destruction radius r, cannot exceed $1 - \exp(-\lambda/\|\mathcal{R}\|)$, where $\|\mathcal{R}\|$ denotes the area of $\mathcal{R}$ and $n\pi r^2 \to \lambda$.

This treatment of the ballistic problem is closely related to work on vacancy in Section 1.5. To draw the parallel more closely, let C denote the amount of the plane $\mathbb{R}^2$ destroyed by at least one of the weapons. In coverage process terminology, C equals the *total coverage* of the random disks placed in the plane. Letting $\mathcal{R}$ increase to $\mathbb{R}^2$ in formula (1.39) we see that

$$E(C) = \int_{\mathbb{R}^2} \left[1 - \left\{ 1 - \int_{|\mathbf{y} - \mathbf{x}| \le r} f(\mathbf{x}) \, d\mathbf{x} \right\}^n \right] d\mathbf{y}.$$

If $n\pi r^2 \to \lambda$ as n increases and r decreases, and if f is uniformly continuous, then

$$E(C) \to \int_{\mathbb{R}^2} [1 - \exp\{-\lambda f(\mathbf{y})\}] \, d\mathbf{y}.$$

Likewise,

$$E(C^2) = \int_{\mathbb{R}^2} \int_{\mathbb{R}^2} \left[1 - \left\{ 1 - \int_{|\mathbf{y}_1 - \mathbf{x}| \le r} f(\mathbf{x}) \, d\mathbf{x} \right\}^n \right.$$
$$- \left\{ 1 - \int_{|\mathbf{y}_2 - \mathbf{x}| \le r} f(\mathbf{x}) \, d\mathbf{x} \right\}^n$$
$$\left. + \left\{ 1 - \int_{|\mathbf{y}_1 - \mathbf{x}| \le r \text{ or } |\mathbf{y}_2 - \mathbf{x}| \le r} f(\mathbf{x}) \, d\mathbf{x} \right\}^n \right] d\mathbf{y}_1 \, d\mathbf{y}_2.$$

Therefore, writing $B(x)$ for the area of the lens of intersection of two unit disks centered $2x$ apart [see (1.23)], we have

$$
\begin{aligned}
\mathrm{var}(C) &= \int_{\mathbb{R}^2} \int_{\mathbb{R}^2} \left[\left\{ 1 - \int_{|\mathbf{y}_1 - \mathbf{x}| \leq r \text{ or } |\mathbf{y}_2 - \mathbf{x}| \leq r} f(\mathbf{x})\, d\mathbf{x} \right\}^n dy_1\, dy_2 \right. \\
&\qquad \left. - \prod_{i=1}^{2} \left\{ 1 - \int_{|\mathbf{y}_i - \mathbf{x}| \leq r} f(\mathbf{x})\, d\mathbf{x} \right\}^n \right] dy_1\, dy_2 \\
&\sim \int_{\mathbb{R}^2} \int_{\mathbb{R}^2} \exp\left\{ -\lambda f(\mathbf{y}_1) - \lambda f(\mathbf{y}_2) \right\} \\
&\qquad \times \left[\exp\left\{ \lambda f(\mathbf{y}_1) \pi^{-1} B(|\mathbf{y}_1 - \mathbf{y}_2|/2r) \right\} - 1 \right] dy_1\, dy_2 \\
&\sim 8\pi r^2 \int_{\mathbb{R}^2} \exp\left\{ -2\lambda f(\mathbf{y}) \right\} d\mathbf{y} \\
&\qquad \times \int_0^1 \left[\exp\left\{ \lambda f(\mathbf{y}) \pi^{-1} B(x) \right\} - 1 \right] x\, dx.
\end{aligned}
$$

Thus, $E(C)$ converges to a finite, positive limit as $n \to \infty$, while $\mathrm{var}(C) \sim n^{-1}\tau^2$ where $0 < \tau < \infty$. A modification of the argument leading to Theorem 1.1 shows that the standardized version of expected total coverage,

$$
\{C - E(C)\} / \{\mathrm{var}(C)\}^{1/2},
$$

is asymptotically normal $N(0,1)$. Analogous results may be proved in any number of dimensions.

A problem which, superficially at least, seems very similar, is that of determining properties of total coverage C when a very large number of overlapping disks of *fixed* radius r are independently and identically centered in the plane according to density f. More generally, we might distribute n k-dimensional fixed-radius spheres in $\mathbb{R}^k$, using a k-variate density f. Moran (1974) has treated the case where $k = 3$ and f is a spherically symmetric normal density. These results are very sensitive to the specific form chosen for f. Little work has been done on other cases. Not surprisingly, the radius of the region in which "most" spheres are distributed depends on tail properties of f, as does the average amount of overlap of spheres. Asymptotic properties of coverage, as n increases, are closely bound up with multivariate extreme value theory for f.

1.7. POISSON AND OTHER POINT PROCESSES

In Section 1.5 we discussed properties of the coverage process created when n spheres are centered at n points $\xi_1, \ldots, \xi_n$ placed independently and uniformly into a cube $\mathcal{R} \subseteq \mathbb{R}^k$. If $\mathcal{R}$ is large then the points ξ_i are essentially those of a Poisson process, as we now show.

For convenience, take $\mathcal{R}$ to be the cube $[-l, l]^k$, having content $\|\mathcal{R}\| = (2l)^k$, and suppose l and n increase together in such a manner that $n/\|\mathcal{R}\| \to \lambda$ where $0 < \lambda < \infty$. Let $\mathcal{S}$ be a bounded Borel subset of $\mathbb{R}^k$. For large l the region $\mathcal{R}$ completely envelops $\mathcal{S}$, and then the chance that $\mathcal{S}$ contains precisely j of the uniformly distributed points ξ_i equals

$$\binom{n}{j} \left(\frac{\|\mathcal{S}\|}{\|\mathcal{R}\|} \right)^j \left(1 - \frac{\|\mathcal{S}\|}{\|\mathcal{R}\|} \right)^{n-j}$$

$$= \left(\frac{n}{\|\mathcal{R}\|} \|\mathcal{S}\| \right)^j \frac{1}{j!} \left(1 - \frac{1}{n} \right) \left(1 - \frac{1}{n} \right)$$

$$\times \ldots \left(1 - \frac{j-1}{n} \right) \left(1 - \frac{1}{n} \frac{n}{\|\mathcal{R}\|} \|\mathcal{S}\| \right)^n$$

$$\to (\lambda \|\mathcal{S}\|)^j (1/j!) \exp(-\lambda \|\mathcal{S}\|),$$

as $\mathcal{R}$ increases. Thus in the limit, the number of points ξ_i in $\mathcal{S}$ is Poisson distributed with mean $\lambda \|\mathcal{S}\|$. The quantity λ equals the mean number, or intensity or density, of points per unit content of $\mathcal{R}^k$.

Similarly we may prove that if $\mathcal{S}_1, \ldots, \mathcal{S}_m$ are disjoint Borel subsets of $\mathbb{R}^k$ then the numbers of points in $\mathcal{S}_1, \ldots, \mathcal{S}_m$ are asymptotically independent. These two properties distinguish a *stationary* (or homogeneous) *Poisson process*, $\mathcal{P} \equiv \{\xi_i, i \geq 1\}$, with intensity λ:

(i) the number of points ξ_i in any Borel subset $\mathcal{S}$ of $\mathbb{R}^k$ is Poisson distributed with mean $\lambda \|\mathcal{S}\|$; and

(ii) the numbers of points in any number of disjoint Borel subsets are independent random variables.

More generally, the process $\mathcal{P}$ is Poisson with intensity *function* $\lambda \geq 0$ if (ii) holds and property (i) is changed to (i)$'$ the number of points ξ_i in S is Poisson distributed with mean

$$\int_S \lambda(\mathbf{x})\,d\mathbf{x}.$$

The process is called stationary if and only if the function λ is constant almost everywhere. Sometimes (mainly in queueing theory, where $k = 1$) λ is called the *rate*. Of course the points in $\mathcal{P}$ are denumerable (countable), and the indexing of ξ_i, $i \geq 1$, is arbitrary. We might take ξ_i to be that point of $\mathcal{P}$ ith closest to the origin.

A stationary Poisson process in space is sometimes referred to simply as a random distribution of points, in much the same way that a point ξ distributed uniformly in a region $\mathcal{R}$ is often said to be randomly placed into $\mathcal{R}$.

From axioms (i) and (ii) follow several important properties of Poisson processes. For example, if $\mathcal{P}$ is a stationary Poisson process then the mean and variance of the number of points of $\mathcal{P}$ lying within S equal $\lambda\|S\|$, and the chance that no points lie in S equals $\exp(-\lambda\|S\|)$, with the obvious changes if axiom (i)$'$ operates in place of (i). Conditional on n points of $\mathcal{P}$ occurring in S, the positions of those points are independently and uniformly distributed over S.

One of the more interesting features of a Poisson process $\mathcal{P}$ is its "forgetfulness." The so-called "lack-of-memory" property may be expressed in a variety of ways, some of which are little more than restatements of axiom (ii) above. For example, if S_1 and S_2 are disjoint Borel subsets of $\mathbb{R}^k$, and $\mathcal{E}_i$ is an event concerned wholly with those points of $\mathcal{P}$ falling within S_i, then the probability of $\mathcal{E}_1$ given $\mathcal{E}_2$ equals the probability of $\mathcal{E}_1$. Other lack-of-memory properties are more subtle. In particular, a stationary Poisson process $\mathcal{P}$ conditional on n points of $\mathcal{P}$ being sited at $\mathbf{x}_1, \ldots, \mathbf{x}_n \in \mathbb{R}^k$ has the properties of $\mathcal{P} \cup \{\mathbf{x}_1, \ldots, \mathbf{x}_n\}$. Thus for all intents and purposes the process "forgets" where it had the n points, and behaves as though it were $\mathcal{P}$ with the n points adjoined. The notion of conditioning in this case is not entirely straightforward, since the event that some point of $\mathcal{P}$ takes the value $\mathbf{x}_i$ has probability 0 [see axiom (i) above]. However we may condition in a formal way by using the notions of Palm probability and Palm distribution [see, e.g., Karr (1986, Section 1.7)].

If $\mathcal{P} \equiv \{\xi_i, i \geq 1\}$ is a stationary Poisson process in $\mathbb{R}^k$ with intensity λ, and $\{\eta_i, i \geq 1\}$ is a sequence of independent and identically distributed random k-vectors, independent also of $\mathcal{P}$, then $\mathcal{P}' \equiv \{\xi_i + \eta_i, i \geq 1\}$ is another stationary Poisson process in $\mathbb{R}^k$ with intensity λ. This result follows easily from axioms (i) and (ii);

see Exercise 1.8. The result will be used several times in later chapters, and is one feature of the stationarity of $\mathcal{P}$. Generally, a point process $\mathcal{P}$ (not necessarily Poisson) is said to be stationary if for each Borel set $\mathcal{S} \subseteq \mathbb{R}^k$ the distribution of the number of points of $\mathcal{P}$ which fall within the set $\mathbf{x} + \mathcal{S} \equiv \{\mathbf{x} + \mathbf{y} : \mathbf{y} \in \mathcal{S}\}$ does not depend on $\mathbf{x}$. See Chapter 4, Section 4.1 for more details. Point processes $\mathcal{P}_1, \mathcal{P}_2, \ldots$ are said to be independent if for each finite collection $\mathcal{P}_{i_1}, \ldots, \mathcal{P}_{i_n}$ and for all Borel sets $\mathcal{S}_{i_1}, \ldots, \mathcal{S}_{i_n}$, the numbers of points from $\mathcal{P}_{i_j}$ in $\mathcal{S}_{i_j}$, $1 \le j \le n$, are independent random variables.

The notion of a *marked point process* will prove useful for deriving various properties of coverage processes. A marked point process may be constructed by taking an ordinary point process $\mathcal{P} \equiv \{\xi_i, i \ge 1\}$ (in $\mathbb{R}^k$, say), and associating with the ith point ξ_i a mark M_i, for $i \ge 1$. The marks can be almost arbitrary; in general they are elements of a complete separable metric space. We delay a detailed discussion until Chapter 4, Section 4.1. However, the following special case is worth mentioning here. Let $\mathcal{P} \equiv \{\xi_i, i \ge 1\}$ be a stationary Poisson process with intensity λ in $\mathbb{R}^k$, and let $\{M_i, i \ge 1\}$ be an independent and identically distributed sequence of marks, independent also of $\mathcal{P}$. Assume that the marks take only a countable range of distinct values $m_1, m_2, \ldots$, with

$$P(M_i = m_j) = p_j, \qquad i \ge 1 \ \text{ and } \ j \ge 1,$$

and

$$\sum_{j=1}^{\infty} p_j = 1.$$

In this case there is no loss of generality in thinking of the random marks M_i as real-valued random variables, although in practice they will often be sets. (In that case the marks $m_1, m_2, \ldots$ represent different configurations of the random sets M_i. For example, m_i might denote an i-sided figure.) The process $\mathcal{Q} \equiv \{(\xi_i, M_i), i \ge 1\}$ is a marked point process, and the individual point processes

$$\mathcal{P}_j \equiv \{\xi_i : M_i = m_j\}, \qquad j \ge 1,$$

are independent stationary Poisson processes, the jth having intensity λp_j (see Exercise 1.9).

Chapter 2 is devoted to studying the coverage pattern formed by centering random-length segments at points of a stationary Poisson process on the real line. Let $\mathcal{P} \equiv \{\xi_i, i \ge 1\}$ be a stationary Poisson process with intensity λ on $\mathbb{R}$. (Some authors prefer to denote this point process by $\{\xi_i, -\infty < i < \infty\}$, where $\ldots < \xi_{-2} < \xi_{-1} < \xi_0 < \xi_1 < \xi_2 < \ldots$.) Spacings between adjacent points ξ_i are independent and exponentially distributed with common mean λ^{-1}. If t is any given real number then the distance from t to that point ξ_i greater than t and nearest to t,

$$\xi_{i(t,+)} \equiv \inf\{\xi_i : \xi_i > t\},$$

is exponentially distributed with mean λ^{-1}. The distance from t to

$$\xi_{i(t,-)} \equiv \sup\{\xi_i : \xi_i < t\}$$

is also exponentially distributed with mean λ^{-1}. These facts are related to the lack-of-memory property of $\mathcal{P}$. They may be interpreted by arguing that, when viewed from the position of any given point t, the process behaves as though t were one of its points. Therefore it comes as no surprise to learn that the variables $\xi_{i(t,+)} - t$ and $t - \xi_{i(t,-)}$ are independent as well as exponentially distributed. In consequence, the length $\xi_{i(t,+)} - \xi_{i(t,-)}$ of that interval between adjacent points ξ_i, containing the given number t, has a gamma distribution with exponent 2 and mean $2\lambda^{-1}$. This property is sometimes seen as a contradiction, and is called the *waiting-time paradox*. It declares that the mean length of that interval between adjacent points of $\mathcal{P}$ which contains a given real number t, is twice the mean length of an "arbitrary" interval between adjacent points of $\mathcal{P}$. We may resolve the paradox by noting that the definition of an arbitrary interval is rather special, and that intervals which contain a given real number tend to be longer (on average) than arbitrary intervals. The latter phenomenon is one aspect of length-biased sampling. Notes at the end of this chapter give details of sources.

When discussing the Poisson centering of random, independent, and identically distributed segments on a line, we should bear in mind that there are many ways of "centering" each segment about its Poisson point. However, remember that independent and identically distributed (i.i.d.) translates of a stationary Poisson process result in another stationary Poisson process with the same intensity as the first; see the discussion three paragraphs above. Since different centerings are tantamount to different i.i.d. translates of the driving Poisson process, then any systematic approach to centering (and also any completely random approach) will result in a coverage pattern with identical properties. In Chapter 2 we shall assume for the sake of definiteness that points of the driving Poisson process are *left-hand* endpoints of random segments. Arguing in the same vein we see that if independent and identically distributed random sets are centered throughout $\mathbb{R}^k$ according to a stationary Poisson process, then the exact definition of "center" does not affect properties of the coverage pattern.

1.8. MOSAICS AND BOOLEAN MODELS IN THE CONTINUUM

The term "mosaic" has entered statistical usage to describe the random pattern, usually in the plane, formed by a process that classifies each point as one of several "types" or "phases." In one very general formulation, a mosaic is just a patchwork of different phases (e.g., Pielou 1965). From this point of view, tessellations and other models for the random division of space (e.g., Miles 1972) come under the heading of "mosaics." Information science literature (e.g., Ahuja 1981a) is more specific concerning the random mechanism by means of which mosaics are constructed, and asks that in some sense, sets generating the mosaic be put down independently and (usually) uniformly. Most of the information science mosaics would be termed binary or two-phase models by biologists, since they classify each point as either marked or unmarked. Examples were discussed in Section 1.3, with more sophisticated generalizations in Section 1.4. These models are usually lattice based, reflecting the digital pixel structure of real images.

The mosaics in which we are most interested are termed *Boolean models*. A Boolean model in the continuum $\mathbb{R}^k$ is a collection of sets $\mathcal{C} \equiv \{\xi_i + S_i, i \geq 1\}$, where

$$\xi_i + S_i \equiv \{\xi_i + \mathbf{x} : \mathbf{x} \in S_i\},$$

$\mathcal{P} \equiv \{\xi_i, i \geq 1\}$ is a *stationary* Poisson process in $\mathbb{R}^k$, and S_i, $i \geq 1$ are independent and identically distributed random subsets of $\mathbb{R}^k$, independent of $\mathcal{P}$. We call ξ_i the *center* of the random set $\xi_i + S_i$. Clearly this definition of the centers depends on the way we constructed the Boolean model, not just on the sets it contains. Further details, particularly of the definition of a random set, will be given in Chapter 3 where we begin discussing Boolean models in some depth. A point $\mathbf{x} \in \mathbb{R}^k$ is said to be *covered* by the Boolean model $\mathcal{C}$ if it is contained in at least one of the sets in $\mathcal{C}$.

Bearing in mind that a stationary Poisson process is a "random distribution of points in space" (see Section 1.7), a Boolean model represents a sequence of random sets randomly (i.e., uniformly) centered in space. However, note the definition of randomness in Markovian terms discussed four paragraphs below.

The random pattern formed from a Boolean model may be constructed in a variety of ways. Most commonly we construct the binary pattern, in which each point in $\mathbb{R}^k$ is either covered or uncovered. An n-phase pattern might be one in which any given point is classified as being of type ν if it is contained in precisely $\nu - 1$ of the sets $\xi_i + S_i$, for $1 \le \nu \le n - 1$, and of type n if it is covered by $n - 1$ or more sets. Clearly there are multiphase patterns that are not well modeled by these processes. Some alternatives will be discussed in Chapter 5, Section 5.1.

The union of all the sets $\xi_i + S_i$ in a Boolean model,

$$B \equiv \bigcup_{i=1}^{\infty} (\xi_i + S_i),$$

is often called a *germ-grain model* (Hanisch 1981) or *Boolean set*. Individual sets $\xi_i + S_i$ are often called *grains* or *Boolean grains*, although this name is occasionally reserved for the special case where the sets S_i are convex with probability 1. The set "centers" ξ_i are often called *germs*. Discussion of Boolean models is sometimes restricted to properties of the set B.

Discrete analogues of Boolean models, defined on lattices, have already been discussed in Sections 1.3 and 1.4. Boolean models in the continuum may be regarded as limits of discrete Boolean models, as the lattices get finer and finer. Discrete Boolean models are also useful for checking properties of continuum Boolean models. For example, it is occasionally suggested that a truly "random" spatial pattern should in some sense be Markovian (e.g., Pielou 1965). While this definition is sometimes convenient, it excludes most Boolean models. Since the latter are very popular as first-order approximations to many random natural phenomena, the Markovian definition of randomness seems rather restrictive. We show below that even very simple binary Boolean models on lattices do not form Markov chains. Similarly, Boolean models in the continuum do not form Markov processes.

Consider the following discrete Boolean model on the integers $\mathbb{Z}$. Each point of $\mathbb{Z}$ is marked with probability p independently of all other points. If i is marked then integers i and $i + 1$ are said to be covered. Thus, an integer j is uncovered if and only if neither j nor $j - 1$ is marked. This pattern of covered and uncovered points may be thought of as arising from a Boolean model in which line segments of length two have their left-hand ends assigned randomly to elements of $\mathbb{Z}$. Notice that

$$P(2 \text{ covered} \mid 1 \text{ covered}, \ 0 \text{ uncovered}) = 1,$$

while

$$P(2 \text{ covered} \mid 1 \text{ covered}) < 1.$$

Therefore

$$P(2 \text{ covered} \mid 1 \text{ covered}, \ 0 \text{ uncovered}) \neq P(2 \text{ covered} \mid 1 \text{ covered}),$$

and so the Markov property fails.

Random function models in the continuum may be defined similarly to their discrete counterparts (see Section 1.4 for the latter). If we are working in $\mathbb{R}^k$ then we take $\{\xi_j, j \geq 1\}$ to be a stationary Poisson process in $\mathbb{R}^k$, with intensity λ. The function $F : \mathbb{R}^k \to \mathbb{R}$ is assumed to be random and measurable,* and analogues of conditions (1.14) and (1.16) are

$$E\left\{ \int_{\mathbb{R}^k} |F(\mathbf{x})| \, d\mathbf{x} \right\} < \infty \qquad (1.44)$$

and

$$E\left\{ \int_{\mathbb{R}^k} |F(\mathbf{x})| \, d\mathbf{x} \right\}^2 < \infty, \qquad (1.45)$$

respectively. Definition (1.15) of total intensity $t(\xi)$, for $\xi \in \mathbb{R}^k$, is still appropriate. Formulas (1.17), (1.18), and (1.19) become

$$\mu_1 \equiv E\{t(0)\} = \lambda \int_{\mathbb{R}^k} E\{F(\mathbf{x})\} \, d\mathbf{x}$$

[assuming (1.44)],

$$a(\xi) \equiv E\{t(0)t(\xi)\} - \mu_1^2$$
$$= \lambda \int_{\mathbb{R}^k} E\{F(\mathbf{y})F(\xi+\mathbf{y})\} \, d\mathbf{y}$$

and

$$\alpha(\theta_1,\ldots,\theta_k) \equiv \int_{\mathbb{R}^k} \exp\left(i \sum_{j=1}^{k} \theta_j x_j \right) a(\mathbf{x}) \, d\mathbf{x}$$
$$= \lambda E\left\{ \left| \int_{\mathbb{R}^k} \exp\left(i \sum_{j=1}^{k} \theta_j x_j \right) F(\mathbf{x}) \, d\mathbf{x} \right|^2 \right\}$$

---

*A useful general definition of "a random function $F : \mathbb{R}^k \to \mathbb{R}$" is difficult to formulate. For many purposes it is reasonable to take F to be continuous and to vanish outside a compact set $K \subseteq \mathbb{R}^k$. Then F may be considered to be a random element of the metric space $C(K, \rho)$ of continuous functions vanishing outside K. Here ρ is Euclidean distance.

[assuming (1.45)]. While these formulas are intuitively clear as analogues of the discrete case, a thoroughly rigorous proof would proceed via the Campbell theorem (see Chapter 4, Section 4.1).

Much of this monograph is devoted to the study of Boolean models in the continuum, and several of our results are readily extended
to random function models. For example, the concept of *vacancy* or
uncovered content within a region $\mathcal{R}$ (see Section 1.5 and Chapter 3)
may be replaced by *total intensity* T, where

$$T(\mathcal{R}) \equiv \int_{\mathcal{R}} t(\mathbf{x})\, d\mathbf{x}.$$

Properties of total intensity may be explored much as in Chapter 3.

1.9. STEREOLOGY

Stereology is the science of inferring properties of a spatial pattern
from a lower dimensional section through it. Typically the original
pattern is in $\mathbb{R}^3$, and the section is in $\mathbb{R}^2$ (e.g., a thin slice viewed
on a microscope slide) or in $\mathbb{R}$ (e.g., a drill core), or the original
pattern is in $\mathbb{R}^2$, and the section is in $\mathbb{R}$ (e.g., a sampling transect).
The theory of stereology is beyond the scope of this monograph, but
we sketch several aspects of it here for the purpose of illustration.

An l-dimensional section through a k-dimensional Boolean model
$(k > l)$ produces an l-dimensional Boolean model. The component
sets in the l-dimensional Boolean model are of course sections of
the component sets in the k-dimensional model. Short proofs of
these properties may be based on the "Poisson-ness" and marked
point process arguments introduced in Chapter 4, Section 4.1, although there are other approaches (e.g., Matheron 1975, p.144). Recall from Section 1.7 that the "center" of a random set in a Boolean
model may be defined somewhat arbitrarily. Centers of sets in the
l-dimensional Boolean model may be defined in almost any way
that is convenient. If the sets in the k-dimensional Boolean model
are convex then so are those in the section, although most other
important features of the original model (such as the intensity of
the driving Poisson process) are transformed. Fortunately, the covered and uncovered proportions are unchanged. Indeed, this is true
for any stationary coverage pattern, not just for Boolean models.
Sometimes this feature is proposed as a means of estimating the
uncovered proportion via a *line transect sample*, as follows. Let $\mathcal{C}$
be a Boolean model in $\mathbb{R}^3$ or $\mathbb{R}^2$, and let $\mathcal{L}$ be a line segment

somewhere in the pattern. Let $\hat{p}$ denote the proportion of $\mathcal{L}$ not covered by $\mathcal{C}$. Then $\hat{p}$ is an unbiased estimator of the proportion p of space not covered by $\mathcal{C}$. The estimator is consistent if the length of $\mathcal{L}$ is allowed to increase indefinitely. (Unbiasedness follows from stationarity of $\mathcal{C}$, and consistency follows via the ergodic argument in Section 3.4.)

Stereological problems frequently involve nonoverlapping sets (the so-called "dilute medium" case) or overlapping sets where the boundary of each set is readily discernible. In such circumstances the spatial character of the section of a coverage process under stereological investigation is of limited importance, being of interest primarily because it may induce edge effects that need correcting. Edge effect problems are discussed in Chapter 4, Section 4.2 and in the notes to that section. If we neglect edge effects then we have essentially a random sample of random sections of independent and identically distributed random sets.

In the remainder of this section we concentrate on the case of a Boolean model $\mathcal{C} \equiv \{\xi_i + S_i, i \geq 1\}$ of isotropic (i.e., randomly and uniformly oriented) convex sets in $\mathbb{R}^3$, sectioned by a plane Π. Assume for the sake of convenience that the centroid (center of gravity) of each random convex shape S_i is located at the origin. Then the centroid of $\xi_i + S_i$ is its center, ξ_i. (Note the definition of "center" in Section 1.8.) Let q_i (a random variable) denote the perpendicular distance from ξ_i to that tangent plane to the boundary of $\xi_i + S_i$ parallel to Π and on the same side of ξ_i as Π. The set $\xi_i + S_i$ intersects Π if and only if ξ_i is distant no more than q_i from Π. If there is an intersection, let η_i be the foot of the perpendicular from ξ_i to Π. (We may take η_i to be the center of the set formed from the sectioning of $\xi_i + S_i$ by Π.) The expected number of points ξ_i per unit volume of $\mathbb{R}^3$ equals λ, and so the expected number of points η_i per unit content of Π equals $2\lambda E(q)$, where q has the distribution of each q_i. (The factor 2 comes from considering points ξ_i on both sides of Π.) Since the convex shapes S_i are randomly and uniformly oriented then $\delta \equiv 2E(q)$ equals the mean distance between two tangent planes parallel to S_i. This is sometimes called the *mean caliper diameter* of S_i. The Poisson process driving the Boolean model in Π has intensity $\delta\lambda$. (A rigorous, less heuristic derivation of this property is readily constructed using "Poisson-ness" and marked point process arguments from Chapter 4. See Exercise 4.14.)

The set sections comprising the Boolean model $\mathcal{C}'$ in Π are isotropic and convex, with mean area α' equal to the expected area of a random section through a random set S_i. We may derive a formula for α' by using the fact that the uncovered proportion is the same for the two-dimensional Boolean model C' and the three-dimensional model C. In fact, we shall show in Chapter 3, Section 3.2 that the chance that an arbitrary point in $\mathbb{R}^3$ is not covered by C equals $e^{-\alpha\lambda}$, where α is the mean volume of a random set from C. Likewise, the chance that an arbitrary point in Π is not covered by C' equals $e^{-\alpha'\delta\lambda}$. Equating these two probabilities we deduce that $\alpha' = \alpha/\delta$.

Now suppose that the random convex sets comprising C are spheres, with positive radii having distribution function F. Assume mean sphere volume is finite, and let

$$\delta \equiv 2 \int_0^\infty x\, dF(x)$$

equal mean sphere diameter. The theorem below presents Wiksell's formula describing the distribution of the radius of the random disks that comprise the Boolean model C' in Π. In the formula for g, and in the proof at the end of the section, we write

$$\int_z^\infty \ldots dF(x)$$

to mean

$$\int_{x>z} \ldots dF(x).$$

Theorem 1.3. *The two-dimensional Boolean model C' resulting from sectioning the Boolean model C by a plane Π is composed of random disks, centered at points of a Poisson process with intensity $\delta\lambda$ in Π and having radii distributed according to the density*

$$g(z) \equiv 2\delta^{-1} z \int_z^\infty (x^2 - z^2)^{-1/2}\, dF(x), \qquad z > 0. \tag{1.46}$$

If F is absolutely continuous with density f then formula (1.46) reduces to

$$g(z) = 2\delta^{-1} z \int_z^\infty (x^2 - z^2)^{-1/2} f(x)\, dx. \tag{1.47}$$

(A related integral equation arises in Chapter 5, Section 5.2, where we discuss point covariance in a Boolean model.) We may not directly sample the distribution with density f, only that with density *g*. However, the Abel-type integral equation (1.47) may be solved to express *f* in terms of *g*:

$$f(x) = -\frac{\delta x}{\pi} \int_x^\infty (y^2 - x^2)^{-1/2} \frac{d}{dy} \left\{ y^{-1} g(y) \right\} dy. \tag{1.48}$$

This procedure is known as *unfolding*. From (1.47) we may deduce a formula for δ in terms of an integral of *g*. First, notice that

$$\begin{aligned}
1 &= \int_0^\infty f(x) \, dx \\
&= -\frac{\delta}{\pi} \int_0^\infty \frac{d}{dy} \left\{ y^{-1} g(y) \right\} dy \int_0^y x (y^2 - x^2)^{-1/2} \, dx \\
&= \delta \pi^{-1} \int_0^\infty \left\{ y^{-1} g(y) - g'(y) \right\} dy \\
&= \delta \pi^{-1} \int_0^\infty y^{-1} g(y) \, dy,
\end{aligned}$$

since $g(0) = 0$. Therefore

$$\delta = \pi \left\{ \int_0^\infty y^{-1} g(y) \, dy \right\}^{-1}.$$

If $R_1, \ldots, R_n$ are observations of radii of disks in Π, then

$$\hat{\delta} \equiv \pi \Big/ \left(n^{-1} \sum_{i=1}^n R_i^{-1} \right)^{-1}$$

is an estimate of mean sphere diameter. (Throughout this discussion we have worked with sphere radius rather than sphere diameter, since the former is a trifle simpler from a theoretical standpoint. However, practical measurements are usually made directly of diameter.)

Sometimes the sizes of sets in a planar Boolean model are of interest in their own right. If individual sets (as opposed to just the pattern of covered and uncovered regions) are visible then features of size distribution, such as the distribution function of diameter of disks in C', may be estimated by methods of *sieving* and *granulometry*. See Ripley (1981, Section 9.4) and Serra (1982, Chapter X). These techniques are designed for application to a random sample of observations of a random set, not to coverage processes per se.

Proof of Theorem 1.3. Suppose initially that sphere radii are fixed and equal to *x*. A sphere whose center is distant $y \le x$ from the plane Π intersects Π in a disk of radius $z \equiv (x^2 - y^2)^{1/2}$. Thus, $x^2 =$

$y^2 + z^2$, whence $0 = y\,dy + z\,dz$, where dz equals the change in z if y is perturbed to $y + dy$. In consequence,

$$|dy| = (z/y)\,dz = z(x^2 - z^2)^{-1/2}\,dz.$$

The expected number of spheres whose centers are perpendicularly distant between y and $y + dy$ from a region of unit area in Π equals

$$2\lambda|dy| = 2\lambda z(x^2 - z^2)^{-1/2}\,dz.$$

Therefore the expected number of disks of radius between z and $z + dz$, per unit area of Π, equals $2\lambda z(x^2 - z^2)^{-1/2}\,dz$. If sphere radius is given distribution function F, rather than being fixed at x, then this expected number changes to

$$\nu(z)\,dz \equiv \int_{x>z} 2\lambda z(x^2 - z^2)^{-1/2}\,dz\,dF(x)$$

$$= 2\lambda z\,dz \int_z^\infty (x^2 - z^2)^{-1/2}\,dF(x).$$

Therefore the mean number of disks per unit area of Π equals

$$\int_0^\infty \nu(z)\,dz = 2\lambda \int_0^\infty dF(x) \int_0^x z(x^2 - z^2)^{-1/2}\,dz$$

$$= 2\lambda \int_0^\infty x\,dF(x)$$

$$= \delta\lambda,$$

where δ is mean sphere diameter. (This serves as a check on the earlier heuristic argument!) In consequence, disk radius has density

$$(\delta\lambda)^{-1} 2\lambda z \int_z^\infty (x^2 - z^2)^{-1/2}\,dF(x) = g(z),$$

as had to be shown. $\square$

1.10. PACKING PROBLEMS

Packing problems, or space-filling problems, are concerned with the "random" distribution of "solid" sets in k-dimensional space, with no overlap of sets permitted. The problems can be of various types. We shall consider here only two, the first being the so-called "complete random packing" of solid sets into a container.

Let $\mathcal{D}$ (the container) be a very large k-dimensional cube, and place into $\mathcal{D}$ a unit radius sphere $\mathcal{T}_1$, its center distributed randomly and uniformly over the set of all points such that the sphere will be contained wholly within $\mathcal{D}$. Next, place into $\mathcal{D}$ another unit sphere $\mathcal{T}_2$, its center distributed uniformly over all points such that $\mathcal{T}_2$ does not intersect $\mathcal{T}_1$ and is contained wholly within $\mathcal{D}$; then place into $\mathcal{D}$ a third unit sphere $\mathcal{T}_3$, its center distributed uniformly over all points such that $\mathcal{T}_3$ intersects neither $\mathcal{T}_1$ nor $\mathcal{T}_2$ and is contained wholly within $\mathcal{D}$, and so on. This process will terminate after a finite number of steps, when there are no gaps in $\mathcal{D}$ large enough to take another unit radius sphere. If the process is allowed to proceed to termination, we say that the spheres are *completely randomly packed* into $\mathcal{D}$. We are interested in characteristics such as the proportion of $\mathcal{D}$ covered by spheres, which of course is simply related to the total number of spheres in $\mathcal{D}$.

It appears that only the case $k = 1$ is amenable to rigorous analysis. Results in that circumstance date back to Rényi (1958). For example, we have the following description of the proportion of a large interval covered by segments of unit length.

Theorem 1.4. *Place unit length segments into an interval of length $x > 1$ according to the scheme discussed above, until there is no room left for further segments. Let $N(x)$ denote the number of segments that the interval contains, and put*

$$\nu(x) \equiv E\{N(x)\}.$$

Then as $x \to \infty$,

$$x^{-1}\nu(x) \to c \equiv \int_0^\infty \exp\left\{-2\int_0^u t^{-1}(1 - e^{-t})\,dt\right\}du$$
$$\simeq 0.7476.$$

In this special case the "cube" $\mathcal{D}$ is an interval of length x, and the "spheres" are of radius $\frac{1}{2}$ (i.e., diameter 1).

A sharpening of the argument used to prove Theorem 1.4 (see the end of the section) provides asymptotic formulas for general moments of $N(x)$. In particular, it may be shown that

$$\text{var}\{N(x)\} \sim 0.0382x$$

as $x \to \infty$ (Mackenzie 1962, Solomon and Weiner 1986), from which it follows that $x^{-1}N(x) \to c$ in probability as $x \to \infty$. Furthermore,

$$\{N(x) - EN(x)\} / \{\operatorname{var} N(x)\}^{1/2} \to N(0,1)$$

in distribution as $x \to \infty$ (Dvoretzky and Robbins 1964; Mannion 1964).

This relatively simple problem is sometimes referred to as the "parking problem," since the unit length segments may be thought of as cars parked beside a kerb of length x. Palásti (1960) conjectured that if unit area squares are packed randomly into a large square of side length x according to the scheme outlined earlier, and if $\nu_2(x)$ denotes the expected number of squares after packing terminates, then

$$x^{-2}\nu_2(x) \to c^2 = 0.5589$$

as $x \to \infty$. Computer simulation suggests that while this is a good approximation, it is not strictly correct. Apparently,

$$x^{-2}\nu_2(x) \to c_2 \simeq 0.562 \quad \text{or} \quad 0.563$$

(e.g., Akeda and Hori 1976; Jodrey and Tory 1980). More generally, it seems that in the k-dimensional problem where cubes of unit side length are completely randomly packed into an $x \times \ldots \times x$ cube, the mean number $\nu_k(x)$ of cubes after packing terminates satisfies

$$x^{-k}\nu_k(x) \to c_k > c^k, \tag{1.49}$$

the "error" $c_k - c^k$ apparently increasing with increasing k.

In formula (1.49), the left-hand side equals

$$\frac{\text{expected amount of container covered after packing terminates}}{\text{total content of container}}.$$

The limit of this ratio as $x \to \infty$ is called the *packing density*. Packing densities for many shapes may be determined by simulation. In the case of disks completely randomly packed into the plane, Monte Carlo studies suggest that the packing density is about 0.55 or 0.56 (Tanemura 1979; Lotwick 1982). Solomon (1967) reports an approximate packing density of 0.28 for spheres completely randomly packed into $\mathbb{R}^3$, based on 10 replicates. Packing densities for closest possible *lattice* packings are $\pi/12^{1/2} \simeq 0.91$ in the case of disks packed into the plane, and $\pi/18^{1/2} \simeq 0.74$ in the case of spheres packed into $\mathbb{R}^3$.

The complete random packing model may be used to describe the distribution of particles in a medium that is midway between "dilute" and "solid packed"—e.g., of cells in blood. Sometimes it is appropriate to simulate a system in which packing ceases early, before all places are taken up, since many physical phenomena exhibit packing densities less than the maximum. See the notes at the end of this chapter for sources. On other occasions one is interested in a more solidly packed system, where each particle touches several of its neighbors. Observe that in the complete random packing model, no two sets ever touch. Such a process could not be used to model the positions of rocks in a bed of gravel, for example.

Packing arrangements in which solid sets touch, yet are placed at random (in some sense) into a region, are sometimes called *random dense packings*. They are easy to simulate mechanically, difficult to simulate numerically, and very hard to describe theoretically. For example, consider a large bucket full of small solid spheres (e.g., marbles). Pour the spheres gently into an empty bucket, being careful not to jiggle the contents. Mechanical simulation of this process suggests that the spheres will occupy about 0.60 of the volume (Scott 1960). If the bucket of spheres is gently shaken for some time then the spheres will settle into a closer packing arrangement, and occupy about 0.64 of the volume (Scott 1960). The former figure is often referred to as the density for random *loose* packing, and the latter as the density for random *close* packing. Recall that the densest packing of spheres is about 16% denser even than this, at 0.74 of total volume. Versions of this random packing experiment exist in two dimensions; consider packing circular coins into a sloping tray. Experiments indicate a random close packing density of about 0.8 for disks in the plane. All three densities are "confirmed" by theoretical analysis; see the notes at the end of the chapter for details.

We close this section with a proof of Theorem 1.4.

Proof of Theorem 1.4.   First we show that the function ν satisfies an integral equation,

$$\nu(x+1) = 2x^{-1} \int_0^x \nu(t)\,dt + 1, \qquad x > 0. \tag{1.50}$$

Suppose a unit-length segment $\mathcal{I}$ has already been placed within the interval $(0, x+1)$, and that the left-hand endpoint of the interval is t, where $0 < t < x$. Conditional on this event, the expected number of segments to the left of $\mathcal{I}$ (i.e., with left-hand endpoints between

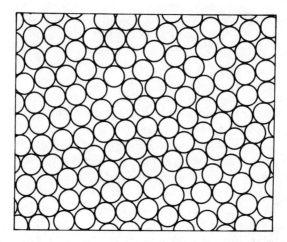

Figure 1.10. Random close packed disks in the plane. [After Cowan (1984).]

0 and $t-1$) equals $\nu(t)$, and the expected number to the right (i.e., with left-hand endpoints between $t+1$ and x) equals $\nu(x-t)$. Since t is uniformly distributed on $(0,x)$ then

$$\nu(x+1) = x^{-1} \int_0^x \{\nu(t) + \nu(x-t)\}\,dt + 1$$
$$= 2x^{-1} \int_0^x \nu(t)\,dt + 1,$$

proving (1.50).

Multiply both sides of (1.50) by x and differentiate, obtaining

$$x\nu'(x+1) + \nu(x+1) = 2\nu(x) + 1. \qquad (1.51)$$

Let

$$\phi(s) \equiv \int_0^\infty \nu(x)e^{-sx}\,dx \qquad (1.52)$$

denote the (ordinary) Laplace transform of ν, finite since $\nu(x) \le x$. Multiply (1.51) by e^{-sx} and integrate over $0 < x < \infty$, obtaining

$$\int_0^\infty \{x\nu'(x+1) + \nu(x+1)\}\, e^{-sx}\,dx = 2\phi(s) + s^{-1}. \qquad (1.53)$$

Since $\nu(x) = 0$ for $0 \le x \le 1$ then

$$-\int_0^\infty x\nu'(x+1)e^{-sx}\,dx = \frac{d}{ds}\left\{\int_0^\infty \nu'(x+1)e^{-sx}\,dx\right\}$$

$$= \frac{d}{ds}\left\{e^s\int_0^\infty \nu'(x)e^{-sx}\,dx\right\}$$

$$= \frac{d}{ds}\{e^s s\phi(s)\}\,.$$

Similarly but more simply we obtain

$$\int_0^\infty \nu(x+1)e^{-sx}\,dx = e^s\phi(s).$$

Substituting into (1.53) we deduce that

$$-\frac{d}{ds}\{e^s s\phi(s)\} + e^s\phi(s) = 2\phi(s) + s^{-1}\,.$$

Put $\psi(s) \equiv e^s\phi(s)$. Then the above equation becomes

$$-s\psi'(s) = 2e^{-s}\psi(s) + s^{-1}\,,$$

or equivalently,

$$2s^{-1}e^{-s}\psi(s) + \psi'(s) = -s^{-2}\,.$$

To solve this differential equation, multiply throughout by the "integrating factor"

$$\exp\left(-2\int_s^\infty t^{-1}e^{-t}\,dt\right),$$

obtaining:

$$\frac{d}{ds}\left\{\exp\left(-2\int_s^\infty t^{-1}e^{-t}\,dt\right)\psi(s)\right\} = -s^{-2}\exp\left(-2\int_s^\infty t^{-1}e^{-t}\,dt\right).$$

Integrating, we get

$$\psi(s) = \exp\left(2\int_s^\infty t^{-1}e^{-t}\,dt\right)\int_s^\infty u^{-2}\exp\left(-2\int_u^\infty t^{-1}e^{-t}\,dt\right)du.$$

(Here we have used the fact that

$$0 \le \psi(s) \le \int_1^\infty xe^{-sx}\,dx \to 0$$

as $s \to \infty$.) Therefore

$$\phi(s) = e^{-s}\psi(s)$$
$$= e^{-s}s^{-2}\int_s^\infty \exp\left\{-2\int_s^u t^{-1}(1-e^{-t})\,dt\right\}\,du.$$

Consequently,

$$\lim_{s\to 0} s^2\phi(s) = \int_0^\infty \exp\left\{-2\int_0^u t^{-1}(1-e^{-t})\,dt\right\}\,du$$
$$= c.$$

We claim that this limit equals the limit of $x^{-1}\nu(x)$ as $x \to \infty$. To see this, notice from (1.52) that

$$c = \lim_{s\to 0} s^2 \int_0^\infty e^{-sx}\,d_x\left\{\int_0^x \nu(t)\,dt\right\},$$

so that by a Tauberian theorem (e.g., Feller 1971, Theorem 2, p. 445),

$$\tfrac{1}{2}c = \lim_{x\to\infty} x^{-2}\int_0^x \nu(t)\,dt.$$

Remember from (1.50) that

$$\int_0^x \nu(t)\,dt = \tfrac{1}{2}x\left\{\nu(x+1)-1\right\},$$

so that

$$c = \lim_{x\to\infty} x^{-1}\left\{\nu(x+1)-1\right\} = \lim_{x\to\infty} x^{-1}\nu(x),$$

as had to be proved. $\square$

1.11. COVERING A SPHERE WITH CIRCULAR CAPS

One of the oldest problems in the theory of coverage processes is that of calculating the chance that a given region is completely covered by a sequence of random sets. Unfortunately there is only a small number of useful circumstances where this probability may be calculated explicitly. One of these is discussed in Exercise 1.1 at the end of this chapter, and another in Theorem 2.6 of Chapter 2 [see formula (2.24)]. In certain other cases the Laplace transform of the

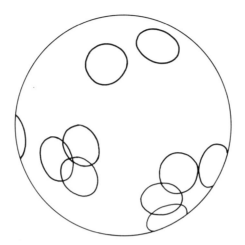

Figure 1.11. Equal circular caps placed on surface of three-dimensional sphere.

probability of complete coverage may be derived [see the formula for $\pi(s)$ in Theorem 2.6], although anything more than formal inversion of the transform can be prohibitively complex. In the present section we investigate the probability of covering the surface of a sphere by a sequence of independently and uniformly centered circular caps. On occasion this probability may be calculated explicitly.

Let $\mathcal{T}$ denote a k-dimensional sphere, which we assume for definiteness to be of unit radius. Let $\mathbf{X}_1,\ldots,\mathbf{X}_n$ be n points independently and uniformly distributed over the surface of $\mathcal{T}$. Each $\mathbf{X}_i$ is taken to be the pole (center) of a circular cap C_i of angular radius θ:

$$C_i = \left\{ \mathbf{x} \in \mathbb{R}^k : |\mathbf{x}| = 1 \text{ and } (\mathbf{x},\mathbf{X}_i) \geq \cos\theta \right\},$$

where $(\mathbf{u},\mathbf{v})$ denotes the usual inner product between vectors $(u_1,\ldots,u_k)$ and $(v_1,\ldots,v_k)$:

$$(\mathbf{u},\mathbf{v}) = \sum_{i=1}^{k} u_i v_i.$$

Figure 1.11 illustrates a typical configuration in the case $k = 3$. The sphere surface is said to be *completely covered* if each point $\mathbf{x}$ on the surface of $\mathcal{T}$ belongs to at least one cap. Let $r_n = r_n(\theta)$ denote the probability of this event.

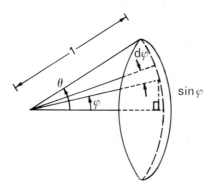

Figure 1.12. Angles used to find area of circular cap. Area of annulus of radius $\sin \phi$ and width $d\phi$ is $2\pi \sin \phi \, d\phi$. Area of entire sphere surface is 4π. Therefore fractional area of cap is $(4\pi)^{-1} \int_0^\theta 2\pi \sin \phi \, d\phi$.

For the time being, take $k = 3$. Each circular cap covers a fraction of the surface area of $\mathcal{T}$ equal to

$$
\begin{aligned}
\rho &\equiv (4\pi)^{-1} \int_0^\theta 2\pi \sin \phi \, d\phi \\
&= \tfrac{1}{2}(1 - \cos\theta) \\
&= \sin^2(\theta/2)
\end{aligned}
\tag{1.54}
$$

(see Figure 1.12). Therefore $r_n(\theta) = 0$ if $n \sin^2(\theta/2) \leq 1$. Less trivially, if $\theta = \tfrac{1}{2}\pi$ then

$$
r_n = 1 - 2^{-n}(n^2 - n + 2),
$$

as will follow from Theorem 1.5 below.

Now return to the case of a k-dimensional sphere. Notice that if $\theta = \tfrac{1}{2}\pi$ then the sphere surface is not completely covered by the caps if and only if all the points $\mathbf{X}_1, \ldots, \mathbf{X}_n$ lie in the same hemisphere. (The sphere surface is not covered if and only if some point $\mathbf{x}$ is not covered, and that means that none of the points $\mathbf{X}_i$ lies in the hemisphere whose pole is $\mathbf{x}$.) Equivalently, the sphere is covered by the caps if and only if the center of the sphere is within the convex hull defined by the points $\mathbf{X}_1, \ldots, \mathbf{X}_n$. (The convex hull defined by n points is the smallest closed convex set containing the n points.)

We now give a formula for $r_n(\tfrac{1}{2}\pi)$.

Theorem 1.5. *If n circular caps with angular radius $\theta = \tfrac{1}{2}\pi$ are distributed independently and uniformly on the surface of a k-dimen-*

sional sphere, then the probability that the sphere is completely covered by the caps equals

$$r_n\left(\tfrac{1}{2}\pi\right) \equiv 1 - 2^{-n+1}\sum_{i=0}^{k-1}\binom{n-1}{i}.$$

As the proof at the end of this section will show, Theorem 1.5 is a straightforward consequence of *Schläfli's formula* for the number of regions formed by n great circles drawn on the surface of a sphere. That formula is discussed in Appendix IV, where it is used to derive the mean content of a convex cell formed by a random collection of lines in space. The mean value is needed in Chapter 3.

We now treat circular caps whose angular radius is less than $\tfrac{1}{2}\pi$, and for simplicity return to the case of $k = 3$ dimensions. (The case $k = 2$ is treated in Exercise 1.1.) A *crossing* between two caps C_i and C_j $(i \neq j)$ is defined to be a point of intersection of the boundaries of C_i and C_j. We say that the crossing is *covered* if it is contained in a third cap C_k $(k \neq i$ or $j)$. Notice that if $\theta \leq \tfrac{1}{2}\pi$ then *the sphere is covered if and only if at least one crossing exists and all crossings are covered.* This simple property is surprisingly useful; we shall encounter it again in Chapter 3, Section 3.7. It is very often the key to distinguishing an uncovered region in computer simulations.

We wish to derive the expected number of uncovered crossings. Now, each cap C_i has two crossings with each cap C_j for which the poles $\mathbf{X}_i$ and $\mathbf{X}_j$ are separated by an angle less than 2θ. For given i, the expected number of C_j's $(j \neq i)$ fulfilling this condition is $(n-1)\rho_1$, where, using the argument leading to (1.54),

$$\begin{aligned}
\rho_1 &\equiv (4\pi)^{-1}\int_0^{2\theta} 2\pi\sin\phi\,d\phi\\
&= \tfrac{1}{2}(1 - \cos 2\theta)\\
&= \sin^2\theta\\
&= 4\sin^2\left(\tfrac{1}{2}\theta\right)\cos^2\left(\tfrac{1}{2}\theta\right)\\
&= 4\rho(1 - \rho).
\end{aligned}$$

The chance that a given crossing between caps C_i and C_j is uncovered equals $(1 - \rho)^{n-2}$. Therefore the expected number of uncovered crossings of C_i with other caps equals

$$2(n-1)4\rho(1-\rho)(1-\rho)^{n-2} = 8(n-1)\rho(1-\rho)^{n-1}.$$

Adding over i, and noting that each uncovered crossing is counted twice in this argument, we see that the expected total number of uncovered crossings is

$$m(n) \equiv 4n(n-1)\rho(1-\rho)^{n-1}. \tag{1.55}$$

Assume that n is sufficiently large for at least one crossing to exist. Let $s_i = s_i(n)$ denote the probability that the n caps leave exactly i crossings uncovered. Then

$$m(n) = \sum_{i=1}^{\infty} i s_i,$$

and $r_n = r_n(\theta) = s_0$. Therefore

$$\begin{aligned} r_n &= 1 - (1 - s_0) \\ &= 1 - m(n)/u(n), \end{aligned} \tag{1.56}$$

where

$$u(n) \equiv (1 - s_0)^{-1} \sum_{i=1}^{\infty} i s_i$$

equals the expected number of crossings given that not all crossings are covered.

It may be proved (Miles 1969) that $u(n) \to 4$ as $n \to \infty$. Heuristically, this follows from the fact that if the sphere surface is not covered then the uncovered region is a single tiny "polygon," with sides made up from boundaries of caps, and after rescaling is approximately one of the polygons formed by a random sequence of lines in the plane. The expected number of sides, and hence vertices (uncovered crossings), of such a polygon equals 4 (see Exercise 1.12). We may now deduce from formulas (1.55) and (1.56) that

$$\begin{aligned} r_n &= 1 - 4n(n-1)\rho(1-\rho)^{n-1}/u(n) \\ &= 1 - \{1 + o(1)\} n^2 \rho(1-\rho)^{n-1} \end{aligned}$$

as $n \to \infty$.

If the sphere surface is not covered then at least three crossings must be uncovered. Therefore $u(n) \geq 3$, and so by (1.56),

$$r_n \geq 1 - m(n)/3 = 1 - \tfrac{4}{3}n(n-1)\rho(1-\rho)^{n-1}.$$

We pause to combine these results into a theorem.

Theorem 1.6. *Suppose n circular caps with angular radius $\theta \leq \frac{1}{2}\pi$ are distributed independently and uniformly on the surface of* *a three-dimensional sphere. Let $\rho \equiv \sin^2(\theta/2)$ be the fraction of the surface covered by a single cap. The probability $r_n = r_n(\theta)$ that the surface is completely covered by the caps satisfies*

$$r_n \geq 1 - \tfrac{4}{3}n(n-1)\rho(1-\rho)^{n-1}$$

and

$$r_n = 1 - \{1 + o(1)\}\, n^2 \rho(1-\rho)^{n-1}$$

as $n \to \infty$.

We close this section with a proof of Theorem 1.5.

Proof of Theorem 1.5. Remember that the event that the sphere is *not* covered by the n circular caps can be interpreted as the event that the n points $\mathbf{X}_1, \ldots, \mathbf{X}_n$ all lie in the same hemisphere. Let p_n denote the probability of this event. Suppose the sphere is centered at the origin, and let $\mathcal{L}_1, \ldots, \mathcal{L}_n$ be n distinct lines passing through the origin. Each line $\mathcal{L}_i$ intersects the surface of the sphere in two points, $\mathbf{x}_i^{(1)}$ and $\mathbf{x}_i^{(2)}$ say. Put

$$q_n \equiv q_n(\mathcal{L}_1, \ldots, \mathcal{L}_n)$$
$$= P\left(\mathbf{X}_1, \ldots, \mathbf{X}_n \text{ all in same hemisphere} \,\Big|\, \mathbf{X}_i = \mathbf{x}_i^{(1)} \text{ or } \mathbf{x}_i^{(2)},\right.$$
$$\left. 1 \leq i \leq n\right).$$

We shall prove that if each subset of size k of $\mathcal{L}_1, \ldots, \mathcal{L}_n$ is linearly independent, and if for each collection $j_1, \ldots, j_n$ of 1's and 2's,

$$P(\mathbf{X}_i = \mathbf{x}_i^{(j_i)}, 1 \leq i \leq n \mid \mathbf{X}_i = \mathbf{x}_i^{(1)} \text{ or } \mathbf{x}_i^{(2)}, 1 \leq i \leq n) = 2^{-n}, \quad (1.57)$$

then

$$q_n = 2^{-n+1} \sum_{i=0}^{k-1} \binom{n-1}{i}. \qquad (1.58)$$

Clearly (1.57) holds almost surely under the hypothesis that $\mathbf{X}_1, \ldots, \mathbf{X}_n$ are independently and uniformly distributed over the surface of the sphere, and so by (1.58), taking expectation E over independently and uniformly distributed orientations of lines $\mathcal{L}_1, \ldots, \mathcal{L}_n$, we obtain

$$p_n = E(q_n) = 2^{-n+1} \sum_{i=0}^{k-1} \binom{n-1}{i},$$

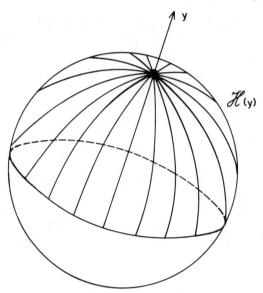

Figure 1.13. Hemisphere $\mathcal{H}(\mathbf{y})$ with outward-pointing normal $\mathbf{y}$.

as had to be shown. Indeed, this method of proof establishes Theorem 1.5 under more general conditions, assuming only that the joint distribution of $\mathbf{X}_1,\ldots,\mathbf{X}_n$ is invariant under reflections through the origin and is such that with probability 1 all subsets of size k of $\{\mathbf{X}_1,\ldots,\mathbf{X}_n\}$ are linearly independent.

Let $\mathbf{y}$ be any vector in $\mathbb{R}^k$ with length equal to the radius of the sphere, and let $\mathcal{H}(\mathbf{y})$ denote that hemisphere whose outward-pointing normal is in the direction of $\mathbf{y}$ (see Figure 1.13). Put

$$s_i(\mathbf{y}) \equiv \begin{cases} 1 & \text{if } \mathbf{X}_i \in \mathcal{H}(\mathbf{y}) \\ 0 & \text{otherwise,} \end{cases}$$

and $\mathbf{s}(\mathbf{y}) \equiv (s_1(\mathbf{y}),\ldots,s_n(\mathbf{y}))$. If $\mathbf{s}_0$ is an n-vector of 0's and 1's, we say that "$\mathbf{s}_0$ occurs" if for *some* $\mathbf{y}$, $\mathbf{s}(\mathbf{y}) = \mathbf{s}_0$. Notice that many different $\mathbf{s}_0$'s may occur simultaneously. Furthermore, q_n equals the conditional probability that $(1,1,1,\ldots,1)$ occurs. By symmetry, the conditional probability that any given $\mathbf{s}_0$ occurs does not depend on $\mathbf{s}_0$. (Any $\mathbf{s}_0$ can be changed into any other by reflecting the appropriate $\mathbf{X}_i$'s through the origin.) There are 2^n different $\mathbf{s}_0$'s. Therefore

$$2^n q_n = E'(N), \tag{1.59}$$

where N equals the total number of s_0's that occur and E' denotes expectation conditional on $\mathbf{X}_i = \mathbf{x}_i^{(1)}$ or $\mathbf{x}_i^{(2)}$, $1 \le i \le n$.

Ostensibly N is a random variable, but in fact it is constant with probability 1. To see this, let Π_i denote the plane through the origin perpendicular to $\mathcal{L}_i$. Then N equals the number of regions formed on the surface of the sphere by the Π_i's. [The intersection of each Π_i with the sphere surface is a great circle, and N equals the number of regions formed by the pattern of great circles. Each region is composed of the set of all vectors $\mathbf{y}$ for which $\mathbf{s}(\mathbf{y})$ has a fixed value.] Schläfli's formula [proved in Appendix IV, part (ii)] demonstrates that

$$N = 2 \sum_{i=0}^{k-1} \binom{n-1}{i}, \qquad (1.60)$$

a constant, provided only that all k-subsets of $\mathcal{L}_1, \dots, \mathcal{L}_n$ are linearly independent. The desired result (1.58) follows from (1.59) and (1.60). $\square$

EXERCISES

1.1 Let $\mathcal{J}$ denote the unit interval $[0,1]$, and place $n-1$ points independently and at random (that is, with the uniform distribution) into $\mathcal{J}$. Together these points divide $\mathcal{J}$ into n consecutive intervals $\mathcal{J}_1, \dots, \mathcal{J}_n$.

(i) Show that for any sequence of events $\mathcal{E}_1, \dots, \mathcal{E}_n$,

$$P(\mathcal{E}_1 \cup \dots \cup \mathcal{E}_n) = \sum_{i_1=1}^{n} P(\mathcal{E}_1) - \sum\sum_{1 \le i_1 < i_2 \le n} P(\mathcal{E}_{i_1} \cap \mathcal{E}_{i_2})$$
$$+ \sum\sum\sum_{1 \le i_1 < i_2 < i_3 \le n} P(\mathcal{E}_{i_1} \cap \mathcal{E}_{i_2} \cap \mathcal{E}_{i_3})$$
$$- \dots + (-1)^{n+1} P(\mathcal{E}_1 \cap \dots \cap \mathcal{E}_n).$$

This is a well-known formula of elementary probability theory. [Hint: Write $I(\mathcal{E}_1 \cup \dots \cup \mathcal{E}_n)$ as $1 - \Pi_{1 \le i \le n}(1 - I_i)$, where $I_i \equiv I(\mathcal{E}_i)$.]

(ii) Let l_i denote the length of interval $\mathcal{J}_i$, and take $\mathcal{E}_i$ to be the event that $l_i > a$, where $0 < a < 1$. Show that

$$P(\mathcal{E}_{i_1} \cap \dots \cap \mathcal{E}_{i_m}) = P(\mathcal{E}_1 \cap \dots \cap \mathcal{E}_m)$$

whenever $1 \leq m \leq n$ and $1 \leq i_1 < \ldots < i_m \leq n$.

(iii) With $\mathcal{E}_1, \ldots, \mathcal{E}_n$ as in (ii) above, show that

$$P(\mathcal{E}_1 \cap \ldots \cap \mathcal{E}_m) = (1 - ma)_+^{n-1},$$

where $x_+ = x$ if $x \geq 0$; $x_+ = 0$ if $x < 0$. [Hint: Use induction over m.]

(iv) Combining the results above, derive a formula for the probability that none of the intervals $\mathcal{J}_i$ is longer than a. Hence show that if n arcs, each of length $a > n^{-1}$, are placed independently and at random on a circle of unit circumference, then the chance that the circle is completely covered by the arcs equals

$$\sum_{i=0}^{n} (-1)^i \binom{n}{i} (1 - ia)_+^{n-1}.$$

1.2   (Here we extend some of the work in Section 1.2 to "oval" curves.) Let f be periodic of period 2π, having four bounded derivatives on $(-\infty, \infty)$ and satisfying $f > 0$ and $f + f'' > 0$. Let Γ_r denote the set of points (x, y) defined by

$$x = x(\theta) = r\{f(\theta)\cos\theta - f'(\theta)\sin\theta\},$$
$$y = y(\theta) = r\{f(\theta)\sin\theta + f'(\theta)\cos\theta\},$$

for $0 \leq \theta < 2\pi$. Then Γ_r is a bounded closed convex curve with no "flat bits." In the special case $f(\theta) \equiv 1$, Γ_r is the circle of unit radius centered at the origin. In general, let $\mathcal{S}_r$ denote the set of all points interior to Γ_r or on Γ_r; then $(0, 0) \in \mathcal{S}_r$. Put

$$I_r(u, v) \equiv \begin{cases} 1 & \text{if } (u, v) \in \mathcal{S}_r \\ 0 & \text{otherwise,} \end{cases}$$

$$N(r; u, v) \equiv \sum_j \sum_k I_r(j - u, k - v)$$

and $N(r) \equiv N(r; U, V,)$, where U and V are independent and uniformly distributed on the interval $(0, 1)$. Then $N(r)$ has the distribution of the number of lattice points contained in a randomly centered copy of Γ_r, with orientation identical to that of Γ_r. Define

$$a_{mn} \equiv \int_0^1 \int_0^1 N(r; u,v) \exp\left\{2\pi i(mu+nv)\right\} du\, dv.$$

(i) Arguing as in Section 1.2, show that

$$E\{N(r)\} = a_{00} = \alpha r^2,$$

where α denotes the area of S_1. Prove that for general (m,n),

$$a_{mn} = r^2 \int\int_{S_1} \exp\left\{2\pi i r(mu+nv)\right\} du\, dv,$$

and that

$$\mathrm{var}\{N(r)\} = \sum_m \sum_n {}' |a_{mn}|^2,$$

where $\sum_m \sum_n {}'$ denotes summation over all pairs (m,n) with $(m,n) \neq (0,0)$.

(ii) Show that

$$|a_{mn}| \leq \mathrm{const.}\, r^{1/2}(m^2+n^2)^{-3/4}$$

uniformly in $r > 1$ and $(m,n) \neq (0,0)$, and hence that $\mathrm{var}\{N(r)\} = O(r)$ as $r \to \infty$.

[Kendall (1948)]

1.3 Let $\mathcal{L}$ be a regular triangular, square, or hexagonal lattice in the plane, and mark each vertex of $\mathcal{L}$ with probability p independently of all other vertices. Delete from the lattice any edge that does not join two marked vertices. Define the "center" vertex of each finite-order clump in some unique way, for example, by using the lowest right-hand vertex notation of Section 1.3. Let N equal the total number of vertices in the clump containing a given vertex P (such as the origin in a square lattice), with $N=0$ if P is not marked. Define p^* to equal the chance that P is the center of some clump of finite order. Show that

$$p^* = E\left\{N^{-1}I(N \geq 1)\right\}.$$

[Hint: Consider the pattern of marked vertices within an increasing sequence of rectangles.]

[Grimmett (1976)]

1.4 Verify the row corresponding to $n = 5$ in Table 1.1.

1.5 Let t be the total intensity function of a random function model on the integer square lattice, as defined in Section 1.4. We may interpret $t(\xi)$ as the "true" intensity at pixel ξ in a digitized photographic image. Recording of the image may result in aberrations of at least two types—blurring, due to lens aberrations or image movement; and noise, due to film grain, etc. Blur may be represented by a *point-spread function* h, which we take to be nonnegative and to satisfy

$$\sum_{\xi \in \mathbb{Z}^2} h(\xi) = 1.$$

The blurred version of t is denoted by b, where

$$b(\xi) \equiv \sum_{\eta \in \mathbb{Z}^2} h(\eta) t(\xi + \eta), \quad \xi \in \mathbb{Z}^2.$$

To the blurred image we add an amount of noise, obtaining the *real image* r given by

$$r(\xi) \equiv b(\xi) + N(\xi),$$

where $\{N(\xi), \xi \in \mathbb{Z}^2\}$ is a sequence of independent and identically distributed random variables, independent of the stochastic process t.

(i) Show that the squared-error fidelity of b to t admits the formula

$$E\{b(\xi) - t(\xi)\}^2$$
$$= (2\pi)^{-2} \int_{(-\pi,\pi)^2} \left\{ |\chi(\boldsymbol{\theta})|^2 - 2\chi(-\boldsymbol{\theta}) + 1 \right\} \alpha(\boldsymbol{\theta}) \, d\boldsymbol{\theta},$$

where

$$\chi(\boldsymbol{\theta}) \equiv \sum_{(\xi_1,\xi_2) \in \mathbb{Z}^2} \exp\{i(\theta_1 \xi_1 + \theta_2 \xi_2)\} h(\xi_1, \xi_2)$$

is the Fourier transform of h, and

$$\alpha(\boldsymbol{\theta}) \equiv \sum_{(\xi_1,\xi_2) \in \mathbb{Z}^2} \exp\{i(\theta_1 \xi_1 + \theta_2 \xi_2)\} a(\xi_1, \xi_2)$$

is the Fourier transform of the covariance function a of t.

(ii) Write down the analogue of this formula if b is replaced by the real image r.

1.6 Prove that the content of the lens of intersection of two k-dimensional unit spheres centered $2x$ apart equals

$$B(x) \equiv 2\pi^{(k-1)/2}\Gamma(\tfrac{1}{2} + \tfrac{1}{2}k)^{-1} \int_x^1 (1 - y^2)^{(k-1)/2}\,dy$$

if $0 \le x \le 1$. Hence show that the content of a single k-dimensional unit sphere equals $\pi^{k/2}/\Gamma(1 + \tfrac{1}{2}k)$. In the cases $k = 2$ and 3, evaluate the integral and thereby provide simple, explicit formulas for $B(x)$.

1.7 Show that the large-region formulas for $E(V)$ and $\mathrm{var}(V)$ stated in Theorem 1.1 remain true *without* the provisor that vacancy is interpreted using the torus convention. [Hint: Go back to the original formulas for $E(V)$ and $\mathrm{var}(V)$ in terms of an integral and a double integral (respectively) over $\mathcal{R}$, and take a little extra care in working out the integrals at points distant one unit or less from the boundary of $\mathcal{R}$. The contributions of these parts of the integrals are asymptotically negligible.]

1.8 Let $\mathcal{P} \equiv \{\boldsymbol{\xi}_i, i \ge 1\}$ be a stationary Poisson process in $\mathbb{R}^k$, with intensity λ, and let $\{\boldsymbol{\eta}_i, i \ge 1\}$ be a sequence of independent and identically distributed random k-vectors, independent also of $\mathcal{P}$. Prove that $\mathcal{P}' \equiv \{\boldsymbol{\xi}_i + \boldsymbol{\eta}_i, i \ge 1\}$ is another stationary Poisson process in $\mathbb{R}^k$ with intensity λ. [Hint: Use axioms (i) and (ii) in Section 1.7.]

1.9 Let $\mathcal{P}$ be as above, and let $\{M_i, i \ge 1\}$ be independently and identically distributed random variables taking only a countable number of distinct values $m_1, m_2, \ldots$:

$$P(M_i = m_j) = p_j, \qquad i \ge 1 \text{ and } j \ge 1,$$

and

$$\sum_{j=1}^{\infty} p_j = 1.$$

Define $\mathcal{P}_j$ to be the point process

$$\mathcal{P}_j \equiv \{\boldsymbol{\xi}_i \in \mathcal{P} : M_i = m_j\}, \qquad j \ge 1.$$

Show that the point processes $\mathcal{P}_j$ are independent, stationary, and Poisson, and that the jth has intensity λp_j. [Hint: Use the axioms again. Independence means that for any finite collection of the $\mathcal{P}_j$'s, the numbers of points from the $\mathcal{P}_j$'s in respective subsets of $\mathbb{R}^k$ are independent random variables.]

1.10 Prove that if $\{\xi_i, i \geq 1\}$ is a stationary Poisson process on $(-\infty, \infty)$ with intensity λ, and g is an integrable function, then

$$E\left\{\sum_{i=1}^{\infty} g(\xi_i)\right\} = \lambda \int_{-\infty}^{\infty} g(x)\, dx$$

and

$$E\left[\prod_{i=1}^{\infty} \{1 + g(\xi_i)\}\right] = \exp\left\{\lambda \int_{-\infty}^{\infty} g(x)\, dx\right\}.$$

If g is nonnegative, these results extend to the case where $\int g = \infty$. (The first of these identities is a simple Campbell theorem; see Chapter 4, Section 4.1.) [Hint: Approximate $\{\xi_i\}$ by a sequence of $2n$ points independently and uniformly distributed on the interval $(-n\lambda^{-1}, n\lambda^{-1})$, and let $n \to \infty$. Alternatively, approximate g by a function that vanishes outside a compact interval, and condition on there being just N points of the Poisson process within that interval.]

1.11 Let $\mathcal{C}$ be a Boolean model of random-radius disks in the plane $\mathbb{R}^2$, and $\mathcal{L}$ be a line (a copy of $\mathbb{R}$) in the plane. Find the intensity and the distribution of segment length of the Boolean model of line segments induced on $\mathcal{C}$ by the section $\mathcal{L}$.

1.12 Suppose n great circles are drawn independently and at random (i.e., uniformly) on the surface of a three-dimensional sphere.

    (i) Prove by induction over n that with probability 1, these great circles divide the surface of the sphere into $n(n-1)+2$ regions, and divide themselves into $2n(n-1)$ arcs. (Here, an arc is a segment of a great circle.)

    (ii) Show that the expected number of sides of each quasi-polygonal region formed on the surface of the sphere by the great circles is

$$\frac{4n(n-1)}{n(n-1)+2}.$$

(iii) By letting $n \to \infty$, show that the expected number of sides of a random polygon formed by a random, uniformly distributed sequence of lines in the plane, equals 4. (Technically, such a collection of lines is called a *Poisson field*; see Chapter 3, Section 3.5. See Appendix IV for further arguments of the type considered in this Exercise.)

1.13 Points are placed at random on the surface of a k-dimensional sphere until they form the vertices of an inscribed polyhedron containing the center of the sphere. Prove that the expected number of faces of the polyhedron equals $2k + 1$. Find the variance of the number of faces. [Hint: Let N equal the number of points placed on the sphere surface. Then $P(N \le n)$ is given by Theorem 1.5.]

[After the contribution of F. G. Schmitt to Silverman (1964).]

NOTES

Section 1.2
The work in Section 1.2 is taken from Kendall (1948). For related results on covering a lattice by random sets, including results for hexagonal lattices, see Hlawka (1950b), Moran (1950), Kendall and Rankin (1953), de Bruin (1965), Russell and Josephson (1965), and Gács and Szász (1975). Similar problems connected with measurement of length and area are discussed by Moran (1966b, 1968a). An excellent account of developments in number theory concerning Hardy's conjecture may be found in the introduction to Kendall (1948). See also Dickson (1952, Chapter VI), Shanks (1962, p. 163ff), and Gelfond and Linnik (1965, Chapter 8).

Sections 1.3 and 1.4
Section 1.3 is based on Roach (1968). The same techniques were used earlier by other authors to obtain approximations to clumping characteristics on various lattices. For example, Domb and Sykes (1961) used enumeration arguments to approximate expected order of the clump containing an arbitrary vertex (different from ν^*, which is expected order of an arbitrary clump). They treated square, triangular, hexagonal (all two-dimensional) and cubic (three-dimensional), lattices. See also Elliott, Heap, Morgan and Rushbrooke (1960).

The technique is common in studies of lattice percolation; see for example Domb (1972) and the references therein. Ahuja (1981a) has surveyed results of this type, and calculated (among other characteristics) measures of dispersedness of random clumps. Note, however, that Ahuja's triangular lattice is our hexagonal lattice, and vice versa. Our notation for lattice type coincides with that used in percolation; see Appendix VI at the end of this monograph and Kesten (1982, pp. 14–15). The reason for the confusion is that our triangular lattice, while it has triangular faces, has six (not three) edges radiating from each vertex; our hexagonal lattice has hexagonal faces but three (not six) edges radiating from each vertex.

The lattice process discussed in Section 1.3 is a discrete version of the Boolean model, to which we devote much of our attention in this monograph. The discrete Boolean model is a simple theoretical model for images, and so finds application in the theory of image processing. Ahuja and Rosenfeld (1982) survey the class of image models, and Ahuja (1981a, 1981b, 1981c) develops theory specifically for discrete Boolean models. See also Ryle and Scheuer (1960), Abend, Harley, and Kanal (1965), Schacter and Ahuja (1979), Ahuja and Dubitzki (1980), Miles (1980), Modestino, Fries, and Vickers (1980), Serra (1980), and Ahuja and Rosenfeld (1981). The random function models outlined in Section 1.4 are special cases of shot noise models, details of which may be found in Daley (1971), Westcott (1976), Rice (1977), and Schmidt (1984, 1985a, 1985b). Random function models have been used by Hall and Titterington (1986) and Hall (1987) to model images. There they were called "Boolean images." See also Feron (1972, 1976a, 1976b).

Section 1.5
There is a very large literature on covering the line by random segments or the circle by random arcs, dating back to the last century; see the notes below on Exercise 1.1. Some problems of this type are included in the case $k = 1$ studied in Section 1.5. As a rule, both the problems and the techniques of solution when $k = 1$ and when the number of segments is nonrandom are related to the statistical theory of spacings of order statistics; see Pyke (1965, 1972) and Holst (1979) for examples of the latter. Deheuvels (1983) and Janson (1984), for example, discuss multidimensional spacings. Amenable characteristics of segment or arc processes include numbers and lengths of spacings (i.e., of gaps between segments or arcs) and vacancy. By adjusting the numbers and lengths of segments or arcs one may model a wide variety of situations, from high to low inten-

sity of sets per unit length of the line or circumference. Vacancy in this range of circumstances will be discussed in Chapter 3, for the case of Poisson-distributed random sets in $\mathbb{R}^k$. Other aspects of the case $k = 1$ will be treated in Chapter 2.

Work on the distribution of a fixed (as opposed to random) number of sets on the line may be found in, for example, Lévy (1939), Darling (1953), Le Cam (1958), Mackenzie (1962), Klamkin (1966), Steutel (1967), Glaz and Naus (1979), and Holst (1980b). References concerning the distribution of a fixed number of arcs on the circle include Stevens (1939), Flatto and Konheim (1962), Flatto (1973), Edens (1975), Rao (1976), Kaplan (1978), Siegel (1978, 1979a), Solomon (1978, p. 75), Holst (1980a, 1980b, 1981, 1983), Hüsler (1982), Siegel and Holst (1982), Janson (1983a), and Holst and Hüsler (1984). Related work concerning sample spacings, the scan statistic, and goodness-of-fit testing may be found in Weiss (1959), Pyke (1965), Cressie (1976, 1979), del Pino (1979), and references therein.

In the case of $k = 3$ dimensions, some of the results in Section 1.5 are available from Moran (1973a, 1973b). Moran's proof of the central limit theorem for vacancy is different from ours; Moran uses techniques of Rényi (1962) concerning sums of conditionally independent random variables. See also Ailam (1966, 1968, 1970) and Stam (1981). The random, nonuniform distribution of spheres in space, in $k \geq 1$ dimensions, has been discussed by Hall (1984a, 1984b, 1985a).

Section 1.6
Part of the work on ballistics problems may be traced to attempts during and shortly after World War II to model the damage caused by bombs (e.g., Garwood 1947; Solomon 1950, 1953; Miser 1951; Rand Corporation 1952; Dishington 1956). The published literature alone runs to hundreds of papers, and cannot be treated in any depth in this monograph. Surveys of the literature include Guenther and Terragno (1964), Eckler (1969), and Eckler and Burr (1972). Section 1.6 gives a taste of the type of coverage problem considered in ballistics. Solutions are generally distinguished by exact numerical results for concise models; for example, Eckler (1969) lists many sources of data on kill probabilities and expected destroyed proportions. This "engineering approach" to coverage problems is in contradistinction to many mathematical and statistical accounts, where emphasis is often on a more qualitative description, frequently using some sort of limit theory in place of exact numerical solution.

These two approaches come closer together in certain treatments of coverage problems (e.g., Groves and Smith 1957; Morgenthaler 1961).

Schroeter (1982, 1984) provides a clear introduction to moment problems connected with ballistic coverage, similar in some respects to the outline given in Section 1.6. He also surveys some of the more recent literature.

The problem of describing the coverage of an increasingly large number of disks or spheres, mentioned briefly at the end of Section 1.6, is related to properties of the convex hull of a multivariate sample, since the latter is connected with the maximum content that the spheres can cover. Moran (1974, Section 7) discusses this relationship in more detail. See also Daniels (1952), Rényi and Sulanke (1963, 1964), Efron (1965), Ling (1971), Eddy (1980), Eddy and Gale (1981), and Jewell and Romano (1982).

Section 1.7
Cox and Miller (1965, pp. 146–153), Karlin and Taylor (1975, pp. 167–175), Snyder (1975, Chapter 2), Cox and Isham (1980, pp. 45–49 and 145–147), Stoyan, Kendall, and Mecke (1987, Section 2.4), and Daley and Vere-Jones (1988, Chapter 2) give detailed discussions of Poisson processes, usually with emphasis to processes on the real line. Cox and Isham (1980) provide a particularly accessible, relatively nontheoretical account of a wide variety of point processes, stationary and otherwise. Daley and Milne (1973) give a point process bibliography.

Section 1.8
Hanisch (1981) was one of the first to discuss coverage processes in great generality and with rigor. However, early accounts of multiphase models, such as those mentioned in Section 1.8, date back to Watt (1947), Kershaw (1957), Matérn (1960), Pielou (1961, 1964, 1965, 1967), Bartlett (1964), and Switzer (1965). Sources for discrete Boolean models are given in the notes to Section 1.3. Diggle (1981) and Serra (1982, Chapters XIII, XIV), for example, fit Boolean models to multiphase data. We shall have much more to say about Boolean models and their properties in Chapter 3, Section 3.1.

Section 1.9
The stereological literature is rich and vast. Monograph-length treatments include Elias (1967), De Hoff and Rhines (1968), Under-

wood (1970), Coleman (1979), Weibel (1979, 1980), Ambartzumian (1982), and Saxl (1986). Recent surveys are by Baddeley (1977), Coleman (1981), Saxl (1981), Jensen, Baddeley, Gundersen, and Sundberg (1985), and Stoyan, Kendall, and Mecke (1987, Chapter 11). Sources describing line transect samples, relevant to the very simple case treated in Section 1.9, include McIntyre (1953), Lucas and Seber (1977), Eberhardt (1978), Stoyan (1979a), and McDonald (1980). The "folding" and "unfolding" formulas for spherical particles given in Section 1.9 date back to Wicksell (1925, 1926). See also Nicholson (1970), Tallis (1970), Watson (1971, 1975), Moran (1972), Ambartzumian (1982, Chapters 9 and 10), Cruz-Orive (1983), Jensen (1984), and Blödner, Mühlig, and Nagel (1984). The latter paper comprises a simulation study of various solutions to Wicksell's problem, and is a source for recent literature.

Wicksell's problem is an example of an *ill-posed problem*, that is, one where small measurement errors can lead to large discrepancies in the value of a "final solution," such as a parameter estimate. Many stereological problems are intrinsically of this form; it is a characteristic of trying to obtain information about k-dimensional properties from data in $k - 1$ or fewer dimensions.

Most stereological formulas involve the concept of random, uniform distribution of particles ("grains," or random sets) in some sense, although not necessarily the full force of the Boolean model hypothesis. In connection with Wicksell's problem, the Poisson assumption about the driving point process is not essential in deriving the integral equation (1.46); see Mecke and Stoyan (1980b). Nevertheless, Boolean models are common in stereology, as is the assumption that the random sets be convex. See the notes to Chapter 4.

Section 1.10
Problems concerning the packing of solid sets into space (Section 1.10) have long intrigued mathematicians. The closest packing (i.e., closest lattice packing) densities for disks in two dimensions and spheres in three dimensions date back to Lagrange and Gauss. Rogers (1964, p. 3) and Thompson (1983, Appendix 1) list these densities and their higher dimensional counterparts, and discuss many aspects of deterministic packing. See also Fejes Tóth (1953). Theoretical investigation of complete random packing has its genesis in Rényi's (1958) now classic paper, from which our Theorem 1.4 is taken. A reprint of that article in Rényi's collected works (Rényi 1976, pp. 173–178) concludes with an interesting bibliographic note. Solomon and Weiner (1986) give an interesting review of the

packing problem, with particular emphasis on Rényi's contribution and the Palásti conjecture.

Rényi's work quickly inspired a number of generalizations and extensions, including Palásti (1960), Bánkövi (1962), Mackenzie (1962), Ney (1962), Dvoretzky and Robbins (1964), Mannion (1964, 1976, 1979, 1983), Widom (1966), Mullooly (1968), and Itoh (1980). Lattice versions of the problem have been considered by Jackson and Montroll (1958), Page (1959), Downton (1961), Mackenzie (1962), Widom (1966), and Runneberg (1982), for example. Reviews and important numerical work appear in Solomon (1967) and Blaisdell and Solomon (1970). The former includes discussion of a one-dimensional system for "shaking" line segments, increasing packing density from 0.7476 to 0.8087. Common applications of packing models are in physics and chemistry (e.g., Bernal 1959, 1960; Bernal and Mason 1960); other applications are to linguistics (Dolby and Solomon 1975) and election theory (Itoh and Ueda 1979).

A large literature has grown up around Palásti's conjecture, with several lively contributions. Blaisdell and Solomon (1970, 1982) give good summaries of the debate and its implications. See also Palásti (1976). Palásti's conjecture is *relatively* easy to check (or "disprove") by simulation, since computation of distances between parallel-sided figures is straightforward. Jodrey and Tory (1980) and Blaisdell and Solomon (1982) investigate the conjecture as far as $k = 4$ dimensions, by simulation. Solomon (1967) is a key source for discussion and early simulation of a variety of packing models.

Simulations of complete random packing models involving disks or spheres exist in the literature, but are time consuming to conduct and fewer in number than those involving squares or cubes (i.e., concerning Palásti's conjecture). Lotwick (1982) simulates complete random packing of disks in $k = 2$ dimensions, and compares his results with those of Tanemura (1979). The aim of Lotwick's work is not so much the determination of packing density, but the simulation of packing models having given packing density. Such processes form the basis for "simple sequential inhibition" point processes (Diggle, Besag, and Gleaves 1976), and have applications in Monte Carlo studies of statistical point patterns. Inhibition point processes are sometimes called hard core point processes. References to the theory and applications of such processes, and a discussion of their approximations, may be found in Westcott (1982). Lotwick (1982) suggests an algorithm for computer simulation of a hard core point process model due to Kelly and Ripley (1976). See Woodcock (1976) and references in Solomon (1967) for examples of the application of complete packing densities to problems in physical chemistry.

Simulation studies suggest that for spheres in $\mathbb{R}^3$, the loose packing density is 0.60 ± 0.02 and the close packing density is 0.6366 ± 0.0004 (e.g., Scott and Kilgour 1969; Finney 1970). A theoretical argument (Gotoh and Finney 1974) based on geometric assumptions, specifically that dense random packings can be thought of as tetrahedral aggregates of spheres, produces densities of 0.6099 and 0.6357 for loose packing and close packing, respectively. Randomness enters the argument in the sense that calculations are based on the "most probable" tetrahedral aggregate. Several authors (e.g., Gamba 1975; Finney and Gotoh 1975; Mackay 1980) have remarked on the nearness of the close packing density to $2/\pi \simeq 0.63662$.

An alternative theoretical approach (Berryman 1983), which makes fewer assumptions about detailed structure of the geometry of the packing, indicates a close packing density of 0.64 ± 0.02 for close packing. When Berryman's argument is applied to disks in $\mathbb{R}^2$, it suggests a density of 0.82 ± 0.02, in good agreement with conclusions of simulation studies (e.g., Stillinger, DiMarzio and Kornegay 1964; Quickenden and Tan 1974).

Visscher and Bolsterli (1972) describe a computer program for numerical simulation of dense packing of spheres and disks, although it does not produce the densities obtained by mechanical simulation. Closer results are obtained by Matheson (1974). See also Adams and Matheson (1972). Tables I and II of Berryman (1983) provide very convenient surveys of simulation results in both the two- and three-dimensional cases. See also Gilbert (1964) and Solomon (1967). Duer, Greenstein, Oglesby, and Millero (1977) survey the literature on three-dimensional packing. Cowan (1984) discusses a "local" version of random dense packing in two dimensions, and indicates further literature sources.

Girling (1982) uses a physical approximation from classical liquid theory (Percus and Yevick 1958) to develop an approximate formula for the variance of the number of three-dimensional spheres packed within a given region, for a given density of spheres per unit volume.

Section 1.11

Theorem 1.5 is due to Wendel (1962), and Theorem 1.6 to Gilbert (1965) and Miles (1969). The simple and elegant crossing-based argument, and indeed much of the development in Section 1.11, come from Gilbert (1965). Miles (1969) gives generalizations of Theorem 1.6, for example to higher dimensions. See too Moran and Fazekas de St. Groth (1962). Solomon and Sutton (1986) use an extensive simulation study to compare approximations to the probability of covering a sphere by circular caps. Related problems are discussed in Chapter 3, Section 3.6, and further references are given in the notes to that section.

Miscellaneous

Exercise 1.1 constructs a solution to one of the oldest and best-known problems connected with probability of complete coverage: the chance of covering a circle by n equal-length arcs placed at random on the circumference. The earliest solution is by Whitworth (1897, 1901). See also Fisher (1929, 1940), Baticle (1933a, 1933b, 1935), Lévy (1939), Stevens (1939), Davis (1941), Garwood (1960), and Votaw (1946). Glaz and Naus (1979) derive a related theorem for multiple coverage, and survey related literature.

Grimmett (1976), Wierman (1978), and Kesten (1982, p. 239ff) discuss problems related to Exercise 1.3.

We have omitted mentioning several topics in the theory of coverage processes, not because they are unimportant but because they do not fit easily into our line of development. One such topic is that of "sequential" coverage of a circle by arcs placed upon it one-by-one. Suppose the circle is of unit circumference, and that arcs $A_1, A_2, \ldots$ of lengths $l_1, l_2, \ldots$ have their centers distributed independently and at random on the circumference. A necessary and sufficient condition for $\bigcup_{n \geq 1} A_n$ to equal almost all (i.e., all but a set of measure zero) of the circumference with probability 1 is $\sum_{n \geq 1} l_n = \infty$ (Dvoretzky 1956). If $l_1 \geq l_2 \geq \ldots$, then a necessary and sufficient condition for $\bigcup_{n \geq 1} A_n$ to equal *all* of the circumference with probability 1 is

$$\sum_{n=1}^{\infty} n^{-2} \exp \left(\sum_{i=1}^{n} l_i \right) = \infty$$

(Shepp 1972a). A version of Dvoretzky's result, in higher dimensions, appears in Chapter 3 (Theorem 3.1). Versions of Shepp's result (Shepp 1972b) are stated in Section 3.2 of Chapter 3 and in Appendix III, where related literature is discussed. See also Flatto and Konheim (1962), Hoffman-Jørgensen (1973), Wschebor (1973a, 1973b), and Cooke (1974).

Chernoff and Daly (1957), Yadin and Zacks (1982, 1985), and Zacks and Yadin (1984) have discussed "shadowing" problems, where random sets illuminated by a light cast shadows upon a region. See also Wieacker (1986).

Properties of clumps and clusters of randomly centered overlapping sets will be discussed in Chapter 4. There are many related problems of clumping and clustering; see, for example, Ambartzumian (1965, 1966, 1971), Whittle (1965), Tate (1969), Eberl and Hafner (1971), and Hafner (1972a, 1972b). Discussions of tessellations and triangulations, which are sometimes regarded as coverings of the plane, may be found in Rogers (1964), Miles (1974b), Green and Sibson (1978), Sibson (1980a, 1980b), Ripley (1981, Section 4.3), and Mecke (1984).

Aldous (1984) shows that ingenious interpretations of Boolean model properties may be used as the basis for a wide variety of heuristic proofs of results about stochastic processes.

CHAPTER TWO
Random Line Segments, Queues, and Counters

2.1. INTRODUCTION

In this chapter we examine one of the simplest examples of coverage processes—the random (i.e., Poisson) distribution of segments on a line. We show that the resulting pattern of clumps of segments can be used to model a variety of physical phenomena. Sections 2.2 and 2.3 provide expressions for the exact distribution of clump length, and Section 2.4 suggests an approximation. These results are used in Section 2.5 to derive the probability that a given interval is completely covered by segments, and in Sections 2.6 and 2.7 to discuss inference for linear coverage processes.

We begin by describing the basic process. Consider an electronic counter in which counting is governed by a charge-storage device such as a capacitor. The arrival of an event at the counter triggers discharge of the capacitor, and if the capacitor was initially fully charged then the discharged current kicks the meter over. It takes a short but nonzero time for the capacitor to fully recharge, and each time an event occurs, any charge in the capacitor is released. Therefore any event occurring between one discharge of the capacitor and the next time at which the capacitor is fully charged is not recorded on the meter.

A counting device with this mode of operation is called a Type II counter. Recharging intervals are called *dead times* or holding times or locked times. The process can be illustrated by repeatedly

79

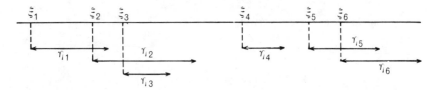

Figure 2.1. Type II counter. Events occur at times ξ_i, and dead times are of durations η_i. Only those events occurring at times ξ_1, ξ_4, and ξ_5 are recorded.

discharging an electronic flash gun, observing that the capacitor will discharge but the flash fail to fully light if the time between two successive discharges is less than the gun's recycling time.

To model this process mathematically, let $\mathcal{P} \equiv \{\xi_i\}$ denote the Poisson process of occurrence times of events, and suppose a capacitor discharging at time ξ_i takes time η_i to fully recharge. An event occurring between times ξ_i and $\xi_i + \eta_i$ goes unrecorded, and initiates its own dead time. Figure 2.1 illustrates the pattern of dead and active times.

The dead times may be thought of as line segments having independent and identically distributed lengths $\{\eta_i, i \geq 1\}$, located at respective points of the stationary Poisson process $\mathcal{P} = \{\xi_i, i \geq 1\}$ on $(-\infty, \infty)$. Of course, $\mathcal{P}$ is taken to be independent of the random variables η_i. In the description above we have placed left-hand ends of segments at points of $\mathcal{P}$, and then the coverage process is the sequence of intervals $\mathcal{C} \equiv \{(\xi_i, \xi_i + \eta_i), i \geq 1\}$. We might call $\mathcal{C}$ a simple linear Boolean model (simple because the component random sets are line segments, linear because it is on $\mathbb{R}^1$). It makes no difference to the distribution of $\mathcal{C}$ if we place segment midpoints, or right-hand endpoints, at points of $\mathcal{P}$; note the discussion of Poisson processes in Chapter 1, Section 1.7. We say that $\mathcal{P}$ *drives* $\mathcal{C}$. Those regions of the real line covered by segments represent times at which the capacitor in our electronic counter has less than maximum charge. A connected set of segments not intersected by any other segments is called a *clump*,* and gaps between successive clumps are termed *spacings*. The number of segments comprising a clump is called its *order*.

Simple linear Boolean models (called Boolean models in the remainder of this chapter) arise in many situations. For example, consider a Poisson stream of customers arriving at a service desk

*Some authors (e.g., Roach 1968) use the term "clump" to denote any connected set of segments, so that a given segment can be part of several different clumps. In our notation, any given segment is part of a unique clump.

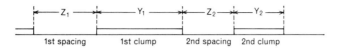

Figure 2.2. Lengths of clumps and spacings.

where there is an unlimited number of servers. Arrival times are represented by left-hand ends of segments, segment lengths represent service times, and clumps are busy periods (periods during which one or more servers are occupied). Such a process is called an $M/G/\infty$ queue—M for Markov, or Poisson, arrivals, G for general service time distribution, ∞ for infinite number of servers. Also, if we draw a line through a higher dimensional random pattern comprised of overlapping convex random sets, then the pattern of covered and uncovered regions on the line is the same as that of clumps and spacings in a random segment process; see Chapter 1, Section 1.9.

Let $Y_1, Y_2, \ldots$ and $Z_1, Z_2, \ldots$ denote successive lengths of clumps and spacings, respectively, counted from the end of an arbitrary clump, such as the first clump starting after time zero (see Figure 2.2). Assume the driving Poisson process has intensity (or rate) λ. The random variables $Y_1, Y_2, \ldots, Z_1, Z_2, \ldots$ are stochastically independent, the variables $Y_1, Y_2, \ldots$ are identically distributed, and the variables $Z_1, Z_2, \ldots$ are identically exponentially distributed with mean λ^{-1} (see Exercise 2.1). This is the prescription for a particular type of *alternating renewal process*, in which independent "lifetimes" having one distribution alternate with independent "lifetimes" having another distribution. Let N_i equal the number of segments comprising the clump whose length is Y_i. The distributions of Y_i and N_i are (defined to be) the distributions of length and order of an arbitrary clump. In general they are different from the distributions of length and order of the clump containing an arbitrary segment. Some aspects of the latter will be discussed in Chapter 4.

Let α denote mean segment length. We claim that the number M of segments that cover the origin has mean $\alpha\lambda$. To see this, let the Boolean model $\mathcal{C}$ be the sequence of intervals $\{(\xi_i, \xi_i + \eta_i), i \geq 1\}$ defined three paragraphs earlier. Notice that

$$M = \sum_{i:\xi_i < 0} I(\xi_i + \eta_i > 0),$$

so that, writing G for the distribution function of segment length, and taking expectations with respect to the η_i's first,

$$E(M) = E\left[\sum_{i:\xi_i<0} E\{I(\xi_i+\eta_i>0)\,|\,\mathcal{P}\}\right]$$

$$= E\left[\sum_{i:\xi_i<0}\{1-G(-\xi_i)\}\right]$$

$$= \lambda\int_0^\infty\{1-G(x)\}\,dx$$

$$= \alpha\lambda,$$

the second-last identity following from the fact that points of $\mathcal{P}$ are distributed with intensity λ per unit length along the negative half-line (see Exercise 1.10, Chapter 1). This is a simple Campbell theorem, and more general results of this type will be discussed in Chapter 4, Section 4.1.

Since segments are placed "at random" onto the real line, then M must have a Poisson distribution. In more detail it may be shown that for any nonnegative function g,

$$E\left[\prod_{i=1}^\infty\{1+g(\xi_i)\}\right] = \exp\left\{\lambda\int_{-\infty}^\infty g(x)\,dx\right\}.$$

(This is a property of Poisson processes; see Exercise 1.10 in Chapter 1.) Using the representation for M above, writing G for the distribution function of segment length, setting $g(x) \equiv (e^\theta-1)\{1-G(-x)\}$ for $x<0$ and $g(x) \equiv 0$ otherwise, and taking expectations with respect to the η_i's first, we obtain

$$E(e^{\theta M}) = E\left[\prod_{\xi_i<0}\exp\{\theta I(\eta_i>-\xi_i)\}\right]$$

$$= E\left(\prod_{\xi_i<0}[1+(e^\theta-1)\{1-G(-\xi_i)\}]\right)$$

$$= E\left[\prod_{i=1}^\infty\{1+g(\xi_i)\}\right]$$

$$= \exp\left\{\alpha\lambda(e^\theta-1)\right\}.$$

The latter is the moment generating function of a Poisson random variable with mean $\alpha\lambda$. Therefore

$$P(M=m) = (\alpha\lambda)^m e^{-\alpha\lambda}/m!.$$

In particular, the probability that the origin is uncovered equals $P(M = 0) = e^{-\alpha\lambda}$. Of course, stationarity of the process $\mathcal{C}$ implies that statements made here and in the previous paragraph apply to segments covering a general point, not necessarily the origin. In particular,

$$P(x \text{ not covered}) = e^{-\alpha\lambda}, \qquad (2.1)$$

for $-\infty < x < \infty$. There are several other ways of deriving this identity; see Chapter 3, Section 3.2, and Chapter 4, Section 4.1.

Suppose we observe the pattern of clumps and spacings over an interval $[0, t]$. In principle, any unknown parameters that determine the distribution of segment length, and the intensity of the driving Poisson process, can be estimated by combining information from several different sources:

(i) total number of clumps in $[0, t]$,
(ii) length of each clump,
(iii) total number of spacings in $[0, t]$,
(iv) length of each spacing.

So as to use information about clumps to the fullest possible extent, we should determine the distribution of clump length. This is most easily done in the case where segment length is constant, and so we start there.

2.2. DISTRIBUTION OF CLUMP LENGTH IN THE CASE OF FIXED SEGMENT LENGTH

Assume that segment length is fixed and equal to a. We begin by deriving the density, f, of the distribution of clump length, given that the clump contains two or more segments. We may suppose that each segment is constructed to the *right* of points of the Poisson process $\mathcal{P}$, whose intensity is constant and equal to λ. Let N denote the number of segments in an arbitrary clump (say, the first clump starting after the origin). Then $N = 1$ if and only if no points of $\mathcal{P}$ lie within the length of the first segment. Therefore

$$P(N \geq 2) = 1 - P(N = 1) = 1 - e^{-a\lambda}. \qquad (2.2)$$

Translate the origin to the beginning of the clump. The clump's total length, C, lies between x and $x + dx$, and it is composed of precisely $N = n$ segments, if and only if

(i) a point of $\mathcal{P}$ lies between $x - a$ and $x - a + dx$;

(ii) no point of $\mathcal{P}$ lies between $x - a$ and x;

(iii) precisely $n - 2$ points of $\mathcal{P}$ lie between 0 and $x - a$; and

(iv) none of the $n - 2$ points lying between 0 and $x - a$ is distant more than a from the nearest adjacent point (including the points 0 and $x - a$).

Clearly, we must have $x \geq a$, and $x = a$ only if the clump consists of a single segment. The probabilities of events (i)–(iii) are, respectively, $\lambda \, dx$, $e^{-a\lambda}$, and

$$\frac{\{\lambda(x - a)\}^{n-2}}{(n - 2)!} e^{-\lambda(x-a)}.$$

The probability of event (iv) equals the chance that, given the unit interval is divided into $n - 1$ parts by $n - 2$ independent uniform variates, none of the parts is of length exceeding $a/(x - a)$. Let this probability be $p_{n-2}\{a/(x - a)\}$.

   Events (i), (ii), and (iii) are independent, and conditional on event (iii), event (iv) is independent of (i) and (ii). Consequently the probability of the intersection of events (i)–(iv), with (iv) conditioned on (iii), equals the product of their probabilities. We have therefore proved that

$$\frac{d}{dx} P(C \leq x; N = n) = \lambda e^{-a\lambda} \{\lambda(x - a)\}^{n-2} e^{-\lambda(x-a)} \{(n - 2)!\}^{-1}$$
$$\times p_{n-2}\{a/(x - a)\}. \tag{2.3}$$

Thus, the density of the distribution of clump length, given that the clump consists of *two or more* segments, is given by

$$f(x) \equiv \left\{ \frac{d}{dx} \sum_{n=2}^{\infty} P(C \leq x; N = n) \right\} \Big/ P(N \geq 2)$$

$$= \lambda(1 - e^{-a\lambda})^{-1} e^{-x\lambda} \sum_{n=2}^{\infty} \{\lambda(x - a)\}^{n-2}$$
$$\times p_{n-2}\{a/(x - a)\}/(n - 2)!, \qquad x > a. \tag{2.4}$$

[(Use (2.2) and (2.3).] The probability that an arbitrary clump is of length precisely equal to a is $P(N = 1) = e^{-a\lambda}$.

   Our final task is to compute $p_n(u)$, equal to the probability that the largest division formed by placing n points independently and uniformly into the interval $(0, 1)$ does not exceed u. Now, the probability that the smallest of n independent uniform variates takes a value lying between v and $v + dv$ equals

$$n(1-v)^{n-1} dv, \qquad 0 < v < 1.$$

Given that the smallest uniform variate equals v, the probability that the smallest division formed by the other $n-1$ points placed between v and 1 does not exceed u equals $p_{n-1}\{u/(1-v)\}$. (The remaining $n-1$ points are placed uniformly between v and 1.) Therefore

$$p_n(u) = \int_0^u p_{n-1}\{u/(1-v)\} n(1-v)^{n-1} dv. \qquad (2.5)$$

Clearly,

$$p_0(u) = \begin{cases} 0 & \text{if } u < 1 \\ 1 & \text{if } u \ge 1, \end{cases}$$

which in that special case agrees with the formula below:

$$p_n(u) = \sum_{j=0}^{[u^{-1}]} (-1)^j \binom{n+1}{j} (1-ju)^n = \sum_{j=0}^{n+1} (-1)^j \binom{n+1}{j} (1-ju)_+^n,$$

$$(2.6)$$

where $[u^{-1}]$ denotes the integer part of u^{-1}, and $x_+ = x$ if $x > 0$, 0 if $x \le 0$. If (2.6) holds for $n-1$ instead of n, then by (2.5),

$$p_n(u) = \sum_{j=0}^{n} (-1)^j \binom{n}{j} n \int_0^u (1-ju-v)_+^{n-1} dv$$

$$= \sum_{j=0}^{n} (-1)^j \binom{n}{j} [(1-ju)_+^n - \{1-(j+1)u\}_+^n]$$

$$= \sum_{j=0}^{n+1} (-1)^j \left\{ \binom{n}{j} + \binom{n}{j-1} \right\} (1-ju)_+^n$$

which is identical to (2.6). Therefore (2.6) holds generally, by induction.

Substituting (2.6) into (2.4), we see that

$$f(x) = \lambda(1-e^{-a\lambda})^{-1} e^{-x\lambda} \sum_{n=2}^{\infty} \lambda^{n-2} \{(n-2)!\}^{-1}$$

$$\times \sum_{j=0}^{k} (-1)^j \binom{n-1}{j} \{x - (j+1)a\}^{n-2}, \qquad (2.7)$$

where $k = [(x/a) - 1]$.

It is convenient to simplify (2.7), to facilitate computation. The double series on the right-hand side may be written as

$$A \equiv \sum_{j=0}^{k} \frac{(-1)^j}{j!} \sum_{n=2}^{\infty} (n-1) \frac{\{\lambda(x-(j+1)a)\}^{n-2}}{(n-j-1)!}$$

$$= \sum_{n=2}^{\infty} \frac{\{\lambda(x-a)\}^{n-2}}{(n-2)!} + \sum_{j=1}^{k} \frac{(-1)^j}{j!} \{\lambda(x-(j+1)a)\}^{j-1}$$

$$\times \sum_{n=j+1}^{\infty} (n-1) \frac{\{\lambda(x-(j+1)a)\}^{n-j-1}}{(n-j-1)!}$$

$$= \sum_{r=0}^{\infty} \frac{\{\lambda(x-a)\}^r}{r!} + \sum_{j=1}^{k} \frac{(-1)^j}{j!} \{\lambda(x-(j+1)a)\}^{j-1}$$

$$\times \sum_{r=0}^{\infty} (r+j) \frac{\{\lambda(x-(j+1)a)\}^r}{r!}.$$

Since $\sum_{r\geq 0} ru^r/r! = ue^u$, and of course $\sum_{r\geq 0} u^r/r! = e^u$, then

$$A = e^{\lambda(x-a)} + \sum_{j=1}^{k} \frac{(-1)^j}{j!} \{\lambda(x-(j+1)a)\}^{j-1}$$

$$\times \{\lambda(x-(j+1)a) + j\} \exp\{\lambda(x-(j+1)a)\}.$$

Therefore

$$f(x) = \lambda(e^{a\lambda} - 1)^{-1} \left[1 + \sum_{j=1}^{[(x/a)-1]} \frac{(-1)^j}{j!} \{\lambda(x-(j+1)a)\}^{j-1} \right.$$

$$\left. \times e^{-ja\lambda} \{\lambda(x-(j+1)a) + j\} \right], \qquad x > a, \quad (2.8)$$

where the series in (2.8) is interpreted as zero if $a < x < 2a$.

The density f is constant on $(a, 2a)$, linear on $(2a, 3a)$, quadratic on $(3a, 4a)$, and so on. It is continuous on $(2a, \infty)$, but has a jump discontinuity at $x = 2a$. The density is monotone decreasing on (a, ∞).

The density is pictured in Figure 2.3. Its skewness suggests that it may be approximated by the exponential density. Indeed, if C denotes the length of a multisegment clump (i.e., a clump made up

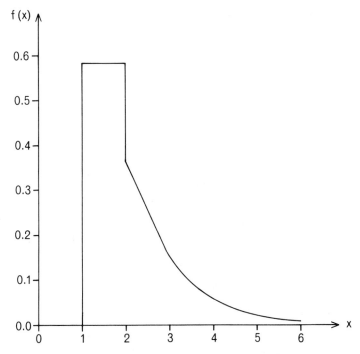

Figure 2.3. Density f, in the case $a = \lambda = 1$.

of two or more segments), then $\lambda e^{-a\lambda}C$ has density given by

$$f^*(x) \equiv \lambda^{-1}e^{a\lambda}f(\lambda^{-1}e^{a\lambda}x)$$

$$= (1 - e^{-a\lambda})^{-1}\left(1 + \sum_{j=1}^{[(xe^{a\lambda}/a\lambda)-1]} \frac{(-1)^j}{j!}\left\{x - (j+1)a\lambda e^{-a\lambda}\right\}^j\right.$$

$$\left. \times \left[1 + j\Big/\left\{e^{a\lambda}x - (j+1)a\lambda\right\}\right]\right).$$

Suppose x is fixed, in the range $(0,\infty)$. If a and λ vary in any manner such that $a\lambda \to \infty$, then clearly

$$f^*(x) \to 1 + \sum_{j=1}^{\infty} \frac{(-1)^j}{j!}x^j = e^{-x}.$$

Therefore, for large values of $a\lambda$, the length of a multisegment clump is approximately exponentially distributed with mean $\lambda^{-1}e^{a\lambda}$. Single-segment clumps are very rare when $a\lambda$ is large [see (2.2)], and so the distribution of clump length (both single- and multisegment) is approximately exponential with mean $\lambda^{-1}e^{a\lambda}$. A slightly better approximation may be obtained using the exact formula for mean clump length, $E(C) = \lambda^{-1}(e^{a\lambda} - 1)$ [see (2.9) below].

At the other extreme, if a and λ vary in any manner such that $a\lambda \to 0$, then the density of $a^{-1} \times$ (length of multisegment clump) converges to the uniform density on $(1,2)$. This may be checked by letting $a\lambda \to 0$ in the density $a f(ax)$, and may be interpreted by noting that for small values of $a\lambda$ there is very little tendency for segments to overlap, so that most multisegment clumps consist of just two overlapping segments. The distribution of the distance between the furthest points of two overlapping segments is uniform on $(a, 2a)$.

To conclude this section we combine the information we have acquired about the distribution of clump length when segment length is fixed.

Theorem 2.1. *Suppose the left-hand ends of segments of fixed length a form a Poisson process of constant intensity λ. Then the distribution of clump length consists of an atom of size $e^{-a\lambda}$ at a, and an absolutely continuous component distributed on (a, ∞) with density*

$$f(x) = \lambda(e^{a\lambda} - 1)^{-1}\left[1 + \sum_{j=1}^{[(x/a)-1]} \frac{(-1)^j}{j!}\{\lambda(x - (j+1)a)\}^{j-1}\right.$$

$$\left. \times e^{-ja\lambda}\{\lambda(x - (j+1)a) + j\}\right], \qquad x > a,$$

pictured in Figure 2.3. For large values of $a\lambda$, clump length is approximately exponentially distributed with mean $\lambda^{-1}(e^{a\lambda} - 1)$, while for small $a\lambda$, the continuous component of clump length is approximately uniform on $(a, 2a)$.

It is easily proved from the formula for f that expected clump length and variance of clump length are given by

$$E(C) = \lambda^{-1}(e^{a\lambda} - 1) \quad \text{and} \quad \text{var}(C) = \lambda^{-2}(e^{2a\lambda} - 2a\lambda e^{a\lambda} - 1),$$
$$(2.9)$$

in the case of fixed segment length.

2.3. DISTRIBUTION OF CLUMP LENGTH IN THE GENERAL CASE

Suppose segments are constructed to the right of points of a Poisson process whose intensity is constant at λ. Let the segment lengths be independently distributed with common distribution function G. We assume that mean segment length is finite and equal to α:

$$\alpha = \int_0^\infty x\,dG(x) = \int_0^\infty \{1 - G(x)\}\,dx < \infty.$$

The case where $\alpha = \infty$ is pathological, since it results in a single, infinite clump covering the whole real line, with probability 1 (see Exercise 2.5). We use techniques from the theory of renewal processes to derive the Laplace-Stieltjes transform of the distribution of clump length.

To simplify notation, place the origin at the end of a clump (start of a spacing). Let $Y_1, Y_2, \ldots$ be the lengths of successive clumps, and $Z_1, Z_2, \ldots$ the lengths of successive spacings (refer to Figure 2.2). Then the number of clumps (or parts of clumps) observed in the interval $[0, t]$ equals

$$N(t) \equiv \sum_{n=1}^\infty I\left(\sum_1^{n-1} Y_i + \sum_1^n Z_i \le t\right),$$

where $I(\mathcal{E})$ denotes the indicator function of an event $\mathcal{E}$. The variable $N(t)$ has mean

$$M(t) \equiv \sum_{n=1}^\infty P\left(\sum_1^{n-1} Y_i + \sum_1^n Z_i \le t\right).$$

Taking Laplace-Stieltjes transforms of both sides, we obtain

$$\mu(s) \equiv \int_0^\infty e^{-st}\,dM(t) = \sum_{n=1}^\infty \int_0^\infty e^{-st}\,dP\left(\sum_1^{n-1} Y_i + \sum_1^n Z_i \le t\right)$$

$$= \sum_{n=1}^\infty \gamma^{n-1}(s)\delta^n(s),$$

where γ and δ are the Laplace-Stieltjes transforms of the distributions of Y_i and Z_i, respectively. Since Z_i has an exponential distribution with mean λ^{-1}, then $\delta(s) = \{1 + (s/\lambda)\}^{-1}$, and so

$$\mu(s) = \{1 + (s/\lambda) - \gamma(s)\}^{-1}. \qquad (2.10)$$

The next step is to deduce an alternative formula for $\mu(s)$. Remember that we have placed the origin at the start of a spacing. Let $p_0(t)$ equal the probability that no segments cover the point $t > 0$. Given that K points of the Poisson process occur between 0 and t, their positions have the distributions of variates distributed independently and uniformly on $[0, t]$. Conditional on the ith uniform variate being situated at x $(0 < x < t)$, the chance that the resulting segment does not cover t is $G(t - x)$, where G is the distribution function of segment length. Therefore the chance that the segment starting at the ith variate does not cover t equals

$$q(t) \equiv t^{-1} \int_0^t G(t - x) \, dx.$$

Conditional on K, the probability that none of the K segments covers t equals $q(t)^K$. Since K has a Poisson distribution with mean λt, then the unconditional probability that none of the segments covers t equals

$$p_0(t) = E\left\{ q(t)^K \right\} = \exp[\lambda t \{q(t) - 1\}] = \exp\left[-\lambda \int_0^t \{1 - G(x)\} \, dx \right].$$

We may express $M(t)$ in terms of $p_0(t)$, as follows. The probability of a clump starting between t and $t + h$ equals

$$P(\text{a spacing exists at } t) \times \{\lambda h + o(h)\} = \lambda p_0(t) h + o(h),$$

as $h \to 0$. The chance of more than one clump starting between t and $t + h$ is $o(h)$. Therefore

$$\begin{aligned} M(t + h) &= (\text{expected number of clumps starting} \\ &\qquad \text{between 0 and } t + h) \\ &= M(t) + \lambda p_0(t) h + o(h), \end{aligned}$$

as $h \to 0$. Consequently, $M'(t) = \lambda p_0(t)$, whence

$$M(t) = \int_0^t \lambda p_0(u) \, du = \lambda \int_0^t \exp\left[-\lambda \int_0^u \{1 - G(x)\} \, dx \right] du.$$

Therefore the Laplace-Stieltjes transform of M is given by

$$\begin{aligned} \mu(s) &= \int_0^\infty e^{-st} \, dM(t) \\ &= \lambda \int_0^\infty \exp\left[-st - \lambda \int_0^t \{1 - G(x)\} \, dx \right] dt. \end{aligned} \qquad (2.11)$$

Let C have the distribution of clump length, and Laplace-Stieltjes transform γ. Equating (2.10) and (2.11), we see that

$$\{1 + (s/\lambda) - \gamma(s)\}^{-1} = \lambda \int_0^\infty \exp\left[-st - \lambda \int_0^t \{1 - G(x)\}\, dx\right] dt,$$

whence

$$\gamma(s) = \int_0^\infty e^{-sx}\, dP(C \le x)$$

$$= 1 + (s/\lambda) - \left(\lambda \int_0^\infty \exp\left[-st - \lambda \int_0^t \{1 - G(x)\}\, dx\right] dt\right)^{-1}.$$

Moments of clump length may be deduced from this transform (see Exercise 2.3). Thus we obtain:

Theorem 2.2. *Suppose the left-hand ends of segments form a Poisson process $\mathcal{P}$ of constant intensity λ, and segment lengths are independent of $\mathcal{P}$ and independent and identically distributed with distribution function G and finite mean α. Then the distribution of clump length C has Laplace-Stieltjes transform*

$$\gamma(s) = \int_0^\infty e^{-sx}\, dP(C \le x)$$

$$= 1 + (s/\lambda) - \left(\lambda \int_0^\infty \exp\left[-st - \lambda \int_0^t \{1 - G(x)\}\, dx\right] dt\right)^{-1},$$

and mean

$$E(C) = \lambda^{-1}(e^{\alpha\lambda} - 1).$$

Clump length has finite variance if and only if segment length has finite variance, and in that case,

$$\text{var}(C) = 2\lambda^{-1} e^{\alpha\lambda} \int_0^\infty \left(\exp\left[\lambda \int_t^\infty \{1 - G(x)\}\, dx\right] - 1\right) dt$$
$$- \lambda^{-2}(e^{\alpha\lambda} - 1)^2.$$

2.4. APPROXIMATIONS TO THE DISTRIBUTION OF CLUMP LENGTH

It is not possible to give a closed, explicit formula for the distribution of clump length in many interesting cases. However, we may

develop approximations to the distribution by permitting parameters governing segment length and Poisson intensity to approach
extreme values. In Section 2.2 we examined an exponential approximation of this type, in the case where mean segment length multiplied by Poisson intensity was large. The need for such approximations is even greater here, since we do not have access to explicit
information about the shape of the distribution of clump length.

We continue to use notation from Section 2.3, where G is the
distribution (function) of segment length, α is mean segment length,
and λ is Poisson intensity. Assume α is finite, which entails

$$\int_0^\infty \{1 - G(x)\}\, dx < \infty.$$

If λ is small then most clumps consist of single segments, and
so clump length has a distribution close to that of segment length.
That is,

$$P(C \leq x) \to G(x)$$

at continuity points x of G, as $\lambda \to 0$. On the other hand, for large
values of λ the distribution of C is often close to exponential, as the
next theorem shows.

Theorem 2.3. *Let C have the distribution of clump length. Assume that G has no atom at the origin, that mean segment length
α is finite, and that*

$$(\log x) \int_x^\infty \{1 - G(y)\}\, dy \to l, \qquad (2.12)$$

*where $0 \leq l \leq \infty$, as $x \to \infty$. Then as $\lambda \to \infty$, $C/E(C)$ has a limit-
ing exponential distribution with mean $e^{-l/\alpha}$:*

$$\lim_{\lambda \to \infty} P\{C \leq xE(C)\} = 1 - \exp(-e^{l/\alpha}x), \quad all \quad x > 0.$$

All proofs are deferred until the end of this section. We pause
here to elucidate the regularity conditions in Theorem 2.3.

If the distribution of segment length, L, has an atom at the origin, let L^* have the distribution of L conditional on $L > 0$. Let C^*
have the distribution of clump length in a Boolean model where
segments are distributed as L^* and where Poisson intensity equals
$\lambda\{1 - P(L = 0)\}$. Then C conditional on $C > 0$, and C^*, have the
same distribution. Therefore the assumption in Theorem 2.3 that G
has no atom at the origin is made without any real loss of generality.

Next we examine condition (2.12). If we insist that $l = 0$ then that restriction is a little more severe than just $E(L) < \infty$ (which of course is true because $\alpha < \infty$), and a little weaker than $E(L^{1+\epsilon}) < \infty$, for any $\epsilon > 0$. Indeed, if $E\{L\log(1+L)\} < \infty$ then (2.12) holds with $l = 0$, which entails a limiting exponential distribution for $C/E(C)$, and "the limit of the means equals the mean of the limit." The case where (2.12) holds for some $l > 0$ is somewhat patholog-ical, in that it applies only when the tails of the distribution are unusually large. If $0 < l < \infty$ then (2.12) holds when

$$1 - G(x) \sim l x^{-1} (\log x)^{-2} \quad \text{as} \quad x \to \infty.$$

If (2.12) holds with $l = \infty$ then by Theorem 2.3, $C/E(C) \to 0$ in probability.

In Theorem 2.3 we kept distribution of segment length fixed as intensity increased. Of course, this led to clump length diverging to $+\infty$ in probability. It is of interest to let segment length distribution shrink as intensity increases, in such a way that clump length has a proper limiting distribution. This model could describe the case where only small numbers of spacings are left after distribution of a very large number of short segments on a line. It will be applied in Section 2.5 to give us an approximation to the probability that a given interval is completely covered. In the next theorem we keep the distribution function G fixed, and in a slight change of notation, assume segments are distributed as δL, where L has distribution function G and $\delta = \delta(\lambda) \to 0$ as $\lambda \to \infty$. The mean of L is denoted by α, and u is a fixed positive number.

Theorem 2.4. *Suppose segment length has distribution function $G(x/\delta)$, where $\delta \to 0$ and $\lambda \to \infty$ together in such a manner that*

$$\alpha\delta\lambda = \log(\lambda/u) + o(1). \tag{2.13}$$

If G has no atom at the origin, and condition (2.12) holds, then clump length has a limiting exponential distribution with mean $\left(ue^{l/\alpha}\right)^{-1}$.

We close this section with proofs of Theorems 2.3 and 2.4.

Proof of Theorem 2.3. Since $E(C) = \lambda^{-1}(e^{\alpha\lambda} - 1) \sim \lambda^{-1}e^{\alpha\lambda}$ as $\lambda \to \infty$, then $C/E(C)$ has a limiting exponential distribution with mean $e^{-l/\alpha}$, if and only if $C/(\lambda^{-1}e^{\alpha\lambda})$ has that weak limit. The limit has Laplace-Stieltjes transform $e^{l/\alpha}/(s+e^{l/\alpha})$, while the Laplace-Stieltjes transform of the distribution of $C/(\lambda^{-1}e^{\alpha\lambda})$ may be deduced from Theorem 2.2. Combining these facts we see that it suffices to show that, under (2.12),

$$\gamma(\lambda e^{-\alpha\lambda}s) \to e^{l/\alpha}/(s+e^{l/\alpha})$$

as $\lambda \to \infty$, or equivalently,

$$\lambda \int_0^\infty \exp\left[-st\lambda e^{-\alpha\lambda} - \lambda \int_0^t \{1 - G(x)\}\, dx\right] dt \to 1 + s^{-1}e^{l/\alpha},$$

$$\text{all} \quad s > 0, \qquad (2.14)$$

as $\lambda \to \infty$.

Let $\tau > 0$ be fixed. Since G has no atom at the origin then

$$t^{-1}\int_0^t \{1 - G(x)\}\, dx \to 1$$

as $t \to 0$. Given $\epsilon \in (0,1)$, choose $\eta \in (0,\tau)$ so small that

$$\left|\int_0^t \{1 - G(x)\}\, dx - t\right| < \epsilon t$$

whenever $0 < t < \eta$. Then

$$J(\tau) \equiv \lambda \int_0^\tau \exp\left[-st\lambda e^{-\alpha\lambda} - \lambda \int_0^t \{1 - G(x)\}\, dx\right] dt$$

$$\leq \lambda \int_0^\eta \exp\{-\lambda(1-\epsilon)t\}\, dt$$

$$+ \lambda \int_\eta^\tau \exp\left[-\lambda \int_0^\eta \{1 - G(x)\}\, dx\right] dt$$

$$= (1-\epsilon)^{-1} + o(1)$$

as $\lambda \to \infty$, while

$$J(\tau) \geq \lambda \exp(-s\eta\lambda e^{-\alpha\lambda}) \int_0^\eta \exp\{-\lambda(1+\epsilon)t\}\, dt$$

$$= (1+\epsilon)^{-1} + o(1).$$

Consequently, $J(\tau) \to 1$ as $\lambda \to \infty$, for each $\tau > 0$. Therefore (2.14) will hold if we prove that for some $\tau > 0$,

$$\lambda s \int_\tau^\infty \exp\left[-st\lambda e^{-\alpha\lambda} - \lambda \int_0^t \{1 - G(x)\}\, dx\right] dt \to e^{1/\alpha}, \quad \text{all} \quad s > 0,$$

$$(2.15)$$

as $\lambda \to \infty$. We shall establish (2.15) for $\tau = 1$.

Let $\epsilon \in (0,1)$, and define $t_1 = t_1(\epsilon)$ as follows. If $0 < l < \infty$, let t_1 be so large that

$$\int_t^\infty \{1 - G(x)\}\, dx$$

lies inbetween $l(1 \pm \epsilon)/\log t$ for all $t > t_1$. If $l = \infty$, choose t_1 so large that

$$\int_t^\infty \{1 - G(x)\}\, dx > \epsilon^{-1}(\log t)^{-1}$$

for $t > t_1$, while if $l = 0$, choose t_1 so large that

$$\int_t^\infty \{1 - G(x)\}\, dx \le \epsilon(\log t)^{-1}$$

for $t > t_1$. We may assume that $t_1 > e^{k/\alpha}$, where k is defined by (2.17) below. Now, the left-hand side of (2.15) equals

$$s\{J(t_1) - J(1)\} + \lambda s \int_{t_1}^\infty \exp\left[-st\lambda e^{-\alpha\lambda} - \lambda \int_0^t \{1 - G(x)\}\, dx\right] dt$$

$$= o(1) + \lambda s e^{-\alpha\lambda} \int_{t_1}^\infty \exp\left[-st\lambda e^{-\alpha\lambda} + \lambda \int_t^\infty \{1 - G(x)\}\, dx\right] dt.$$

Therefore if we show that for each ϵ and s,

$$\lambda s e^{-\alpha\lambda} \int_{t_1}^\infty \exp(-st\lambda e^{-\alpha\lambda} + \lambda k/\log t)\, dt \to e^{k/\alpha} \qquad (2.16)$$

as $\lambda \to \infty$, where

$$k = k(\epsilon) = \begin{cases} \epsilon & \text{if } l = 0 \\ l(1 \pm \epsilon) & \text{if } 0 < l < \infty \\ \epsilon^{-1} & \text{if } l = \infty, \end{cases} \qquad (2.17)$$

then (2.15) will be proved.

Changing variable from t to $u = t\lambda e^{-\alpha\lambda}$, we see that the left-hand side of (2.16) equals

$$s \int_{t_1\lambda e^{-\alpha\lambda}}^\infty \exp\{-su + \lambda k/(\alpha\lambda + \log u - \log \lambda)\}\, du.$$

For each $\epsilon > 0$,

$$\int_\epsilon^\infty \exp\{-su + \lambda k/(\alpha\lambda + \log u - \log\lambda)\}\, du \to e^{k/\alpha} \int_\epsilon^\infty e^{-su}\, du,$$

and so (2.16) will follow if we prove that

$$\limsup_{\lambda\to\infty} \int_{t_1\lambda e^{-\alpha\lambda}}^\epsilon \exp\{\lambda k/(\alpha\lambda + \log u - \log\lambda)\}\, du \le \epsilon e^{k/\alpha}. \quad (2.18)$$

Changing variable from u to $v = \log(\lambda^{-1}e^{\alpha\lambda}u)$, we obtain

$$\int_{t_1\lambda e^{-\alpha\lambda}}^\epsilon \exp\{\lambda k/(\alpha\lambda + \log u - \log\lambda)\}\, du$$
$$= \lambda e^{-\alpha\lambda} \int_{\log t_1}^{\log(\epsilon\lambda^{-1}e^{\alpha\lambda})} \exp(v + \lambda k/v)\, dv.$$

Since $v + \lambda k/v$ is convex in v with minimum at $(\lambda k)^{1/2}$, then for any $c > 1$ and large λ,

$$\lambda e^{-\alpha\lambda} \int_{\log t_1}^{\log(\epsilon\lambda^{-1}e^{\alpha\lambda})} \exp(v + \lambda k/v)\, dv$$
$$\le \lambda e^{-\alpha\lambda} \int_{\log t_1}^{c(\lambda k)^{1/2}} \exp(\log t_1 + \lambda k/\log t_1)\, dv$$
$$+ \lambda e^{-\alpha\lambda} \int_{c(\lambda k)^{1/2}}^{\log(\epsilon\lambda^{-1}e^{\alpha\lambda})} (1 - \lambda k v^{-2})^{-1} d_v\{\exp(v + \lambda k/v)\}$$
$$= o(1) + \lambda e^{-\alpha\lambda}\{1 + o(1)\}$$
$$\times \exp\left\{\log(\epsilon\lambda^{-1}e^{\alpha\lambda}) + \lambda k/\log(\epsilon\lambda^{-1}e^{\alpha\lambda})\right\}$$
$$+ 2\lambda^2 k e^{-\alpha\lambda} \int_{c(\lambda k)^{1/2}}^{\log(\epsilon\lambda^{-1}e^{\alpha\lambda})} (1 - \lambda k v^{-2})^{-2} v^{-3} \exp(v + \lambda k/v)\, dv$$
$$\le \epsilon e^{k/\alpha} + 2\lambda^2 k e^{-\alpha\lambda}(1 - c^{-2})^{-2}$$
$$\times \exp\left\{\log(\epsilon\lambda^{-1}e^{\alpha\lambda}) + \lambda k/\log(\epsilon\lambda^{-1}e^{\alpha\lambda})\right\}$$
$$\times \int_{c(\lambda k)^{1/2}}^\infty v^{-3}\, dv + o(1)$$
$$= \epsilon e^{k/\alpha} + \epsilon e^{k/\alpha}(1 - c^{-2})^{-2}c^{-2} + o(1).$$

(Note that $k/\log t_1 < \alpha$.) This holds for each $c > 1$, and so (2.18) is proved. $\square$

Proof of Theorem 2.4. The claimed limit distribution of clump length has Laplace-Stieltjes transform $\{1 + s(ue^{1/\alpha})^{-1}\}^{-1}$, while the Laplace-Stieltjes transform of the distribution of clump length may be deduced from Theorem 2.2 on replacing $G(x)$ by $G(x/\delta)$. Comparing these results we see that it suffices to prove that

$$\lambda \int_0^\infty \exp\left[-st - \delta\lambda \int_0^{t/\delta} \{1 - G(x)\} \, dx\right] dt \to 1 + s^{-1} u e^{1/\alpha}$$

as $\lambda \to \infty$, for each $s > 0$. In view of the relationship (2.13) connecting δ and λ, this is equivalent to showing that

$$\int_0^\infty \exp\left[-st + \delta\lambda \int_{t/\delta}^\infty \{1 - G(x)\} \, dx\right] dt \to u^{-1} + s^{-1} e^{1/\alpha}. \quad (2.19)$$

For any $\tau > 0$,

$$\int_0^{\delta\tau} \exp\left[-st + \delta\lambda \int_{t/\delta}^\infty \{1 - G(x)\} \, dx\right] dt$$

$$\sim \int_0^{\delta\tau} \exp\left[\delta\lambda \int_{t/\delta}^\infty \{1 - G(x)\} \, dx\right] dt$$

$$= \delta \int_0^\tau \exp\left[\delta\lambda \int_t^\infty \{1 - G(x)\} \, dx\right] dt$$

$$= \delta e^{\alpha\delta\lambda} \int_0^\tau \exp\left[-\delta\lambda \int_0^t \{1 - G(x)\} \, dx\right] dt$$

$$\sim \delta\lambda u^{-1} \int_0^\tau \exp(-\delta\lambda t) \, dt \to u^{-1} \quad (2.20)$$

as $\lambda \to \infty$, the second last asymptotic relation following from the argument preceding (2.15). Let k and t_1 have the meaning they did in the proof of Theorem 2.3. Using (2.20) and the argument succeeding (2.15) we see that (2.19) will follow if we show that

$$\delta s \int_{t_1}^\infty \exp(-\delta st + \delta\lambda k/\log t) \, dt \to e^{k/\alpha}$$

as $\lambda \to \infty$. This result is almost identical to (2.16), and may be proved in the same way. $\square$

2.5. PROBABILITY OF COMPLETE COVERAGE

One of the oldest problems in the theory of coverage processes is that of determining the probability that a given interval is completely covered by random segments. In addition to its obvious aesthetic appeal, it has potential applications. For example, given that an interval $[0, t]$ is observed to be completely covered by segments we may in principle determine a one-sided confidence interval for Poisson intensity per unit of mean segment length; see Table 2.1 below, and our discussion of that table.

Let $p(t)$ denote the probability that interval $[0, t]$ is completely covered. As usual, the case where segment length is fixed is by far the easiest, although it is possible to derive the Laplace transform of coverage probability in the general case. However we shall begin by applying Theorem 2.4 from Section 2.4, to deduce an approximate formula for $p(t)$ in the circumstances of that theorem.

Suppose segments are distributed as δL, where L has distribution function G, $\alpha = E(L) < \infty$, $\delta = \delta(\lambda) \to 0$ as $\lambda \to \infty$,

$$(\log x) \int_x^\infty \{1 - G(y)\}\, dy \to l$$

as $x \to \infty$,

$$\alpha \delta \lambda = \log(\lambda/u) + o(1)$$

as $\lambda \to \infty$, and $0 \le l \le \infty$ and $0 < u < \infty$ are fixed constants. Then according to Theorem 2.4, the distribution of clump length is approximately exponential with mean $\nu^{-1} \equiv (ue^{l/\alpha})^{-1}$. The spacings between clumps are exactly exponential with mean λ^{-1}, which of course converges to zero as $\lambda \to \infty$. Therefore as λ increases, the spacings shrink to points and those points are separated by clumps whose lengths are independent and very nearly exponentially distributed with mean ν^{-1}. That is, the pattern of clumps and spacings resembles a Poisson process $\mathcal{Z}$ of intensity ν, with very small spacings located at the points of $\mathcal{Z}$. The probability that the interval $[0, t]$ is entirely covered converges to the probability that none of the points of $\mathcal{Z}$ lies within $[0, t]$, that is, to $e^{-\nu t}$. In summary:

Theorem 2.5. *Suppose segment length has distribution function* $G(x/\delta)$, *where* $\delta \to 0$ *and* $\lambda \to \infty$ *together in such a manner that*

$$\alpha \delta \lambda = \log(\lambda/u) + o(1).$$

Assume the distribution function G has no atom at the origin, and

$$(\log x) \int_x^\infty \{1 - G(y)\}\, dy \to l$$

as $x \to \infty$. Then the probability $p(t)$ that the interval $[0,t]$ is completely covered, converges to $\exp(-ute^{l/\alpha})$ as $\lambda \to \infty$.

For a practical interpretation of this result, observe from the discussion following Theorem 2.3 that $l = 0$ in most regular cases. Also, since $\alpha\delta\lambda \simeq \log(\lambda/u)$ then $u \simeq \lambda e^{-\alpha\delta\lambda}$, and $\alpha\delta$ equals mean segment length. Therefore in a Boolean model where mean segment length equals α and Poisson intensity equals λ,

$$P([0,t] \text{ completely covered}) \simeq \exp(-\lambda t e^{-\alpha\lambda}). \qquad (2.21)$$

Of course, we have only proved this formula in the special case where segment length shrinks as Poisson intensity increases. But an argument based on monotonicity and Theorem 2.5 shows that in a great many cases which we have not treated explicitly, the left-hand and right-hand sides of (2.21) are either both close to 0 or both close to 1. In particular, both sides equal 1 if $\alpha = \infty$, even though we have hitherto assumed that $\alpha < \infty$.

We now revert to exact rather than approximate methods, and derive the Laplace transform of $p(t)$. Here it is convenient to introduce the concept of a renewal process. Consider a sequence $\varsigma_1, \varsigma_2, \ldots$ of independent and identically distributed, nonnegative random variables, which we might interpret as lifetimes of successive machine parts. As soon as a part fails it is replaced by the next, and so on ad infinitum. Let $N(t)$ equal the number of parts replaced during time interval $[0,t]$. Then N is a *renewal process*. In our case we assume that the machine began operating infinitely far back in time; then N is a *stationary* renewal process. The intervals between successive part failures are called *renewal intervals*. For the present we shall take the common distribution of the renewal intervals (i.e., of the ς_i's) to be that of clump length, although a later application in Section 2.6 will employ a different distribution. (Thus, we are here considering the renewal process obtained from the coverage process by taking out all the spacings, and closing up all the gaps so that adjacent clumps touch.)

The interval $[0,t]$ is completely covered if and only if the point 0 is covered by a clump, and the same clump continues until at least t. The probability that 0 is covered by a clump equals 1 minus the probability that no segments cover 0, i.e., $1 - e^{-\alpha\lambda}$, where α is mean segment length [see identity (2.1)]. Let $r(t)$ be the probability that the clump continues at least until time t, given that 0 is covered by a clump. Now, $r(t)$ equals the chance that in our renewal process, the machine part in operation at time 0 is still operating by time t. In renewal process jargon, $r(t)$ equals the probability that the excess life at time 0 in a stationary renewal process exceeds t. Consider the renewal process as a pattern of intervals on the line, each interval being independently distributed as clump length. Color blue each interval whose length does not exceed t. If an interval is longer than t, color the *last* t units of it blue, and the remainder of it red. Then

$r(t) =$ probability that 0 falls during a red interval

$$= \frac{\text{(expected amount of red per renewal interval)}}{\text{(expected length of renewal interval)}}$$

$$= \left\{ \int_t^\infty (x - t)\, dP(C \le x) \right\} \Big/ E(C)$$

$$= \left\{ \int_t^\infty P(C > x)\, dx \right\} \Big/ E(C),$$

where C has the distribution of clump length. Since $E(C) = \lambda^{-1}(e^{\alpha\lambda} - 1)$ (see Theorem 2.2) then the probability that interval $[0, t]$ is completely covered by segments equals

$$p(t) = (1 - e^{-\alpha\lambda}) r(t) = \lambda e^{-\alpha\lambda} \int_t^\infty P(C > x)\, dx. \qquad (2.22)$$

Alternatively, formula (2.22) could be derived using the *renewal theorem*; see Karlin and Taylor (1975, pp. 189–195).

It follows from (2.22) that $p(t)$ has (ordinary) Laplace transform

$$\pi(s) \equiv \int_0^\infty e^{-st} p(t)\, dt$$

$$= \lambda e^{-\alpha\lambda} s^{-1} \int_0^\infty (1 - e^{-sx}) P(C > x)\, dx$$

$$= -\lambda e^{-\alpha\lambda} s^{-2} \int_0^\infty (1 - sx - e^{-sx})\, dP(C \le x)$$

$$= \lambda e^{-\alpha\lambda} s^{-2} \{\gamma(s) + s E(C) - 1\},$$

where $\gamma(s)$ is the Laplace-Stieltjes transform of clump length distribution. Using the formulas for $\gamma(s)$ and $E(C)$ given in Theorem 2.2 we see that

$$\pi(s) = s^{-1} - \left(s^2 e^{\alpha\lambda} \int_0^\infty \exp\left[-st - \lambda \int_0^t \{1 - G(x)\}\, dx \right] dt \right)^{-1}.$$

The problem of deriving an explicit formula for $p(t)$ in terms of easily calculable functions is similar to that of deriving a formula for distribution of clump length. In most cases it appears to be impossible. However, the special case where segment length is fixed and equal to a is more accessible. Let f denote the density of length of a multisegment clump, given in Theorem 2.1. Since $P(C > a) = 1 - e^{-a\lambda}$, and $f'(x) = -\lambda e^{-a\lambda} f(x - a)$ for $x/a > 1$ and not an integer (verified by differentiating the formula for f), then for $t > a$,

$$P(C > t) = (1 - e^{-a\lambda}) \int_t^\infty f(x)\,dx$$
$$= -\lambda^{-1} e^{a\lambda}(1 - e^{-a\lambda}) \int_t^\infty f'(x + a)\,dx$$
$$= \lambda^{-1}(e^{a\lambda} - 1)f(t + a),$$

and

$$\int_t^\infty P(C > x)\,dx = \lambda^{-1}(e^{a\lambda} - 1) \int_t^\infty f(x + a)\,dx$$
$$= \lambda^{-2} e^{a\lambda}(e^{a\lambda} - 1)f(t + 2a).$$

Substituting this result into (2.22), and noting the formula for f in Theorem 2.1, we see that for $t > a$,

$$p(t) = \lambda^{-1}(e^{a\lambda} - 1)f(t + 2a)$$
$$= 1 + \sum_{j=1}^{[(t/a)+1]} \frac{(-1)^j}{j!} \{\lambda(t - (j-1)a)\}^{j-1} e^{-ja\lambda}$$
$$\times \{\lambda(t - (j-1)a) + j\}. \tag{2.23}$$

In the case $0 < t \le a$ we have from (2.22) that

$$p(t) = \lambda e^{-a\lambda} \int_t^a P(C > x)\,dx + p(a+)$$
$$= \lambda e^{-a\lambda}(a - t) + p(a+)$$
$$= 1 - (1 + \lambda t)e^{-a\lambda},$$

which agrees with (2.23) in the range $0 < t \le a$. Therefore $p(t)$ is given by (2.23) for *all* $t > 0$.

Collecting together the results above, we obtain the following theorem.

Theorem 2.6. *Suppose segment lengths are independent and identically distributed with distribution function G. Then the probability*

$p(t)$ that the entire interval $[0, t]$ is covered by segments has ordinary
Laplace transform

$$\pi(s) = \int_0^\infty e^{-st} p(t)\, dt$$

$$= s^{-1} - \left(s^2 e^{\alpha\lambda} \int_0^\infty \exp\left[-st - \lambda \int_0^t \{1 - G(x)\}\, dx \right] dt \right)^{-1}.$$

When segment length is fixed and equal to a,

$$p(t) = 1 + e^{-a\lambda} \sum_{j=0}^{[t/a]} \frac{(-1)^{j+1}}{(j+1)!} \{\lambda(t - ja)\}^j e^{-ja\lambda}$$

$$\times \{\lambda(t - ja) + j + 1\} \qquad (2.24)$$

for all $t > 0$, where $[t/a]$ denotes the integer part of t/a.

Suppose an interval of length t is observed to be completely covered by a simple linear Boolean model in which segments have constant length, a. Writing $\mu = a\lambda$ for the intensity of the observed
process per unit of segment length, and $u = t/a$, we may deduce
from (2.24) that the function $q(u \mid \mu) \equiv p(au)$ is a function of u and
μ alone. Table 2.1 gives values of μ for which $q(u \mid \mu) = 0.10$, 0.05,
or 0.01, for various values of u. If $q(t/a \mid \mu) = 0.05$, then $(\mu/a, \infty)$ is
a 95% confidence interval for λ, conditional on the interval of length
t being completely covered.

The exact formula for $p(t)$ given by (2.24) gives us an opportunity
to assess the performance of the approximation defined by (2.21).
Table 2.2 compares $p(t)$ with

$$\tilde{p}(t) \equiv \exp(-\lambda t e^{-a\lambda})$$

in the case $a = 1$, for various values of λ and t.

2.6. LIKELIHOOD FOR A SIMPLE LINEAR BOOLEAN MODEL

In this and the next section we consider the possibility of conducting
inference using data garnered by observing a simple linear Boolean
model over the interval $[0, t]$. As noted in Section 2.1, this information comes in a variety of forms. One of the first steps in conducting
inference for more standard problems is often to write down the likelihood of all the information, and we show here how to do that. We

TABLE 2.1 Values of μ for which $q(u \mid \mu) = \alpha$, for Various Values of u and for $\alpha = 0.10$, 0.05, or 0.01

μ	$u(\alpha = 0.10)$	$u(\alpha = 0.05)$	$u(\alpha = 0.01)$
0.2	0.4963	0.8017	1.2920
0.4	0.8566	1.1385	1.9508
0.6	1.1563	1.6278	2.6598
0.8	1.5649	2.1220	3.4297
1.0	1.9935	2.6886	4.2983
1.2	2.4949	3.3363	5.2905
1.4	3.0683	4.0808	6.4318
1.6	3.7308	4.9409	7.7505
1.8	4.4986	5.9378	9.2794
2.0	5.3905	7.0962	11.0567
3.0	12.6532	16.5350	25.5484
4.0	28.8222	37.5620	57.8551
5.0	65.8207	85.6898	131.8243

TABLE 2.2 Comparison of $p(t)$ with $\tilde{p}(t)$, for $a = 1$ and Various Values of λ and t

λ	\multicolumn{6}{c}{t}
	1	2	4	8	16	32
1	.264, .692	.099, .479	.013, .230	.000, .053	.000, .003	.000, .000
2	.594, .763	.397, .582	.176, .339	.035, .115	.001, .013	.000, .000
3	.801, .861	.670, .742	.469, .550	.230, .303	.055, .092	.003, .008
4	.908, .929	.839, .864	.716, .746	.521, .556	.276, .310	.078, .096
5	.960, .967	.927, .935	.864, .874	.752, .764	.569, .583	.325, .340
6	.983, .985	.968, .971	.939, .942	.884, .888	.783, .788	.615, .621

First entry in each cell is $p(t)$, second is $\tilde{p}(t)$.

assume that only the pattern of covered and uncovered regions is observed. Segment length is given an arbitrary distribution, determined up to a set of unknown parameters. The entire stochastic process is governed by a vector θ of parameters, consisting of segment length parameters and Poisson intensity.

Our aim is to conduct asymptotic inference as $t \to \infty$, and so we are not particularly concerned about remnants of clumps or spacings which arise at the ends of the interval $[0, t]$. They provide an

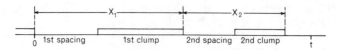

Figure 2.4. Renewal process of spacing/clump pairs.

asymptotically negligible amount of information. Therefore we shall simplify matters by moving the origin to the end of a clump (start of a spacing).

Pair the ith spacing with the ith clump, $i \geq 1$. In this manner we derive a renewal process whose ith renewal life is of length X_i, equal to the sum of the lengths of the ith spacing and the ith clump (see Figure 2.4). Since X_i has the distribution of clump length convolved with an independent exponential distribution, then X_i is absolutely continuous with a bounded density h, no matter what the distribution of segment length. Our first task is to derive the likelihood of the renewal process.

Recall that in the classical theory of inference for a vector parameter $\boldsymbol{\theta}$, the likelihood of an independent sample $\{Y_1, \ldots, Y_n\}$ in which each Y_i has density $f(\cdot \,|\, \boldsymbol{\theta})$ is given by

$$\mathcal{L} \equiv \prod_{i=1}^{n} f(Y_i \,|\, \boldsymbol{\theta}).$$

That is, the likelihood equals the (Radon-Nikodym) derivative of the probability measure P_1, describing the distribution of $(Y_1, \ldots, Y_n)$, with respect to n-variate Lebesgue measure P_2. Let P_3 be any other probability measure that does not depend on the unknown parameters $\boldsymbol{\theta}$ and is such that $P_1 << P_3 << P_2$. ($P << Q$ means that P is absolutely continuous with respect to Q.) Then

$$\mathcal{L} = \frac{dP_1}{dP_2} = \frac{dP_1}{dP_3} \frac{dP_3}{dP_2}.$$

Since dP_3/dP_2 does not depend on $\boldsymbol{\theta}$ then maximizing $\mathcal{L}$ with respect to $\boldsymbol{\theta}$ is equivalent to maximizing dP_1/dP_3 with respect to $\boldsymbol{\theta}$. Therefore the results of maximum likelihood estimation are unchanged if we replace the likelihood dP_1/dP_2 by dP_1/dP_3.

This argument is valid in a very general setting. In the case of our renewal process, take P_1 to be the measure describing the joint distributions of N (the number of renewals in $[0, t]$), of $X_1, \ldots, X_N$ (the first N renewal lifetimes), and of $\tilde{X}_t \equiv t - \sum_{1 \leq i \leq N} X_i$ (the current life, or remnant of renewal life, at time t). Then P_1 depends on $\boldsymbol{\theta}$. Let P_3 be the version of P_1 for any fixed, known value of $\boldsymbol{\theta}$, say

$\boldsymbol{\theta} = \boldsymbol{\theta}_0$. The *direct Radon-Nikodym theorem* (see Appendix I) implies that

$$\frac{dP_1}{dP_3} = \left\{ \prod_{i=1}^{N} \frac{h(X_i \mid \boldsymbol{\theta})}{h(X_i \mid \boldsymbol{\theta}_0)} \right\} \frac{1 - \displaystyle\int_0^{\tilde{X}_t} h(x \mid \boldsymbol{\theta}) \, dx}{1 - \displaystyle\int_0^{\tilde{X}_t} h(x \mid \boldsymbol{\theta}_0) \, dx},$$

where $h(\cdot \mid \boldsymbol{\theta})$ is the density of renewal life. Therefore the likelihood of the observed renewal process is proportional to

$$\left\{ \prod_{i=1}^{N} h(X_i \mid \boldsymbol{\theta}) \right\} \left\{ 1 - \int_0^{\tilde{X}_t} h(x \mid \boldsymbol{\theta}) \, dx \right\}, \qquad (2.25)$$

where the "constant" of proportionality does not depend on $\boldsymbol{\theta}$.

Suppose clump length has a continuous distribution with density $g(\cdot \mid \boldsymbol{\theta})$. Denote the lengths of the ith spacing and ith clump by Z_i and Y_i, respectively. Then $X_i = Y_i + Z_i$. Conditional on $X_i = x$, the spacing Z_i has density

$$\frac{\lambda e^{-\lambda z} g(x - z \mid \boldsymbol{\theta})}{h(x \mid \boldsymbol{\theta})}, \qquad 0 < z < x,$$

using the fact that spacing lengths are exponentially distributed with mean λ^{-1} and are independent of clump lengths. Therefore the conditional likelihood of lengths of the first N spacings, given N and $X_1, \ldots, X_N$, equals

$$\prod_{i=1}^{N} \left\{ \frac{\lambda \exp(-\lambda Z_i) g(Y_i \mid \boldsymbol{\theta})}{h(X_i \mid \boldsymbol{\theta})} \right\}.$$

Combining this result with (2.25) we see that the joint likelihood of N, $Y_1, \ldots, Y_N$, $Z_1, \ldots, Z_N$, and $\tilde{X}_t$ is proportional to

$$\left\{ \prod_{i=1}^{N} h(X_i \mid \boldsymbol{\theta}) \right\} \left\{ 1 - \int_0^{\tilde{X}_t} h(x \mid \boldsymbol{\theta}) \, dx \right\} \left\{ \prod_{i=1}^{N} \frac{\lambda \exp(-\lambda Z_i) g(Y_i \mid \boldsymbol{\theta})}{h(X_i \mid \boldsymbol{\theta})} \right\}$$

$$= \left\{ \prod_{i=1}^{N} \lambda \exp(-\lambda Z_i) \right\} \left\{ \prod_{i=1}^{N} g(Y_i \mid \boldsymbol{\theta}) \right\} \left\{ 1 - \int_0^{\tilde{X}_t} h(x \mid \boldsymbol{\theta}) \, dx \right\}.$$

$$(2.26)$$

A version of this result holds true when clump length does not have a continuous distribution; we shall shortly treat the example where segment length is fixed.

An estimator of $\boldsymbol{\theta}$ could be derived by minimizing the quantity in (2.26). However, provided segment length has finite variance, the current life $\tilde{X}_t$ makes an asymptotically negligible (as $t \to \infty$) contribution to the maximum likelihood estimator. In particular it has negligible effect on variance. Therefore we remove the last factor in (2.26), and work with the *approximate likelihood*, proportional to

$$\tilde{\mathcal{L}} = \left\{ \prod_{i=1}^{N} \lambda \exp(-\lambda Z_i) \right\} \left\{ \prod_{i=1}^{N} g(Y_i \mid \boldsymbol{\theta}) \right\}$$

$$= \lambda^N \exp\left(-\lambda \sum_{1}^{N} Z_i \right) \prod_{i=1}^{N} g(Y_i \mid \boldsymbol{\theta}). \tag{2.27}$$

This minor change simplifies the analysis.

In the case where segment length is fixed and equal to a, the distribution of clump length is not continuous, being made up of an atom at a and a continuous component on (a, ∞) (see Theorem 2.1). In this case the "density" of clump length is given by

$$g(y \mid \lambda) = (e^{-a\lambda})^I \left\{ (1 - e^{-a\lambda}) f(y \mid \lambda) \right\}^{1-I},$$

where $f(y \mid \lambda) = f(y)$ is given in Theorem 2.1, and I equals 1 if $y = a$, 0 if $y > a$. Substituting this formula into (2.27) we find that

$$\tilde{\mathcal{L}} = \lambda^N \exp\left(-\lambda \sum_{1}^{N} Z_i - Ma\lambda \right) (1 - e^{-a\lambda})^{N-M} \prod_{i=1}^{N-M} f(Y_i \mid \lambda),$$

where M equals the number of single-segment clumps and $Y_1, \ldots,$ Y_{N-M} are the lengths of multisegment clumps.

Of course, the total number of complete spacings observed within $[0, t]$ (N_1, say) will equal either the total number of complete clumps (N_2, say), or $N_2 + 1$. Estimation could be conducted using the approximate likelihoods,

$$\tilde{\tilde{\mathcal{L}}} = \lambda^{N_1} \exp\left(-\lambda \sum_{1}^{N_1} Z_i \right) \prod_{i=1}^{N_2} g(Y_i \mid \boldsymbol{\theta}) \tag{2.28}$$

in the continuous case, or

$$\tilde{\tilde{\mathcal{L}}} = \lambda^{N_1} \exp\left(-\lambda \sum_{1}^{N_1} Z_i - Ma\lambda \right) (1 - e^{-a\lambda})^{N_2-M} \prod_{i=1}^{N_2-M} f(Y_i \mid \lambda)$$

$$\tag{2.29}$$

in the case of fixed segment length. These changes have an asymptotically negligible effect on first-order efficiency.

2.7. ESTIMATION OF INTENSITY IN A SIMPLE LINEAR BOOLEAN MODEL

Assume for the time being that clump length has a continuous distribution with density g. Then the approximate log-likelihood of N_1 complete spacings Z_i and N_2 complete clumps Y_i within the interval $[0, t]$ may be found from (2.28):

$$\log \tilde{\mathcal{L}} = N_1 \log \lambda - \lambda \sum_1^{N_1} Z_i + \sum_1^{N_2} \log g(Y_i \mid \boldsymbol{\theta}). \qquad (2.30)$$

As we noted in Section 2.3, it is usually not possible to write down the distribution of clump length in simple, closed form, and so the method of maximum likelihood cannot be considered a practical means of estimating unknown parameters. (An exception is the case where segment length is fixed, which we shall consider shortly.) However the variance of the maximum likelihood estimator provides an asymptotic lower bound to the variance of any other asymptotically unbiased estimator, and so provides a convenient benchmark for the comparison of simpler but less efficient techniques.

We consider only the problem of estimating unknown Poisson intensity λ when distribution of segment length is known, since that case depends least on special properties of segment length distribution. The problem of estimating parameters in a more general setting will be treated in Chapter 5. The argument below is somewhat heuristic, but may be tightened under specific assumptions about the distribution of segment length. We assume the distribution of segment length has finite variance.

A theory of inference based on $\tilde{\mathcal{L}}$ may be developed along classical lines (e.g., as in Kendall and Stuart 1979, Chapter 18). In particular it may be shown that under appropriate conditions on segment length distribution, the maximum likelihood estimator $\hat{\lambda}$ of Poisson intensity is asymptotically normally distributed with mean λ and variance

$$\left\{ -E(\partial^2 \log \tilde{\mathcal{L}} / \partial \lambda^2) \right\}^{-1}.$$

As $t \to \infty$,

$$- E(\partial \log \tilde{\tilde{\mathcal{L}}}/\partial \lambda^2) = \lambda^{-2} E(N_1)$$

$$+ E\left[\sum_1^{N_2} \left\{ g_\lambda^2(Y_i)/g^2(Y_i) - g_{\lambda\lambda}(Y_i)/g(Y_i) \right\} \right]$$

$$\sim \lambda^{-2} E(N_1) + E(N_2) E\left\{ g_\lambda^2(C)/g^2(C) - g_{\lambda\lambda}(C)/g(C) \right\}$$

$$= \lambda^{-2} E(N_1) + E(N_2) \int_0^\infty g_\lambda^2(x) \{ g(x) \}^{-1} \, dx,$$

where $g_\lambda \equiv \partial g/\partial \lambda$, $g_{\lambda\lambda} \equiv \partial^2 g/\partial \lambda^2$, and C has the distribution of clump length. Since $|N_1 - N_2| \leq 1$ then $|E(N_1) - E(N_2)| \leq 1$. Furthermore,

$$E(N_1) \sim t/(\text{expected length of spacing plus clump})$$

$$= t \Big/ \left\{ \lambda^{-1} + \lambda^{-1}(e^{\alpha\lambda} - 1) \right\} = t\lambda e^{-\alpha\lambda},$$

as $t \to \infty$. (A detailed proof uses the renewal theorem.) Consequently,

$$- E(\partial^2 \log \tilde{\tilde{\mathcal{L}}}/\partial \lambda^2) \sim t\lambda^{-1} e^{-\alpha\lambda} \left[1 + \lambda^2 \int_0^\infty g_\lambda^2(x) \{ g(x) \}^{-1} \, dx \right].$$

$$(2.31)$$

A very simple estimator of λ may be constructed using the concept of *vacancy*. Assume for the time being that mean segment length, α, is known. Let $V = V(t)$ denote the total length of the uncovered portions of $[0, t]$. In view of formula (2.1),

$$E(V) = \int_0^t P(x \text{ not covered}) \, dx = te^{-\alpha\lambda}.$$

We shall prove in Chapter 3, Section 3.4 that $t^{-1/2}\{V - E(V)\}$ is asymptotically normally distributed with 0 mean and variance equal to

$$\sigma^2 \equiv 2e^{-2\alpha\lambda} \int_0^\infty \left(\exp\left[\lambda \int_t^\infty \{ 1 - G(x) \} \, dx \right] - 1 \right) dt.$$

Therefore the estimator

$$\hat{\lambda}_1 \equiv \alpha^{-1} \log(t/V)$$

is consistent and has an asymptotic normal distribution with mean λ and variance $t^{-1}\alpha^{-2}e^{2\alpha\lambda}\sigma^2$. The asymptotic efficiency of $\hat{\lambda}_1$ relative to the maximum likelihood estimator $\hat{\lambda}$ is given by

$$
e_1(\lambda) \equiv \lim_{t \to \infty} \left[(t^{-1}\alpha^{-2}e^{2\alpha\lambda}\sigma^2) \left\{ -E(\partial^2 \log \tilde{\mathcal{L}}/\partial\lambda^2) \right\} \right]^{-1}
$$

$$
= \alpha^2 \lambda e^{\alpha\lambda} \left\{ 2 \left[1 + \lambda^2 \int_0^\infty g_\lambda^2(x) \left\{ g(x) \right\}^{-1} dx \right] \right.
$$

$$
\left. \times \int_0^\infty \left(\exp\left[\lambda \int_t^\infty \{1 - G(x)\} \, dx \right] - 1 \right) dt \right\}^{-1},
$$

$$
(2.32)
$$

using (2.31). Provided only that the integrals in the denominator are finite and continuous in λ, $e_1(\lambda)$ is bounded away from 0 on any interval $[\lambda_1, \lambda_2]$ having $0 < \lambda_1 < \lambda_2 < \infty$. As $\lambda \to 0$, $g(x)$ converges to the density of segment length, and

$$
1 + \lambda^2 \int_0^\infty g_\lambda^2(x) \left\{ g(x) \right\}^{-1} dx \to 1.
$$

Furthermore,

$$
\int_0^\infty \left(\exp\left[\lambda \int_t^\infty \{1 - G(x)\} \, dx \right] - 1 \right) dt \sim \lambda \int_0^\infty dt \int_t^\infty \{1 - G(x)\} \, dx
$$

$$
= (\lambda/2) \int_0^\infty x^2 \, dG(x)
$$

as $\lambda \to 0$. Therefore

$$
e_1(\lambda) \to \alpha^2 \bigg/ \int_0^\infty x^2 \, dG(x) = (EL)^2 / E(L^2)
$$

as $\lambda \to 0$, where L has the distribution of segment length.

Next we work out the behavior of $e_1(\lambda)$ as $\lambda \to \infty$. When λ is large the distribution of clump length is approximately exponential with mean $\lambda^{-1}e^{\alpha\lambda}$ (see Theorem 2.3). Therefore as $\lambda \to \infty$,

$$
\int_0^\infty g_\lambda^2(x) \left\{ g(x) \right\}^{-1} dx \sim \left\{ (\partial/\partial\lambda)\lambda e^{-\alpha\lambda} \right\}^2
$$

$$
\times (\lambda e^{-\alpha\lambda})^{-2} \int_0^\infty (1-x)^2 e^{-x} \, dx
$$

$$
\to \alpha^2.
$$

$$
(2.33)
$$

The next theorem, proved at the end of this section, describes high-intensity behavior of the other main component of $e_1(\lambda)$.

Theorem 2.7. *Let G be a distribution function with support confined to $(0, \infty)$, corresponding to a distribution with mean α and finite variance. Then*

$$\int_0^\infty \left(\exp\left[\lambda \int_t^\infty \{1 - G(x)\}\, dx \right] - 1 \right) dt \sim \lambda^{-1} e^{\alpha \lambda} \qquad (2.34)$$

as $\lambda \to \infty$.

Combining (2.32), (2.33), and (2.34) we see that as $\lambda \to \infty$,

$$e_1(\lambda) \sim \alpha^2 \lambda e^{\alpha \lambda} \left\{ 2(\lambda^2 \alpha^2)(\lambda^{-1} e^{\alpha \lambda}) \right\}^{-1} = 1/2.$$

Therefore under conditions of high intensity the vacancy-based estimator $\hat\lambda_1$ is approximately 50% efficient.

From the results of the last two paragraphs we see that a graph of $e_1(\lambda)$ against λ consists of a curve bounded away from 0, asymptotic to $(EL)^2/E(L^2)$ as $\lambda \to 0$ and asymptotic to $1/2$ as $\lambda \to \infty$. A more detailed analysis requires extensive numerical computation using the exact distribution of clump length, and only in the case of constant segment length is that distribution reasonably accessible. There, $(EL)^2/E(L^2) = 1$. Figure 2.5 depicts $e_1(\lambda)$ graphed against $e^{-a\lambda}$, the proportion of single-segment clumps, in the case where segment length is constant and equal to a. The value of $e_1(\lambda)$ increases rapidly to 1 as $e^{-a\lambda}$ increases, exceeding 80% whenever $e^{-a\lambda} > 0.05$ (i.e., $a\lambda < 3$).

A second vacancy-based estimator is given by

$$\hat\lambda_2 = N/V,$$

where N denotes the number of spacings (or alternatively, and almost equivalently, the number of clumps) observed within $[0, t]$. This estimator does not require prior knowledge of mean segment length, α. Observe that

$$V = \sum_1^N Z_i + R,$$

where the spacings $Z_1, Z_2, \ldots$ are independent exponential random variables with mean λ^{-1}, and the remainder term R accounts for any fractional spacings that may make up part of the vacancy. Now,

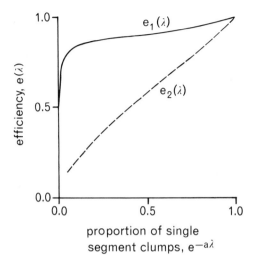

Figure 2.5. Efficiency of estimators of intensity in the case of fixed segment length.

$R = O_p(1)$ as $t \to \infty$, and

$$(V/N) - \lambda^{-1} = N^{-1} \sum_{i=1}^{N} (Z_i - \lambda^{-1}) + (R/N)$$

$$= N^{-1} \sum_{1}^{N} (Z_i - \lambda^{-1}) + O_p(t^{-1}).$$

The sum $N^{-1} \sum_{1}^{N} (Z_i - \lambda^{-1})$ is asymptotically normally distributed (as $t \to \infty$) with zero mean and with variance

$$(EN)^{-1} \text{var}(Z_i) \sim (t\lambda e^{-\alpha\lambda})^{-1} \lambda^{-2} = t^{-1} \lambda^{-3} e^{\alpha\lambda};$$

see Appendix II, or apply the general theory in Chapter 3, Section 3.4. Therefore $\hat{\lambda}_2$ is asymptotically normal $N(\lambda, t^{-1}\lambda e^{\alpha\lambda})$, and has efficiency

$$e_2(\lambda) \equiv \left[1 + \lambda^2 \int_0^\infty g_\lambda^2(x) \{g(x)\}^{-1} \, dx \right]^{-1}$$

relative to the maximum likelihood estimator. Note that $e_2(\lambda) \to 1$ as $\lambda \to 0$, while $e_2(\lambda) \to 0$ as $\lambda \to \infty$, the latter result following from (2.33). Consequently, $\hat{\lambda}_1$ is more efficient than $\hat{\lambda}_2$ under conditions of high intensity. However, if segment length is not constant, meaning that $e_1(0+) = (EL)^2/E(L^2) < 1$, then $\hat{\lambda}_2$ is more efficient than $\hat{\lambda}_1$ under conditions of low intensity. The efficiency of $\hat{\lambda}_2$ in the case of fixed segment length is represented by the broken line in Figure 2.5.

When segment length is constant the approximate likelihood $\tilde{\tilde{\mathcal{L}}}$ is given by (2.29), in which f is defined as in Theorem 2.1. Observe that

$$f(x \mid \lambda) = \lambda(e^{a\lambda} - 1)^{-1}k(x \mid \lambda),$$

where

$$k(x \mid \lambda) = 1 + \sum_{j=1}^{[(x/a)-1]} \left\{(-1)^j/j!\right\} \left\{\lambda(x - (j+1)a)\right\}^{j-1} e^{-ja\lambda}$$

$$\times \left\{\lambda(x - (j+1)a) + j\right\}, \quad x > a.$$

Therefore the equation $(\partial/\partial\lambda)\log\tilde{\tilde{\mathcal{L}}} = 0$, defining the maximum likelihood estimator $\hat{\lambda}$, has the form

$$N_1 + N_2 - M + \hat{\lambda}\sum_1^{N_2-M} k_\lambda(Y_i \mid \hat{\lambda})/k(Y_i \mid \hat{\lambda}) = N_2 a\hat{\lambda} + \hat{\lambda}\sum_1^{N_1} Z_i,$$

where $Z_1, \ldots, Z_{N_1}$ are the lengths of spacings, $Y_1, \ldots, Y_{N_2-M}$ are the lengths of multisegment clumps, and M is the number of single-segment clumps.

We conclude this section with a derivation of formula (2.34).

Proof of Theorem 2.7. An integration by parts reveals that the left-hand side of (2.34) equals

$$\lambda e^{a\lambda} \int_0^\infty t\{1 - G(t)\}\exp\left[-\lambda\int_0^t \{1 - G(x)\}\,dx\right]dt \equiv J(\lambda),$$

say. Given $\epsilon \in (0,1)$, choose $\eta > 0$ so small that

$$\left|\int_0^t \{1 - G(x)\}\,dx - t\right| < \epsilon t$$

whenever $0 < t < \eta$. Then $J(\lambda)$ lies inbetween the lower bound,

$$\lambda e^{a\lambda} \int_0^\eta t\{1 - G(t)\}\exp\left\{-\lambda(1+\epsilon)t\right\}dt \sim (1+\epsilon)^{-2}\lambda^{-1}e^{a\lambda},$$

and the upper bound,

$$\lambda e^{\alpha\lambda} \int_0^\eta t\{1 - G(t)\} \exp\{-\lambda(1-\epsilon)t\}\, dt$$

$$+ \lambda \exp\left[\lambda \int_\eta^\infty \{1 - G(x)\}\, dx\right] \int_\eta^\infty t\{1 - G(t)\}\, dt$$

$$\sim (1-\epsilon)^{-2}\lambda^{-1}e^{\alpha\lambda},$$

as $\lambda \to \infty$. Consequently, $J(\lambda) \sim \lambda^{-1}e^{\alpha\lambda}$ as $\lambda \to \infty$, as had to be proved. $\square$

EXERCISES

2.1 Let $Z_1, Y_1, Z_2, Y_2, \ldots$ denote the successive lengths of spacings and clumps in the alternating pattern depicted in Figure 2.2. Suppose the driving Poisson process $\mathcal{P}$ has intensity λ. Show that the variables $Y_1, Y_2, \ldots, Z_1, Z_2, \ldots$ are stochastically independent, that the Y_i's are identically distributed, and that the Z_i's are exponentially distributed with mean λ^{-1}. [Hint: Use the lack-of-memory property of $\mathcal{P}$.]

2.2 By integrating $xf(x)$, with f defined in Theorem 2.1, and adding in an amount for the atom at the point a, show that clump length C has expected value

$$E(C) = \lambda^{-1}(e^{a\lambda} - 1)$$

when segment length is fixed and equal to a. Similarly, prove that

$$\mathrm{var}(C) = \lambda^{-2}(e^{2a\lambda} - 2a\lambda e^{a\lambda} - 1).$$

2.3 By differentiating the Laplace-Stieltjes transform γ in Theorem 2.2, show that if segment lengths are independent and identically distributed with distribution function G and finite mean α, then

$$E(C) = \lambda^{-1}(e^{\alpha\lambda} - 1)$$

and

$$\mathrm{var}(C) = 2\lambda^{-1}e^{\alpha\lambda} \int_0^\infty \left(\exp\left[\lambda \int_t^\infty \{1 - G(x)\}\, dx\right] - 1\right) dt$$
$$- \lambda^{-2}(e^{\alpha\lambda} - 1)^2.$$

(This is not easy!) Use the second formula to prove that clump length has finite variance if and only if segment length has finite variance.

2.4 By differentiating the formula for f in Theorem 2.1, prove that

$$f'(x) = -\lambda e^{a\lambda} f(x - a), \qquad (2.A)$$

provided x/a is not an integer. Therefore f is monotone decreasing on $(2a, \infty)$; it is constant on $(a, 2a)$. Use formula (2.A) to interpret the exponential approximation to the distribution of clump length.

2.5 Prove that if expected segment length is infinite (i.e., $\alpha = \infty$), then with probability one every point on the real line is covered an infinite number of times by random segments. In particular, this result implies that not only is expected clump length infinite, but $P(C = \infty) = 1$. [Hint: Show that without loss of generality, segments may be taken to have integer-valued lengths.]

2.6 Deduce from Theorem 2.3 that a necessary and sufficient condition for $C/E(C)$ to have a limiting exponential distribution with unit mean is

$$(\log x) \int_x^\infty \{1 - G(y)\}\, dy \to 0$$

as $x \to \infty$.

2.7 Suppose the distribution function G of segment length satisfies

$$(\log x) \sup_{y \geq x} y\, \{1 - G(y)\} \to 0$$

as $x \to \infty$. Show that for large λ, there exists a unique number $\nu = \nu(\lambda)$ exceeding $e^{\lambda/2}$ and satisfying

$$\lambda \int_\nu^\infty \{1 - G(y)\}\, dy + \log(\lambda \nu) = \lambda.$$

Prove that the distribution of clump length C obeys the limit theorem

$$P(C \leq \nu x) \to 1 - e^{-x}, \quad \text{all} \quad x > 0,$$

as $\lambda \to \infty$.

2.8 Verify the comment made after Theorem 2.3: If L and C are random variables having the distributions of segment length and clump length, respectively, in a certain simple linear Boolean model driven by a Poisson process of intensity λ, and if L^* and C^* represent segment length and clump length when segment length has the distribution of L given $L > 0$ and when the driving Poisson process has intensity $\lambda\{1 - P(L = 0)\}$, then C given $C > 0$, and C^*, have the same distribution.

2.9 Consider a simple linear Boolean model in which the driving Poisson process has intensity λ, and segment length is random with distribution function G. Suppose the first clump to start to the right of the origin, starts at $X > 0$ and finishes at $Y \geq X$. Let $C \equiv Y - X$ denote the length of the clump and let N equal the number of segments comprising the clump. Let $N_1(t)$ and $N_2(t)$ be, respectively, the number of segments contained wholly within $[0, t]$ and the number of segments that cover t. Put

$$H(t, n) \equiv P\{N_1(t) = n, \ N_2(t) = 0\}$$

and

$$F_n(x) \equiv P(C \leq x, \ N = n).$$

(i) Notice that the probability of precisely n segments starting during $[0, t]$ equals

$$(t\lambda)^n e^{-t\lambda} / n!,$$

and that if the left-hand endpoint of a segment is uniformly distributed within $[0, t]$ then the chance of that segment not covering t equals

$$t^{-1} \int_0^t G(t - x) \, dx.$$

Hence find an explicit formula for $H(t, n)$, and prove that

$$H^*(s_1, s_2) \equiv \sum_{n=0}^{\infty} \int_0^{\infty} \exp(-s_1 t - s_2 n) \, H(t, n) \, dt$$

$$= \int_0^{\infty} \exp\Big\{-s_1 t - t\lambda$$

$$+ \exp(-s_2)\lambda \int_0^t G(x) \, dx\Big\} \, dt.$$

(ii) By considering the position of the first clump, show that for $n \geq 1$,

$$H(t,n) = \sum_{m=1}^{n} \lambda \int_{y=0}^{t} \int_{x=0}^{y} \exp(-x\lambda)$$
$$\times H(t-y, n-m) \, dF_n(y-x) \, dx.$$

It should follow from your answer to (i) that $H(t,0) = e^{-t\lambda}$.

(iii) Take Laplace transforms of the formula for H in (ii), to obtain

$$F^*(s_1, s_2) = \lambda^{-1}(\lambda + s_1) - \{\lambda H^*(s_1, s_2)\}^{-1},$$

where

$$F^*(s_1, s_2) \equiv \sum_{n=0}^{\infty} \int_{0}^{\infty} \exp(-s_1 x - s_2 n) \, dF_n(x).$$

(iv) By combining (i) and (iii), show that the pair (C, N) has joint Laplace-Stieltjes transform

$$F^*(s_1, s_2) = 1 + \lambda^{-1} s_1 - \left[\lambda \int_{0}^{\infty} \exp\left\{-s_1 t - \lambda t\right.\right.$$
$$\left.\left. + \exp(-s_2)\lambda \int_{0}^{t} G(x) \, dx \right\} dt \right]^{-1}.$$

Compare with Theorem 2.2.

[After Shanbhag (1966).]

2.10 [Continuation of Exercise 2.9(iv)] Write down the Laplace-Stieltjes transform of the distribution of N, the number of segments comprising an arbitrary clump. Prove by differentiation or otherwise that

$$E(N) = e^{\alpha\lambda} = \lambda E(C) + 1$$

and

$$\text{var}(N) = \lambda^2 \text{var}(C) + 2\alpha\lambda e^{\alpha\lambda} - e^{\alpha\lambda} + 1,$$

where C has the distribution of clump length. In particular it follows that $\text{var}(N)$ is finite if and only if variance of segment length is finite.

NOTES

The theory of simple linear Boolean models appears in many guises in the literature—as $M/G/\infty$ queues (Takács 1962, Chapter 3; Gnedenko and König 1984, Chapter 2.3), as Type II counters (Karlin and Taylor 1975, p. 179ff; Takács 1962, p. 210ff), and as renewal processes and alternating renewal processes (Cox 1962; Cox and Miller 1965, p. 340ff; Karlin and Taylor 1975, Chapter 5). Theorem 2.1 is well known in the queueing context. Roach (1968, Chapter 3) treats linear Boolean models in their own right, although directing almost all attention at the case of fixed segment length. Shanbhag (1966) derives the Laplace-Stieltjes transform of the joint distribution of the duration and number of services in a busy period in an $M/G/\infty$ queue. That result immediately gives Theorem 2.2. See also Connolly (1971). The approximation theory in Section 2.4 is based on Hall (1985c). References on the problem of approximating and bounding busy periods of $M/G/\infty$ queues (i.e., clump lengths) include Dvurečenskij (1984), Sathe (1985), and Stadje (1985). See Gnedenko and König (1984, Chapter 2.3) for further citations.

The probability of complete coverage, discussed in Section 2.5, is derived by Domb (1947) in the case of fixed segment length. Domb starts from an unusual form of the Laplace transform of vacancy distribution. Portions of the work on inference in Sections 2.6 and 2.7 are drawn from Hall (1985c, 1985d). The material on likelihood ratios and the direct-Radon Nikodym theorem in Section 2.6 and Appendix I is based on Brémaud (1981, Chapter VI). The classical theory of likelihood may of course be found in many places—for example, Cox and Hinkley (1974, Chapters 8 and 9) and Kendall and Stuart (1979, Chapter 18). Cox and Lewis (1966), Snyder (1975), and Karr (1986, Chapters 5–10) discuss statistical analysis of point processes.

There is a large and growing literature on inference in queueing theory. As a guide only, we mention White, Schmidt, and Bennett (1975, Chapters 7–9), Descloux (1976), Grassman (1981), and Whitt (1983). Problems encountered there are usually distinctly different from those arising in applications of Boolean model theory, since in the queueing case one usually has access to data on arrival and/or departure times. In the Boolean model case it is common to have information only about beginnings and ends of "busy periods" (clumps).

CHAPTER THREE
Vacancy

3.1. INTRODUCTION

In this chapter we use the concept of vacancy to introduce and motivate some of the main features of a Boolean model in k-dimensional Euclidean space. Section 3.1 presents notation and defines Boolean models. Section 3.2 defines vacancy and discusses its basic properties, such as mean and variance. Vacancy within a region is just the uncovered content of that region. Various facets of the distribution of vacancy can be worked out exactly in the case of $k = 1$ dimension, when random sets are fixed-length line segments, and Section 3.3 is devoted to that problem. Section 3.4 discusses properties of vacancy within large regions, including limit-theoretic approximations such as the central limit theorem and law of large numbers. Sections 3.5 and 3.6 derive properties of vacancy in high-intensity Boolean models, where a large fraction of space is covered by random sets. Other cases, such as low-intensity Boolean models, are treated briefly at the end of Section 3.6. Section 3.7 illustrates the use of properties of vacancy to bound the probability that a given region is completely covered by random sets. In Section 3.8 we briefly discuss the effects on vacancy of non-Poisson models for centers of sets. This is the only place in the present chapter where coverage processes other than Boolean models are mentioned.

The remainder of this section introduces necessary definitions and notation, beginning with the concept of a random set. A practical (as opposed to abstract) definition of a random set is often confined only to closed sets, or only to open sets. We now indicate why. Let $\mathcal{B}$ be the class of Borel subsets of the real line, $\mathbb{R}$, that is, the σ-field generated by the open sets. Recall that a (scalar) random variable on a probability space $(\Omega, \mathcal{F}, P)$ is a measurable function $X : \Omega \to \mathbb{R}$, that is, a function such that $X^{-1}(B)$ is an element of the σ-field $\mathcal{F}$ for each $B \in \mathcal{B}$. This measurability is crucial, for without it we cannot ascribe probabilities to even simple events related to X. By analogy, if $\mathcal{X}$ is a class of subsets of $\mathbb{R}$ then a random set S, which is a function $S : \Omega \to \mathcal{X}$, will have to satisfy a property such as $S^{-1}(G) \in \mathcal{F}$ for all $G \in \mathcal{G}$, where $\mathcal{G}$ is a σ-field of sets of elements from $\mathcal{X}$. We could specify $\mathcal{G}$ in some abstract manner, but from a practical point of view we really have to introduce a metric on sets in $\mathcal{X}$ and define $\mathcal{G}$ in terms of that. Recall that $\mathcal{B}$ was generated by the open sets induced by the Euclidean metric on $\mathbb{R}$; if our definition of a random set is to be more than an abstraction, it will have to relate to a meaningful measure of distance on $\mathcal{X}$.

Let ρ be such a distance function, and assume for the sake of simplicity that the sets in $\mathcal{X}$ are subsets of the real line, such as intervals. Since the distribution of pattern in a one-dimensional Boolean model is translation-invariant (see Chapter 2), then ρ should be translation-invariant. Therefore

$$\rho([0, 1], (0, 1)) = \rho([x, x + 1], (x, x + 1))$$

for all x. The intervals $[x, x + 1]$ and $(x, x + 1)$ both converge to $(0, 1]$ as $x \downarrow 0$, and so we would expect that

$$\rho([x, x + 1], (x, x + 1)) \to \rho((0, 1], (0, 1]) = 0$$

as $x \downarrow 0$. In consequence,

$$\rho([0, 1], (0, 1)) = 0,$$

implying (if ρ is a metric) that $[0, 1] = (0, 1)$!

These properties indicate that a distance function ρ on the class of all bounded, nonempty intervals will have to treat intervals having the same endpoints as though they were the same set. In effect, we have to work with *equivalence classes* of intervals, such as $\mathcal{E}(a, b) \equiv \{(a, b), (a, b], [a, b), [a, b]\}$ (assuming $a \neq b$). We may choose a representative from each class—say the closed interval, or the open interval—and call ρ a metric on the space of all nonempty closed intervals, or on the space of all nonempty open intervals. Of course, convergence in this space means convergence in the sense of ρ, and of equivalence classes, so the interval $[x, x+1]$ would converge to $[0,1]$ [the closed representative of $\mathcal{E}(0,1)$] as $x \downarrow 0$, not to $(0,1]$.

We now return to k-dimensional space $\mathbb{R}^k$. Given $\mathcal{S}, \mathcal{S}_1, \mathcal{S}_2 \subseteq \mathbb{R}^k$ and $r > 0$, define

$$\mathcal{S}^r \equiv \left\{ \mathbf{y} \in \mathbb{R}^k : |\mathbf{x} - \mathbf{y}| < r \text{ for some } \mathbf{x} \in \mathcal{S} \right\} \tag{3.1}$$

(the set $\mathcal{S}$ expanded to include points less than r from its borders), and

$$\rho(\mathcal{S}_1, \mathcal{S}_2) \equiv \inf \left\{ t > 0 : \mathcal{S}_1 \subseteq \mathcal{S}_2^t \text{ and } \mathcal{S}_2 \subseteq \mathcal{S}_1^t \right\}. \tag{3.2}$$

The second formula defines the *Hausdorff metric*, ρ; it is a proper metric on the class of all nonempty compact sets, on the class of all nonempty bounded open sets, and in many other cases. It is *not* a metric on the *union* of the classes of nonempty compact sets and nonempty bounded open sets, for reasons which are clear from the example two paragraphs above.

Let $\mathcal{K}$ be the class of compact subsets of $\mathbb{R}^k$, $\mathcal{K}' \equiv \mathcal{K} \setminus \{\phi\}$ the class of nonempty compact sets, and $\mathcal{G}$ the Borel field (σ-field) of subsets of $\mathcal{K}'$ generated by the open sets induced by ρ. The metric space $(\mathcal{K}', \rho)$ is complete and separable (Federer 1969, Section 2.10.21). The topology on $\mathcal{K}'$ generated by ρ is called the *myope topology*. A random, nonempty compact set S defined on a probability space $(\Omega, \mathcal{F}, P)$ is a function $S : \Omega \to \mathcal{K}'$ such that $S^{-1}(G) \in \mathcal{F}$ for all $G \in \mathcal{G}$.

To define a random closed set (not necessarily bounded) we argue as follows. Let $\mathcal{X}$ be the class of all closed subsets of $\mathbb{R}^k$. Given $K \in \mathcal{K}$, put

$$\mathcal{X}^K \equiv \{E \in \mathcal{X} : E \cap K = \phi\}.$$

Let $\mathcal{G}^*$ be the smallest σ-field containing all sets $\mathcal{X}^K$, $K \in \mathcal{K}$. A *random closed set* (RACS) defined on the probability space $(\Omega, \mathcal{F}, P)$ is a function $S : \Omega \to \mathcal{X}$ such that $S^{-1}(G) \in \mathcal{F}$ for each $G \in \mathcal{G}^*$. If each realization of such an S is a bounded nonempty set then S is a random compact set in the sense given earlier. *Random open sets* may be defined in a similar manner to random closed sets: let $\mathcal{X}$ be the class of all open subsets of $\mathbb{R}^k$, replace $\mathcal{X}^K$ by

$$\mathcal{X}_K \equiv \{E \in \mathcal{X} : K \subseteq E\},$$

and change $\mathcal{G}^*$ to the smallest σ-field containing all sets $\mathcal{X}_K$, $K \in \mathcal{K}$ [see Matheron (1975, pp. 27ff, 48)]. A random, nonempty, *bounded* open set is easily defined in terms of the Hausdorff metric ρ.

In this chapter we do not require the topological niceties associated with random sets, except briefly in Section 3.5 when we introduce the concept of "convergence in distribution" for random compact sets. The main significance of the definitions as far as we are concerned is that they ensure that important scalar functions of random sets, such as content, are well-defined, measurable random variables taking values on the extended real line. But we should stress that henceforth, a random set will be assumed either closed with probability one or open with probability one.

There are many ways of generating a random set S. For example, S could be a sphere in $\mathbb{R}^k$ whose radius and center are random variables, or S could be the cell containing the origin formed by a collection of random $(k-1)$-dimensional planes, or S could be equal to one type of set (a sphere, say) with probability p and another type (a cube, say) with probability $1-p$. In each case, stipulating that S be closed or open amounts only to specifying whether the surface (boundary) of S be contained in S or outside S.

If S_1 and S_2 are both random bounded open sets, or random compact sets, nonempty with probability 1, we say that S_1 and S_2 have the *same distribution* if

$$E\{g(S_1)\} = E\{g(S_2)\}$$

for all bounded, real-valued functions g continuous with respect to the Hausdorff metric. If this identity holds for all bounded, real-valued, continuous, and *translation-invariant* g, we say that S_1 and S_2 have *essentially the same distribution*. These definitions are readily extended to more general random sets. For example, if S_1 and S_2 are both random closed sets, let

$$T(r) \equiv \left\{ \mathbf{x} \in \mathbb{R}^k : |\mathbf{x}| \leq r \right\}$$

be the closed sphere centered at the origin and of radius r, and define

$$S_i(r, \mathbf{x}) \equiv \begin{cases} S_i \cap T(r) & \text{if } S_i \cap T(r) \neq \phi \\ \{\mathbf{x}\} & \text{if } S_i \cap T(r) = \phi . \end{cases}$$

Then S_1 and S_2 have the same distribution if

$$E[g\{S_1(r, \mathbf{x})\}] = E[g\{S_2(r, \mathbf{x})\}]$$

for all $r > 0$, $\mathbf{x} \in \mathbb{R}^k$ and all bounded, real-valued, continuous g. Given a random set S we may construct a sequence $S_1, S_2, \ldots$ of independent "copies" of it by defining a product probability space and the associated product probability measure.

The *content*, or Lebesgue measure, $\|S\| = \|S\|_k$ of a Borel set $S \subseteq \mathbb{R}^k$, is of course its area if $k = 2$, its volume if $k = 3$, and so on. If S is a random subset of $\mathbb{R}^k$ then with probability 1, S is a Borel subset of $\mathbb{R}^k$, and $\|S\|$ is a random variable taking values on $[0,\infty]$. Given $\mathbf{x} = (x_1,\ldots,x_k) \in \mathbb{R}^k$, define $|\mathbf{x}| \equiv (x_1^2 + \ldots + x_k^2)^{1/2}$ — the usual Euclidean distance. For $-\infty < c < \infty$, let $cS \equiv \{c\mathbf{x} : \mathbf{x} \in S\}$.

A subset $\mathcal{R}$ of $\mathbb{R}^k$ is said to be *Riemann measurable* if its indicator function,

$$I(\mathcal{R})(\mathbf{x}) \equiv \begin{cases} 1 & \text{if } \mathbf{x} \in \mathcal{R} \\ 0 & \text{otherwise}, \end{cases}$$

is (directly) Riemann integrable. That is, for each $\epsilon > 0$ there exist finite sequences of k-dimensional cubes $\{\mathcal{D}_{i1}\}$ and $\{\mathcal{D}_{i2}\}$ such that

$$\bigcup_i \mathcal{D}_{i1} \subseteq \mathcal{R} \subseteq \bigcup_i \mathcal{D}_{i2}$$

and

$$\left\|\bigcup_i \mathcal{D}_{i2}\right\| - \epsilon \le \left\|\bigcup_i \mathcal{D}_{i1}\right\| + \epsilon.$$

This means that a Riemann measurable set $\mathcal{R}$ is always bounded, and we shall adhere to that convention. However, most of our proofs continue to work if boundedness is relaxed to $\|\mathcal{R}\| < \infty$. That is, we can get away with the weaker assumption that $i_\mathcal{R}$ is Riemann integrable as a limit, or equivalently, that $\|\mathcal{R}\| < \infty$ and

$$I(\mathcal{R} \cap [-n,n]^k)$$

is Riemann integrable for each n.

The *Minkowski sum* of two sets $S_1, S_2 \subseteq \mathbb{R}^k$ is given by

$$S_1 + S_2 \equiv \{\mathbf{x}_1 + \mathbf{x}_2 : \mathbf{x}_1 \in S_1, \mathbf{x}_2 \in S_2\}.$$

(Some authors write the sum as $S_1 \oplus S_2$.) If one of the sets is a singleton, $S_1 \equiv \{\mathbf{x}\}$ say, we write simply

$$\mathbf{x} + S \equiv \{\mathbf{x} + \mathbf{y} : \mathbf{y} \in S\},$$

which is just the translation of S by $\mathbf{x}$. Put also

$$\mathbf{x} - S \equiv \{\mathbf{x} - \mathbf{y} : \mathbf{y} \in S\}.$$

Recall that $T(r)$ is the closed sphere of radius r centered at the origin. The *radius* of a set $S \subseteq \mathbb{R}^k$ is given by

$$\text{rad}(S) \equiv \inf \left\{ r > 0 : \text{for some } \mathbf{x} \in \mathbb{R}^k, \ S \subseteq \mathbf{x} + T(r) \right\}.$$

If S is a random set then $\text{rad}(S)$ is a random variable taking values on the extended positive half-line. The content of a unit k-dimensional sphere is

$$v_k \equiv \|T(1)\| = \pi^{k/2}/\Gamma(1 + \tfrac{1}{2}k).$$

The surface of $T(1)$ is the set $\Omega = \Omega_k$ of all unit vectors in $\mathbb{R}^k$. Its $(k-1)$-dimensional content is

$$s_k = 2\pi^{k/2}/\Gamma(k/2).$$

Occasionally in this chapter we discuss the effect of "randomly and uniformly rotating" a random set S. A set that has undergone such a motion is said to be *isotropic*, or *randomly and uniformly oriented*. Varying the center of such a transformation has the overall effect only of translation, and so we assume that the rotation is about the origin. (We show three paragraphs below that random translations of the sets comprising a Boolean model have no effect on the distribution of the pattern.)

In two dimensions, the notion of a "random, uniformly distributed rotation" about the origin is trivial. Simply pick a random number Φ uniformly distributed on $(0, 2\pi)$ and stochastically independent of S, and rotate S about $\mathbf{O}$ until a vector $\mathbf{U}$ fixed to S is inclined at angle Φ to a fixed direction in the plane. In k dimensions a rotation about the origin is just an orthogonal transformation, given by an orthogonal matrix. Therefore a "random rotation" is representable by a random orthogonal matrix $\mathbf{A}$, determined by the joint distribution of its components. Isotropy requires that for any fixed orthogonal matrix $\mathbf{B}$, $\mathbf{AB}$ has the same distribution as $\mathbf{A}$. This invariance condition determines the distribution of $\mathbf{A}$ uniquely. The same distribution is determined by the condition that $\mathbf{BA}$ be distributed as $\mathbf{A}$ for all orthogonal $\mathbf{B}$ [see Loomis (1953, pp. 115 and 117) and Santaló (1976, Chapter 12)]. Apart from $k = 2$, the case $k = 3$ is the only one that we require with any degree of explicitness, and there the invariant transformation may be represented by a rotation through an angle Φ $(0 < \Phi < 2\pi)$ about an axis in direction Θ. Isotropy requires that Θ have a uniform distribution on Ω (the space of all unit vectors in $\mathbb{R}^3$), that Φ have the distribution with

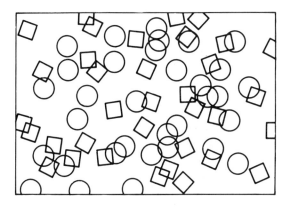

Figure 3.1. Boolean model in the case $k = 2$. Shapes are either disks or square laminae, the sizes of each being fixed and each type occurring with probability $\frac{1}{2}$.

density $\pi^{-1}\sin^2(\frac{1}{2}\phi)$ on $(0, 2\pi)$, and that Θ and Φ be independent (see, e.g., Kendall and Moran 1963, p. 93ff). Heuristic arguments suggesting that Φ should be uniform are therefore invalid.

A Boolean model in k-dimensional Euclidean space is just the coverage pattern created by a Poisson-distributed sequence of random sets. Specifically, let

$$\mathcal{P} \equiv \{\xi_i, i \geq 1\}$$

be a stationary Poisson process of intensity λ in $\mathbb{R}^k$, the points ξ_i being indexed in any systematic order. Let $S_1, S_2, \ldots$ be independent and identically distributed random sets, independent of $\mathcal{P}$. Then

$$\mathcal{C} \equiv \{\xi_i + S_i, i \geq 1\}$$

is a *Boolean model* (see Figure 3.1). We call the sets S_i *shapes*, to help distinguish them from the sets $\xi_i + S_i$, and say that $\xi_i + S_i$ represents the shape S_i *centered at* the point ξ_i. Thus, $\xi_i + S_i$ has "center" ξ_i. In this notation, the random sets (line segments) in the simple linear Boolean models of Chapter 2 have as their "centers" their left-hand endpoints. We say that the Poisson process $\mathcal{P}$ *drives* the Boolean model, and that the shapes S_i *generate* the model. We always assume that the shapes S_i are nonempty. A set or a point is said to be *covered* by the Boolean model if it is contained in the union $\bigcup_{i \geq 1}(\xi_i + S_i)$.

Let $\mathcal{C}$ denote the Boolean model just defined, and let $\mathcal{C}'$ be a Boolean model in which the driving Poisson process is the same but

the random shapes S_i are replaced by $\eta_i + S_i$, where η_i are random
vectors independent of $\mathcal{P}$ and such that $\eta_i + S_i$ are independent and
identically distributed random sets. Then $\mathcal{C}$ and $\mathcal{C}'$ have the same
properties. To see this it is necessary only to note that conditional
on $\{S_i, i \geq 1\}$, the point process $\mathcal{P}' \equiv \{\xi_i + \eta_i\}$ is stationary Poisson
with the same intensity as $\mathcal{P}$ (see Chapter 1, Section 1.7). Therefore
$\mathcal{C}'$, like $\mathcal{C}$, is obtained by centering shapes S_i at points of a stationary
Poisson process of intensity λ. This property motivates the following
definition. Two Boolean models, driven by Poisson processes $\mathcal{P}$ and
$\mathcal{P}'$ and with shapes distributed as S and S', respectively, are said to
have the same distribution if $\mathcal{P}$ and $\mathcal{P}'$ have the same intensity and
if for some random vector ξ, S' has the same distribution as $\xi + S$.

In the case $k = 1$, the Boolean models defined here are more general than the simple Boolean models treated in Chapter 2, since we
no longer insist that the shapes S_i be line segments—they may be
unions of disjoint line segments, for example. Recall that a simple
linear Boolean model can be thought of as an infinite-server queue
in which each line segment represents a service time. By analogy,
a general linear Boolean model is an infinite-server queue in which
each customer can keep coming back for additional services.

In a small number of places in this chapter we require the notion
of an "arbitrary," or "typical," or "randomly chosen" region having
certain specific properties, such as an "arbitrary cell" formed by a
Poisson field of hyperplanes. Chapter 4, Section 4.1, will investigate
aspects of this notion, and it is inconvenient to encumber ourselves
with details here. Fortunately the arbitrariness that we require is
conceptually quite straightforward. If explicitness is required, properties of an arbitrary region can be formally and rigorously defined
by an ergodic argument, as in Miles (1972, pp. 246–247). For example, suppose we wish to define the probability that an "arbitrary"
set in the Boolean model $\mathcal{C}$ intersects at least one other set from
$\mathcal{C}$. Let $I_i = 1$ if $(\xi_i + S_i) \cap (\xi_j + S_j) \neq \phi$ for some $j \neq i$, and $I_i = 0$
otherwise. Then

$$p(r) \equiv \left(\sum_{i:\xi_i \in \mathcal{T}(r)} I_i \right) \Big/ \left(\sum_{i:\xi_i \in \mathcal{T}(r)} 1 \right)$$

equals the proportion of sets centered within the sphere $\mathcal{T}(r)$ that
intersects at least one other set. By the ergodic theorem (Appendix
II), the limit

$$\lim_{r \to \infty} p(r)$$

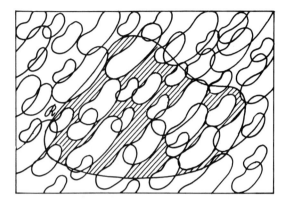

Figure 3.2. Vacancy within region $\mathcal{R}$, for sets with fixed orientation. The random shapes are random scale changes of a fixed sausage-shaped set $\mathcal{S}$, of fixed orientation. Uncovered region is cross-hatched; its area equals V.

exists and is almost surely constant. We define this constant to be the chance that an arbitrary set in $\mathcal{C}$ intersects some other set in $\mathcal{C}$. Of course, the value of the limit does not depend on our decision to concentrate on sets centered within $\mathcal{T}(r)$; many other large regions could have been used instead of $\mathcal{T}(r)$.

3.2. BASIC PROPERTIES OF VACANCY

Let $\mathcal{R}$ be a Borel subset of $\mathbb{R}^k$, and let $\mathcal{C}$ denote our coverage process (a Boolean model in $\mathbb{R}^k$). The vacancy V within $\mathcal{R}$ is defined to be the k-dimensional content of that part of $\mathcal{R}$ not covered by any random sets from $\mathcal{C}$:

$$V = V(\mathcal{R}) \equiv \int_{\mathcal{R}} \chi(\mathbf{x})\,d\mathbf{x}, \qquad (3.3)$$

where

$$\chi(\mathbf{x}) \equiv \begin{cases} 1 & \text{if for all } i, \ \mathbf{x} \notin \xi_i + S_i \\ 0 & \text{otherwise.} \end{cases}$$

Figure 3.2 illustrates the region whose area equals V in the case where random shapes are simply scale changes of a fixed set S, with fixed orientation, and where $k = 2$. Uncovered parts of a region are often referred to as the *pores*.

To find the expected value of vacancy we use Fubini's theorem and take expectations under the integral sign in (3.3). Therefore we must calculate

$$
\begin{aligned}
E\{\chi(\mathbf{x})\} &= P(\mathbf{x}\text{ not covered}) \\
&= P(\text{for all } i,\ \mathbf{x} \notin \xi_i + S_i) \\
&= P(\text{for all } i,\ \xi_i \notin \mathbf{x} - S_i) \\
&= P(\text{for all } i,\ \xi_i \notin S_i),
\end{aligned}
$$

the last equality following from symmetry and homogeneity of the Poisson process. There is a variety of ways of calculating this probability. We give here a direct argument; see Chapter 4, Section 4.1 for a different approach. Let $\alpha \equiv E(\|S\|) < \infty$ denote the expected content of a random shape. Suppose initially that for some $r > 0$, and with probability 1, $|\mathbf{x}| \le r$ for each $\mathbf{x} \in S$. Let the region $\mathcal{A}$ be at least as large as the sphere of radius r centered at the origin. If a point ξ is placed at random into $\mathcal{A}$ then the probability that the point does not lie within S equals $1 - (\alpha/\|\mathcal{A}\|)$. If N points $\xi^{(1)}, \dots, \xi^{(N)}$ are placed independently and uniformly into $\mathcal{A}$, then (conditional on N) the chance that $\xi^{(i)} \notin S_i$ for $1 \le i \le N$ equals $\{1 - (\alpha/\|\mathcal{A}\|)\}^N$. If N is Poisson-distributed with mean $\lambda\|\mathcal{A}\|$, as it is if $\xi^{(1)}, \dots, \xi^{(N)}$ denote those points of the Poisson process that fall within $\mathcal{A}$, then

$$
E[\{1 - (\alpha/\|\mathcal{A}\|)\}^N] = \exp\{-(\alpha/\|\mathcal{A}\|)E(N)\} = e^{-\alpha\lambda}.
$$

Therefore

$$
P(\mathbf{x}\text{ not covered}) = P(\text{for all } i,\ \xi_i \notin S_i) = e^{-\alpha\lambda}. \qquad (3.4)
$$

An argument based on taking limits shows that this formula remains valid when S is unbounded, even when $\alpha = E(\|S\|) = \infty$. Substituting into (3.3) we see that

$$
\begin{aligned}
E(V) &= \int_{\mathcal{R}} E\{\chi(\mathbf{x})\}\, d\mathbf{x} \\
&= \|\mathcal{R}\| P(\mathbf{O}\text{ not covered}) \\
&= \|\mathcal{R}\| \exp\{-\lambda E(\|S\|)\}. \qquad (3.5)
\end{aligned}
$$

Thus, the expected vacancy within a region depends on the distribution of random shapes only through their expected content. In particular, it is not affected by random rotations of the shapes, even though this might have a significant effect on the pattern of covered and uncovered areas. Compare Figures 3.2 and 3.3.

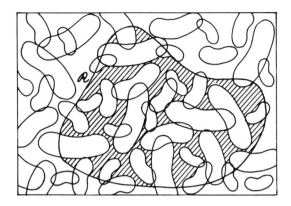

Figure 3.3. Vacancy within region $\mathcal{R}$, for sets with random orientation. (As in Figure 4.2, except the shapes have been given a random rotation uniformly distributed on $[0, 2\pi)$.)

The expected proportion of $\mathcal{R}$ that is not covered,

$$p \equiv E(V)/\|\mathcal{R}\| = \exp\{-\lambda E(\|S\|)\},$$

is often called *porosity.*

The variance of V may be calculated similarly. Indeed,

$$
\begin{aligned}
E\{\chi(\mathbf{x}_1)\chi(\mathbf{x}_2)\} &= P(\text{for all } i, \ \mathbf{x}_1 \notin \boldsymbol{\xi}_i + S_i \text{ and } \mathbf{x}_2 \notin \boldsymbol{\xi}_i + S_i) \\
&= P(\text{for all } i, \ \boldsymbol{\xi}_i \notin \mathbf{x}_1 - \mathbf{x}_2 + S_i \text{ and } \boldsymbol{\xi}_i \notin S_i) \\
&= P\{\text{for all } i, \ \boldsymbol{\xi}_i \notin (\mathbf{x}_1 - \mathbf{x}_2 + S_i) \cup S_i\} \\
&= \exp\left[-\lambda E\{\|(\mathbf{x}_1 - \mathbf{x}_2 + S) \cup S\|\}\right] \\
&= \exp\left[-2\lambda E(\|S\|) + \lambda E\{\|(\mathbf{x}_1 - \mathbf{x}_2 + S) \cap S\|\}\right].
\end{aligned}
$$

Therefore

$$
\begin{aligned}
\operatorname{cov}\{\chi(\mathbf{x}_1), \chi(\mathbf{x}_2)\} &= \exp\{-2\lambda E(\|S\|)\} \\
&\quad \times (\exp[\lambda E\{\|(\mathbf{x}_1 - \mathbf{x}_2 + S) \cap S\|\}] - 1),
\end{aligned}
$$

and

$$
\begin{aligned}
\operatorname{var}(V) &= \int\!\!\int_{\mathcal{R}^2} \operatorname{cov}\{\chi(\mathbf{x}_1), \chi(\mathbf{x}_2)\} \, d\mathbf{x}_1, d\mathbf{x}_2 \\
&= \exp\{-2\lambda E(\|S\|)\} \\
&\quad \times \int\!\!\int_{\mathcal{R}^2} (\exp[\lambda E\{\|(\mathbf{x}_1 - \mathbf{x}_2 + S) \cap S\|\}] - 1) \, d\mathbf{x}_1 \, d\mathbf{x}_2. \quad (3.6)
\end{aligned}
$$

It is clear from this formula that variance of vacancy depends on much more than expected shape content.

The formulas for mean and variance of vacancy simplify in the case of $k = 1$ dimension. For example, suppose the shape S is a line segment of length L and with distribution function G. Put $\alpha = E(L)$. Then the vacancy V within the interval $[0, t]$ has mean $E(V) = te^{-\alpha\lambda}$ and variance

$$\text{var}(V) = e^{-2\alpha\lambda} \int_0^t \int_0^t (\exp[\lambda E \{(L - |x_1 - x_2|)$$
$$\times I(L > |x_1 - x_2|)\}] - 1) \, dx_1 \, dx_2$$
$$= 2e^{-2\alpha\lambda} \int_0^t (t - x) \left(\exp \left[\lambda \int_x^\infty \{1 - G(y)\} \, dy \right] - 1 \right) dx.$$

When segment length is fixed and equal to a, $E(V) = te^{-a\lambda}$ and

$$\text{var}(V) = 2\lambda^{-2} e^{-a\lambda} [\lambda t - 1 + e^{-\lambda \min(a,t)} - \tfrac{1}{2} \{\lambda \min(a,t)\}^2 e^{-a\lambda}$$
$$- \lambda \max(0, t - a)(1 + a\lambda) e^{-a\lambda}].$$

Mean and variance of vacancy may be used to bound the probability of (almost) complete coverage, as follows. By the Cauchy-Schwarz inequality,

$$E(V) = E\{V I(V > 0)\} \leq \left\{ E(V^2) P(V > 0) \right\}^{1/2},$$

whence

$$P(V > 0) \geq (EV)^2 / E(V^2) = 1 - \text{var}(V) / E(V^2)$$

and

$$P(V = 0) \leq \text{var}(V) / E(V^2). \tag{3.7}$$

Notice that $0 \leq V \leq \|\mathcal{R}\|$, and so if $\mathcal{R}$ has finite content then all moments of $V = V(\mathcal{R})$ are finite. Furthermore, $E(V) = 0$ implies $V = 0$ with probability 1. We now see from formula (3.5) that if $\lambda > 0$, $\|\mathcal{R}\| < \infty$, and expected shape content is infinite, then almost all of $\mathcal{R}$ is covered with probability 1. Our first theorem generalizes this result.

Theorem 3.1. *If $\lambda > 0$ and $E(\|S\|) = \infty$ then with probability 1, almost all of $\mathbb{R}^k$ is covered by random sets. Conversely, if $E(\|S\|) < \infty$ then with probability 1 the uncovered portion of $\mathbb{R}^k$ has infinite Lebesgue measure.*

Proof. Suppose $E(\|S\|) = \infty$. Partition $\mathbb{R}^k$ into a countable disjoint union of unit cubes $\mathcal{D}_i$. Then

$$P(\text{vacancy in } \mathbb{R}^k > 0) \leq \sum_i P(\text{vacancy in } \mathcal{D}_i > 0) = 0.$$

Conversely, suppose $E(\|S\|) < \infty$. Fix $n \geq 2$ and let $\mathcal{D}_{ni}$, $i \geq 1$, be any collection of unit cubes in $\mathbb{R}^k$ with adjacent cubes distant at least $n - 1$ apart. For example,

$$\mathcal{D}_{ni} \equiv (0, 1)^{k-1} \times (ni, \, ni + 1), \qquad i \geq 1,$$

will do. Then by Markov's inequality,

$$\begin{aligned}
P(\text{almost all of } \mathcal{D}_{ni} \text{ covered}) &= P(\text{almost all of } \mathcal{D}_1 \text{ covered}) \\
&= P(1 - V_1 \geq 1) \\
&\leq E(1 - V_1) \\
&= 1 - \exp\{-\lambda E(\|S\|)\} \\
&< 1,
\end{aligned}$$

and so for some $\epsilon > 0$,

$$p \equiv P(\text{measure of uncovered portion of } \mathcal{D}_i \text{ exceeds } \epsilon) > 0.$$

Fix j and m, $1 \leq j \leq m < \infty$, and let $n \to \infty$. Then

$$\begin{aligned}
&P(\text{at least } j \text{ of } \mathcal{D}_{n1}, \ldots, \mathcal{D}_{nm} \text{ have uncovered portions} \\
&\quad \text{whose measure exceeds } \epsilon) \\
&\to \sum_{i=j}^{m} \binom{m}{i} p^i (1 - p)^{m-i},
\end{aligned}$$

since $V(\mathcal{D}_{n1}), \ldots, V(\mathcal{D}_{nm})$ are asymptotically independent as $n \to \infty$ (see Exercise 3.1). The left-hand side is less than the probability that the vacancy in $\mathbb{R}^k$ exceeds $j\epsilon$, and the right-hand side converges to 1 as $m \to \infty$, for each j. Therefore $P\{V(\mathbb{R}^k) \geq j\epsilon\} = 1$ for each j, proving that $P\{V(\mathbb{R}^k) = \infty\} = 1$. $\square$

It is not true that if $E(\|S\|) = \infty$ then $\mathbb{R}^k$ is completely covered with probability 1. The problem of determining conditions under which $\mathbb{R}^k$ is almost surely completely covered is a delicate one, and is related to the notion of random Cantor-like sets, as we shall explain in a short while. We treat only the case $k = 1$, and assume shapes have a specific, fixed form.

Let $l_1, l_2, \ldots$ be positive numbers converging to 0 and satisfying

$$l^* \equiv \max_{1 \leq i < \infty} l_i < 1,$$

and put either

$$\mathcal{S} \equiv \bigcup_{i=1}^{\infty} (i, i + l_i) \quad \text{or} \quad \mathcal{S} \equiv \bigcup_{i=1}^{\infty} [i, i + l_i].$$

The former set is open, the latter closed. Let $l_1' \geq l_2' \geq \ldots$ denote the sequence $\{l_i\}$ arranged in decreasing order, and let $\mathcal{C}$ be the linear Boolean model in which each shape equals $\mathcal{S}$ and the driving Poisson process has intensity $\lambda > 0$. The next theorem gives necessary and sufficient conditions for such a Boolean model to completely cover $\mathbb{R}$ with probability 1; they are strictly stronger than the condition $\|\mathcal{S}\| < \infty$.

Theorem 3.2. *The probability that the Boolean model $\mathcal{C}$ defined above completely covers $\mathbb{R}$ equals 0 or 1 according as the series*

$$\sum_{n=1}^{\infty} n^{-2} \exp\left\{\lambda(l_1' + \ldots + l_n')\right\} \tag{3.8}$$

converges or diverges.

Proof. It suffices to consider the case where $\mathcal{S}$ is open, for if $\mathcal{S}$ is closed we may proceed by treating separately the cases where

$$\mathcal{S} \equiv \bigcup_{i=1}^{\infty} (i, i + l_i + 2^{-i}(1 - l^*)) \quad \text{or} \quad \mathcal{S} \equiv \bigcup_{i=1}^{\infty} (i, i + l_i - 2^{-i}l_i).$$

Let $\epsilon \equiv 1 - l^* > 0$, and study the pattern of intervals intersecting $\mathcal{I} \equiv [0, \epsilon]$. Any interval intersecting $\mathcal{I}$ is of the form $\xi_i + (j, j + l_j)$, for some $j \geq 1$ and some point ξ_i from the Poisson process $\mathcal{P}$ driving $\mathcal{C}$. The intersection is nonempty if and only if both

$$\xi_i + j < \epsilon \quad \text{and} \quad \xi_i + j + l_j > 0;$$

that is, $\xi_i \in \mathcal{J}_j \equiv (-j-l_j, -j+\epsilon)$. Since each $l_j \leq 1-\epsilon$ then the intervals $\mathcal{J}_j$, $j \geq 1$ are disjoint, and so we may consider those points ξ_i lying within $\mathcal{J}_j$ to come from *independent* Poisson processes $\mathcal{P}_j$, each of intensity λ. Thus, the distribution of the pattern of intervals intersecting $\mathcal{I}$ is the same as that which would have occurred had we placed the points of an infinite sequence $\mathcal{P}_1, \mathcal{P}_2, \ldots$ of independent Poisson processes, each of intensity λ, along the real line $(-\infty, \infty)$, and at the points $\{\xi_{ji}, i \geq 1\}$ of $\mathcal{P}_j$ placed the interval $(0, l'_j)$, forming the doubly infinite sequence of intervals $\{\xi_{ji} + (0, l'_j), i \geq 1, j \geq 1\}$. Shepp's theorem (Shepp 1972b; see Appendix III) treats this type of process, and implies that $\mathcal{I}$ is completely covered with probability 1 if and only if the series at (3.8) diverges. In consequence, the whole real line is covered with probability 1 if and only if that series diverges. (Note that $\mathbb{R}$ is contained within the union of a countable number of translates of $\mathcal{I}$, that any given translate is covered with probability 1 if and only if $\mathcal{I}$ is covered with probability 1, and that the intersection of a countable number of almost sure events occurs almost surely.)

If the series at (3.8) converges then the probability p that $\mathcal{I}$ is completely covered is strictly less than 1. We prove finally that $\mathbb{R}$ is almost surely not completely covered. Suppose the points ξ_i in $\mathcal{P}$ are reindexed such that $\ldots < \xi_{-1} < 0 < \xi_0 < \xi_1 < \ldots$. Put

$$U_n \equiv \bigcup_{1 \leq |i| \leq n} (\xi_i + \mathcal{S}).$$

The sets $\mathcal{I} \cap \tilde{U}_n$, $n \geq 1$, are closed and decreasing, and so their intersection

$$\bigcap_{n=1}^{\infty} (\mathcal{I} \cap \tilde{U}_n) = \{\text{points in } \mathcal{I} \text{ not covered by } \mathcal{C}\}$$

is empty if and only if $\mathcal{I} \cap \tilde{U}_n$ is empty for all sufficiently large n. In consequence, $\mathcal{I}$ is completely covered by the Boolean model if and only if it is completely covered by a finite number of sets $\xi_i + \mathcal{S}$, and

$$P(\mathcal{I} \text{ completely covered}) = \lim_{r \to \infty} P\left\{\mathcal{I} \subseteq \bigcup_{i:|\xi_i| \leq r} (\xi_i + \mathcal{S})\right\}.$$

It follows that for each $m \geq 1$,

$$P(\mathbb{R} \text{ completely covered})$$

$$\leq \lim_{n \to \infty} P(in + \mathcal{I} \text{ completely covered}, 1 \leq i \leq m)$$

$$= \lim_{n \to \infty} P\left\{ in + \mathcal{I} \subseteq \bigcup_{j:|in-\xi_j| \leq \frac{1}{2}n} (\xi_j + \mathcal{S}), 1 \leq i \leq m \right\}$$

$$= \lim_{n \to \infty} \prod_{i=1}^{m} P\left\{ in + \mathcal{I} \subseteq \bigcup_{j:|in-\xi_j| \leq \frac{1}{2}n} (\xi_j + \mathcal{S}) \right\}$$

$$= p^m.$$

Therefore, since $p < 1$,

$$P(\mathbb{R} \text{ completely covered}) = 0. \quad \square$$

Cases where the series at (3.8) converges for all $\lambda > 0$ include

$$l_{n-2} \equiv n^{-1}(\log n)^{-\epsilon}, \qquad n \geq 3,$$

for any $\epsilon > 0$. If we choose $0 < \epsilon \leq 1$ then we shall have

$$E(\|\mathcal{S}\|) = \|\mathcal{S}\| = \sum_{i=1}^{\infty} l_i = \infty,$$

but

$$P(\mathbb{R} \text{ completely covered}) = 0.$$

It may be proved that if the series at (3.8) converges for all $\lambda > 0$ then with probability 1, the number of points in $\mathbb{R}$ that are not covered is uncountably infinite for each $\lambda > 0$ (see Exercise 3.2). In this case, and assuming $\sum_{i \geq 1} l_i = \infty$, the set of uncovered points is almost surely both uncountable and of measure 0, a property shared by the Cantor set. The Cantor ternary set C is an uncountably infinite, measure-zero subset of the interval $[0, 1]$, and has the curious distinction that its difference set,

$$C - C \equiv \{x - y : x, y \in C\},$$

equals the entire interval $[-1, 1]$ (Gelbaum and Olmsted 1964, p. 87ff). This means that in the Boolean model where each shape S_i is the repeated Cantor set

$$\bigcup_{-\infty < i < \infty} (i + C),$$

which has measure 0, each random set $\xi_i + S_i$ has an infinite number of points in common with *each* other random set. This event occurs not just almost surely, but absolutely certainly!

Examples such as those above are pathological, and are not often encountered in practice. As a general rule, the property $E(\|S\|) = \infty$ in a k-dimensional Boolean model does indicate that all of $\mathbb{R}^k$ is covered with probability 1. In particular, this conclusion is valid if S contains a sphere whose mean content is infinite (see Exercise 3.4).

As far as the relationship between vacancy and coverage is concerned, the case where $E(\|S\|) < \infty$ is relatively straightforward. Before demonstrating this fact we mention two classes of pathological example. First, take $\mathcal{R}'$ to be a closed set and $\mathbf{x}$ any point not in $\mathcal{R}$, and put $\mathcal{R} \equiv \mathcal{R}' \cup \{\mathbf{x}\}$. Then $V(\mathcal{R}) = V(\mathcal{R}')$, but $\mathcal{R}$ and $\mathcal{R}'$ can have quite different probabilities of complete coverage. Second, take the shapes generating the Boolean model to be identical to the unit sphere $\mathcal{T}(1)$, but excluding all rational k-tuples $\mathbf{x} = (x_1, \ldots, x_k)$. A subset $\mathcal{R}$ of $\mathbb{R}^k$ can be completely covered by the spheres and so have zero vacancy, but not be completely covered by the Boolean model, because of the "holes" in the spheres. The following theorem shows that these pathologies vanish if we insist that $\mathcal{R}$ be *open* and S be *closed*.

Theorem 3.3. *Let $\mathcal{C}$ be a Boolean model in $\mathbb{R}^k$ in which shapes are distributed as S. If $\mathcal{R}$ is an open subset of $\mathbb{R}^k$, and S is a random closed set with $E(\|S\|) < \infty$, then the event*

$$\{V(\mathcal{R}) = 0; \ \mathcal{R} \text{ not completely covered}\}$$

has probability 0.

This result may be generalized to other types of coverage processes. The proof is essentially topological, and makes relatively little use of Boolean model properties.

Proof. Assume that the points $\xi_1, \xi_2, \ldots$ of the driving Poisson process $\mathcal{P}$ are indexed such that ξ_i is ith closest to the origin. Put

$$U_n \equiv \bigcup_{i=1}^{n} (\xi_i + S_i), \qquad U^{(n)} \equiv \bigcup_{i=n+1}^{\infty} (\xi_i + S_i),$$

$$U \equiv \bigcup_{i=1}^{\infty} (\xi_i + S_i) = U_n \cup U^{(n)},$$

and $W_n \equiv \{\xi_1, \ldots, \xi_n; S_1, \ldots, S_n\}$. In this notation we must prove that

$$P\{V(\mathcal{R}) = 0; \mathcal{R} \cap \tilde{U} \neq \phi\} = 0. \tag{3.9}$$

If $\mathcal{R}_1, \mathcal{R}_2, \ldots$ are open regions then

$$P\left\{V\left(\bigcup_{i=1}^{\infty} \mathcal{R}_i\right) = 0; \left(\bigcup_{i=1}^{\infty} \mathcal{R}_i\right) \cap \tilde{U} \neq \phi\right\}$$

$$\leq \sum_{i=1}^{\infty} P\left\{V(\mathcal{R}_i) = 0; \mathcal{R}_i \cap \tilde{U} \neq \phi\right\},$$

and so it suffices to prove (3.9) when $\mathcal{R}$ is a bounded open set. In fact, by making the obvious scale and location change we may assume that $\mathcal{R}$ is a subset of $\mathcal{T}(1)$ (the closed sphere of unit radius centered at the origin.) This assumption is made throughout the proof below.

Since $\{\mathcal{R} \subseteq U\} \subseteq \{V(\mathcal{R}) = 0\}$ then result (3.9) is equivalent to

$$P(\mathcal{R} \subseteq U) = P\{V(\mathcal{R}) = 0\}. \tag{3.10}$$

Notice that, with $V = V(\mathcal{R})$,

$$P(\mathcal{R} \subseteq U) = P(\mathcal{R} \cap \tilde{U}_n = \phi) + P(\phi \neq \mathcal{R} \cap \tilde{U}_n \subseteq U^{(n)})$$

and

$$P(V = 0) = P(\|\mathcal{R} \cap \tilde{U}_n\| = 0) + P(0 \neq \|\mathcal{R} \cap \tilde{U}_n\|; V = 0).$$

Since $\mathcal{R}$ is open and U_n is closed then $\mathcal{R} \cap \tilde{U}_n$ is open, and so $\mathcal{R} \cap \tilde{U}_n = \phi$ if and only if $\|\mathcal{R} \cap \tilde{U}_n\| = 0$. Therefore

$$P(\mathcal{R} \cap \tilde{U}_n = \phi) = P(\|\mathcal{R} \cap \tilde{U}_n\| = 0), \qquad n \geq 1,$$

and so (3.10) will be proved if we show that each of

$$P(\phi \neq \mathcal{R} \cap \tilde{U}_n \subseteq U^{(n)}), \qquad P(0 \neq \|\mathcal{R} \cap \tilde{U}_n\|; V = 0)$$

converges to 0 as $n \to \infty$. The first of these probabilities equals

$$E\left\{P(\phi \neq \mathcal{R} \cap \tilde{U}_n \subseteq U^{(n)} \mid W_n)\right\}$$

$$\leq E\left\{\sup_{\mathbf{x} \in \mathcal{R}} P(\mathbf{x} \in U^{(n)} \mid W_n)\right\},$$

while the second equals

$$E[P\{\text{for some } \mathbf{x} \in \mathcal{R} \text{ and } 0 < \epsilon < 1, \text{ the sphere}$$
$$\mathbf{x} + \mathcal{T}(\epsilon) \subseteq \mathcal{R} \cap \tilde{U}_n; V = 0 \mid W_n\}]$$
$$\leq E\left[\sup_{\mathbf{x} \in \mathcal{R}, 0 < \epsilon < 1} P\left\{ \|(\mathbf{x} + \mathcal{T}(\epsilon)) \cap \tilde{U}^{(n)}\| = 0 \,\Big|\, W_n \right\} \right].$$

If $\mathbf{x} \in \mathcal{R}$ and $0 < \epsilon < 1$ then

$$P\left[\| \{\mathbf{x} + \mathcal{T}(\epsilon)\} \cap \tilde{U}^{(n)} \| = 0 \,\Big|\, W_n \right]$$
$$= P\left[\| \{\mathbf{x} + \mathcal{T}(\epsilon)\} \cap U^{(n)} \| \geq \|\mathbf{x} + \mathcal{T}(\epsilon)\| \,\Big|\, W_n \right]$$
$$\leq \|\mathcal{T}(\epsilon)\|^{-1} E\left[\| \{\mathbf{x} + \mathcal{T}(\epsilon)\} \cap U^{(n)} \| \,\Big|\, W_n \right]$$
$$= \|\mathcal{T}(\epsilon)\|^{-1} \int_{\mathbf{x} + \mathcal{T}(\epsilon)} P(\mathbf{y} \in U^{(n)} \mid W_n) \, d\mathbf{y}$$
$$\leq \sup_{\mathbf{y} \in \mathcal{T}(2)} P(\mathbf{y} \in U^{(n)} \mid W_n).$$

Consequently it suffices to prove that

$$E\left\{ \sup_{\mathbf{x} \in \mathcal{T}(2)} P(\mathbf{x} \in U^{(n)} \mid W_n) \right\} \to 0$$

as $n \to \infty$, or equivalently, that

$$\inf_{\mathbf{x} \in \mathcal{T}(2)} P(\text{for each } i \text{ with } |\boldsymbol{\xi}_i| > r, \mathbf{x} \notin \boldsymbol{\xi}_i + S_i) \to 1$$

as $r \to \infty$. Using the argument leading to (3.4) we see that the left-hand side here equals

$$\inf_{\mathbf{x} \in \mathcal{T}(2)} \exp[-\lambda E\{\|(\mathbf{x} - S) \cap \mathcal{T}(r)\tilde{\ }\|\}] \geq \exp[-\lambda E\{\|S \cap \mathcal{T}(r-2)\tilde{\ }\|\}]$$
$$\to 1,$$

as required. $\square$

In view of Theorem 3.3 we may assume for most practical purposes that the event $V(\mathcal{R}) = 0$ is equivalent to $\mathcal{R}$ being completely covered. Although Theorem 3.3 is stated only for open regions $\mathcal{R}$ and closed shapes S, one may readily prove versions of it for sufficiently regular closed $\mathcal{R}$ and (or) open S; see Exercise 3.4, for example. The major requirement is that we exclude pathological cases such as those described just prior to the theorem.

Exact distribution of vacancy is usually prohibitively complex, although some progress can be made in working out its properties for the case of one dimension. We treat next the simplest situation.

3.3. DISTRIBUTION OF VACANCY IN ONE DIMENSION, FOR FIXED SEGMENT LENGTH

Consider a linear Boolean model in which Poisson intensity equals λ, shapes are segments distributed along a line, and segment length is fixed and equal to a. Exactly the same model was treated in Chapter 2, Section 2.2. Assume the segments are placed to the *right* of points of the driving Poisson process $\mathcal{P}$. There is a positive probability that no segments overlap $[0, t]$, and also a positive probability $p(t)$, given in Theorem 2.6, that the interval is completely covered. Thus, the distribution of V has atoms at the points 0 and t. Let n denote the integer part of t/a. If $1 \leq j \leq n$ then there is a positive probability that precisely j nonoverlapping segments lie within $[0, t]$, and that no other segments overlap into $[0, t]$. Thus, V has atoms at the points $t - ja$ for $1 \leq j \leq n$. The $n + 2$ points $0, t - na, \ldots, t - a$ and t ($n + 1$ points if t/a is an integer) represent the only atoms of V.

We shall determine the amount of mass at the atom ja, for $0 \leq j \leq n$. Observe that $V = t - ja$ if and only if

(i) no points of the Poisson process $\mathcal{P}$ lie within $[-a, 0]$ or $[t - a, t]$;

(ii) precisely j points of $\mathcal{P}$ lie between 0 and $t - a$; and

(iii) none of these j points is distant less than a from the nearest adjacent point (*not* including the endpoints 0 and $t - a$).

Events (i) and (ii) have probabilities $e^{-2a\lambda}$ and $\{\lambda(t - a)\}^j e^{-\lambda(t-a)} /j!$, respectively. The probability of event (iii), conditional on (ii), equals the chance that the smallest of the $j - 1$ spacings between j independent variables $U_1, \ldots, U_j$, each distributed uniformly on $[0, 1]$, is at least $a/(t - a)$. Let this probability be $q_j\{a/(t - a)\}$. We shall prove that if $0 \leq (j-1)u \leq 1$, then

$$q_j(u) = \{1 - (j-1)u\}^j .$$

The order statistics $U_{(1)} \leq \ldots \leq U_{(j)}$ of a sample of size j from the uniform distribution on $[0,1]$ have joint density

$$f_U(u_{(1)}, \ldots, u_{(j)}) = j!, \qquad 0 < u_{(1)} < \ldots < u_{(j)} < 1.$$

Define $U_{(0)} \equiv 0$, and let $W_i \equiv U_{(i)} - U_{(i-1)}$, $1 \leq i \leq j$. By a simple change of variable we deduce the density of the W_i's:

$$f_W(w_1, \ldots, w_j) = j!, \quad \text{each } w_i > 0 \quad \text{and} \quad \sum_1^j w_i < 1.$$

Therefore

$$q_j(u) = P(W_2 > u, \ldots, W_j > u)$$

$$= j! \int_u^{1-(j-2)u} dw_j \int_u^{1-(j-3)u-w_j} dw_{j-1} \ldots$$

$$\ldots \int_u^{1-u-\sum_4^j w_i} dw_3 \int_u^{1-\sum_3^j w_i} dw_2 \int_0^{1-\sum_2^j w_i} dw_1$$

$$= \{1 - (j-1)u\}^j ,$$

as required.

Combining the results of the previous two paragraphs we see that for $0 \leq j \leq n \equiv [t/a]$,

$$P(V = t - ja) = e^{-2a\lambda} \frac{\{\lambda(t-a)\}^j}{j!} e^{-\lambda(t-a)} q_j \left(\frac{a}{t-a} \right)$$

$$= \frac{\{\lambda(t-a)\}^j}{j!} e^{-\lambda(t+a)} \left(\frac{t-ja}{t-a} \right)^j$$

$$= \{\lambda(t-ja)\}^j e^{-\lambda(t+a)} / j!.$$

The distribution of V is absolutely continuous on any interval between two adjacent atoms. We shall work out the entire distribution of V in the case $0 < t \leq a$, where the only atoms of V are the points 0 and t. Let $X > 0$ denote the distance from the origin to the nearest point of the Poisson process $\mathcal{P}$ on the positive half-line, and $Y > 0$ the distance from 0 to the nearest point of $\mathcal{P}$ on the negative half-line. Then X and Y are independent, each having an exponential distribution with mean λ^{-1}. Thus,

$$P(V = t) = P(X > t, Y > a) = P(X > t)P(Y > a) = e^{-\lambda(a+t)},$$
$$P(V = 0) = P(X > t, Y \le a - t) + P(X \le t, X + Y \le a)$$
$$= e^{-\lambda t}(1 - e^{-\lambda(a-t)}) + \int_0^t \lambda e^{-\lambda x}(1 - e^{-\lambda(a-x)})\, dx$$
$$= 1 - (1 + \lambda t)e^{-\lambda a},$$

and if $0 < v < t$,

$$P(v < V \le v + dv) = P(X > t, a + v - t < Y \le a + v - t + dv)$$
$$+ P(v < X \le v + dv, Y > a)$$
$$+ P(X \le t, Y \le a, a + v < X + Y \le a + v + dv)$$
$$= e^{-\lambda t}\lambda e^{-\lambda(a+v-t)}dv$$
$$+ \lambda e^{-\lambda v}dv e^{-\lambda a}$$
$$+ \{(d/dv)P(X \le t, Y \le a, X + Y \le a + v)\}\, dv.$$

But

$$P(X \le t, Y \le a, X + Y \le a + v)$$
$$= \int_0^t P\{Y \le \min(a, a + v - x)\}\lambda e^{-\lambda x}\, dx$$
$$= \int_0^v (1 - e^{-\lambda a})\lambda e^{-\lambda x}\, dx + \int_v^t (1 - e^{-\lambda(a+v-x)})\lambda e^{-\lambda x}\, dx$$
$$= 1 - e^{-\lambda a} - e^{-\lambda t} + \{1 - \lambda(t - v)\}e^{-\lambda(a+v)},$$

and so

$$P(v < V \le v + dv) = \lambda\{2 + \lambda(t - v)\}e^{-\lambda(a+v)}\, dv.$$

Thus, conditional on $0 < V < t$, V has density

$$b(v) \equiv \lambda\{2 + \lambda(t - v)\}e^{-\lambda(a+v)} \Big/ \int_0^t \lambda\{2 + \lambda(t - u)\}e^{-\lambda(a+u)}\, du$$
$$= \lambda\{2 + \lambda(t - v)\}e^{\lambda(t-v)} \Big/ \{(1 + \lambda t)e^{\lambda t} - 1\}, \qquad 0 < v < t.$$

This density is independent of a (except of course for our initial constraint that $a \ge t$).

3.4. PROPERTIES OF VACANCY IN LARGE REGIONS

Suppose we observe a coverage process in the plane $\mathbb{R}^2$ from a vantage point a short distance away from the plane. This position might be that of a camera in an aircraft flying over a pattern of leaf covers in a forest, or of a microscope surveying the pattern of dust particles on a glass slide. Suppose we are initially placed so that we view a square of area a. Let v equal the expected vacancy within the square. If we wished to estimate v, we could move the vantage point around so that it took in n different squares, all of area a and separated by very large distances. Our estimate of v would equal the arithmetic mean of the observed vacancies within the n squares, and would have variance of order $1/n$. An alternative approach would be to move the vantage point further away from the plane, so that in one sweep we could take in a single large square of area na. In the latter case, our estimate of v would equal the vacancy within the large square, divided by n. We might expect the second technique to also give an estimate with variance of order $1/n$. However, this is not necessarily the case. Note that the vacancy within the large square cannot be represented as the sum of n *independent* vacancies within regions of area a, since some shapes may overlap two or more regions. If the shapes tend to have very large radii then the amount of overlap can be substantial. In this case the variance of the second estimate may be of larger order than $1/n$.

A necessary and sufficient condition for the second sampling scheme to give an estimate whose variance is of order $1/n$, is that the content of each random shape have finite second moment. We shall arrive at this conclusion by modeling the "increase in perspective" of the previous paragraph, as follows. Suppose our vantage point recedes some distance from $\mathbb{R}^2$, so that we take in a larger region. If we rescale so that the new region acquires the same dimensions as the original, we must multiply the linear dimensions of each random shape by a factor $\delta < 1$, and increase Poisson intensity by a factor δ^{-2}. In k dimensions we should study the Boolean model $\mathcal{C}(\delta, \lambda)$ in which shapes are distributed as δS, where $\delta = \delta(\lambda)$ is such that as Poisson intensity λ diverges, $\delta^k \lambda \to \rho$ and $0 < \rho < \infty$.

Suppose that the region $\mathcal{R}$ is Riemann measurable (that is, the indicator function of $\mathcal{R}$ is Riemann integrable) and $0 < \|\mathcal{R}\| < \infty$.

Theorem 3.4. *Assume $E(\|S\|) < \infty$, and let V denote vacancy within the region $\mathcal{R}$ arising from the Boolean model $\mathcal{C}(\delta, \lambda)$ defined* *just above. If $\delta \to 0$ as $\lambda \to \infty$, in such a manner that $\delta^k \lambda \to \rho$ where $0 \le \rho < \infty$, then*

$$E(V) \to \|\mathcal{R}\| \exp\{-\rho E(\|S\|)\} \qquad (3.11)$$

and

$$E|V - E(V)| \to 0 \qquad (3.12)$$

as $\lambda \to \infty$. A necessary and sufficient condition for

$$\lambda \operatorname{var}(V) = O(1) \qquad (3.13)$$

as $\lambda \to \infty$, is that $E(\|S\|^2) < \infty$. If this condition holds then

$\lambda \operatorname{var}(V) \to \sigma^2(S)$

$$\equiv \rho\|\mathcal{R}\| \exp\{-2\rho E(\|S\|)\} \int_{\mathbb{R}^k} (\exp[\rho E\{\|(\mathbf{x}+S) \cap S\|\}] - 1)\, d\mathbf{x},$$

$$(3.14)$$

as $\lambda \to \infty$.

Proofs of Theorems 3.4–3.6 are deferred until the end of the section.

Of course, result (3.12) implies that $V - E(V) \to 0$ in probability. Since $0 \le V \le \|\mathcal{R}\|$ then by dominated convergence, $V - E(V) \to 0$ in L^p norm for each $1 \le p < \infty$. The second part of the theorem declares that when $E(\|S\|^2) < \infty$,

$$E|V - E(V)|^2 = O(\lambda^{-1}), \qquad (3.15)$$

thereby strengthening (3.12).

It is illuminating to interpret (3.12) and (3.15) in terms of the weak law of large numbers, which states that if $X_1, X_2, \ldots$ are independent and identically distributed with finite mean,

$$E\left| n^{-1} \sum_{i=1}^n X_i - E(X_1) \right| \to 0, \qquad (3.16)$$

while if $E(X_1^2) < \infty$,

$$E\left| n^{-1} \sum_{i=1}^n X_i - E(X_1) \right|^2 = n^{-1} \operatorname{var}(X_1) = O(n^{-1}) \qquad (3.17)$$

as $n \to \infty$. See Appendix II. We may view vacancy within a large region (or vacancy of a scaled-down Boolean model within a fixed region) as a sum of "almost" independent random variables, equal to the vacancies within smaller component parts of the region. Rescaling the model is analogous to dividing by n in (3.16) and (3.17), and the role of n is played by $\delta^{-k} \approx \lambda$. The condition $E(X_1^2) < \infty$ becomes $E(\|S\|^2) < \infty$.

In the case of $k = 1$ dimension, when shapes are line segments with length having distribution function G and mean α,

$$\sigma^2(S) = 2\rho\|\mathcal{R}\|e^{-2\alpha\rho} \int_0^\infty \left(\exp\left[\rho \int_x^\infty \{1 - G(y)\} \, dy\right] - 1 \right) dx.$$

When the line segments are of fixed length a,

$$\sigma^2(S) = 2\|\mathcal{R}\|e^{-a\rho} \left\{ 1 - (1 + a\rho)e^{-a\rho} \right\}.$$

In higher dimension it can be tedious to compute actual values of $\sigma^2(S)$, although upper and lower bounds are easy to derive. Since $e^x - 1 \geq x$ for all $x \geq 0$ then

$$\int_{\mathbb{R}^k} \left(\exp[\rho E\{\|(\mathbf{x} + S) \cap S\|\}] - 1 \right) d\mathbf{x}$$

$$\geq \rho \int_{\mathbb{R}^k} E\{\|(\mathbf{x} + S) \cap S\|\} \, d\mathbf{x}$$

$$= \rho E\left\{ \int_{\mathbb{R}^k} \|(\mathbf{x} + S) \cap S\| \, d\mathbf{x} \right\}. \tag{3.18}$$

Now, for any two Lebesgue measurable sets $\mathcal{A}$ and $\mathcal{B}$ (subsets of $\mathbb{R}^k$), we have

$$\int_{\mathbb{R}^k} \|(\mathbf{x} + \mathcal{A}) \cap \mathcal{B}\| \, d\mathbf{x} = \int_{\mathbb{R}^k} d\mathbf{x} \int_{(\mathbf{x} + \mathcal{A}) \cap \mathcal{B}} d\mathbf{y}$$

$$= \int_{\mathcal{B}} d\mathbf{y} \int_{\mathbf{y} - \mathcal{A}} d\mathbf{x}$$

$$= \int_{\mathcal{B}} \|\mathcal{A}\| \, d\mathbf{y}$$

$$= \|\mathcal{A}\|\|\mathcal{B}\|. \tag{3.19}$$

In consequence the right-hand side of (3.18) equals $\rho E(\|S\|^2)$, and so

$$\sigma^2(S) \geq \rho^2\|\mathcal{R}\|E(\|S\|^2)\exp\{-2\rho E(\|S\|)\}. \tag{3.20}$$

A similar argument based on the inequality $e^x - 1 < xe^x$, $x > 0$, shows that

$$\sigma^2(S) \leq \rho^2\|\mathcal{R}\|E(\|S\|^2)\exp\{-\rho E(\|S\|)\}.$$

Given a random shape S, let S^* denote an "isotropic version" of S, that is, the image of S after applying a "random, uniformly distributed rotation," as discussed in Section 3.1. If each shape generating the Boolean model is oriented independently and uniformly in this way, the effect is to *reduce* (or at least, not increase) the asymptotic variance σ^2 defined at (3.14). To see this, observe that the integral in the definition of σ^2 may be written in polar coordinates as follows:

$$\int_{\mathbb{R}^k} (\exp[\rho E\{\|(\mathbf{x}+S)\cap S\|\}] - 1)\, d\mathbf{x} = \int_0^\infty J(r\mid S) r^{k-1}\, dr,$$

where

$$J(r\mid S) \equiv \int_\Omega (\exp[\rho E\{\|(r\omega+S)\cap S\|\}] - 1)\, d\omega$$

and Ω is the set of all unit vectors ω in $\mathbb{R}^k$. Suppose S is carried into S^* by the random orthogonal transformation T, which represents a rotation of S about the origin (and is stochastically independent of S). If S were replaced by S^* then $J(r\mid S)$ would change to

$$J(r\mid S^*) = \int_\Omega \left(\exp\left[\rho E\{\|(r\omega+S^*)\cap S^*\|\}\right] - 1\right) d\omega$$

$$= \int_\Omega \left(\exp\left[\rho E\{\|(rT^{-1}\omega+S)\cap S\|\}\right] - 1\right) d\omega$$

$$= \int_\Omega \left(\exp\left[\rho E_T\{E(\|(rT^{-1}\omega+S)\cap S\|\mid T)\}\right] - 1\right) d\omega$$

$$\le E_T\left\{\int_\Omega \left(\exp\left[\rho E\{\|(rT^{-1}\omega+S)\cap S\|\mid T\}\right] - 1\right) d\omega\right\},$$

where E_T denotes expectation in the distribution of the components of the orthogonal matrix defining T, and the inequality follows by Jensen's inequality (e.g., Billingsley 1979, p. 240; Chung 1974, p. 47) since the function $e^{\rho x} - 1$ is convex in x. Changing variable to $\omega^* = T^{-1}\omega$ in the integral over Ω, we see that

$$J(r\mid S^*) \le E_T\left\{\int_\Omega (\exp[\rho E\{\|(r\omega^*+S)\cap S\|\mid T\}] - 1)\, d\omega^*\right\}$$

$$= \int_\Omega (\exp[\rho E\{\|(r\omega+S)\cap S\|\}] - 1)\, d\omega$$

$$= J(r\mid S).$$

Writing $\sigma^2(S)$ and $\sigma^2(S^*)$ for the respective versions of σ^2 for random shapes S and S^*, we conclude that

$$\sigma^2(S^*) = \rho\|\mathcal{R}\| \int_0^\infty r^{k-1} J(r \mid S^*)\, dr \exp\{-2E(\|S^*\|)\}$$

$$\leq \rho\|\mathcal{R}\| \int_0^\infty r^{k-1} J(r \mid S)\, dr \exp\{-2E(\|S\|)\}$$

$$= \sigma^2(S).$$

That is, a random, uniformly distributed rotation of each shape has the effect of reducing the large-region variance of vacancy. In this sense, rotations tend to stabilize the coverage process. (Of course, rotations exist only in $k \geq 2$ dimensions!) Random reflections also have a stabilizing effect, as may be proved in the same manner.

The infimum of $\sigma^2(S)$ over all shapes S with given mean content $E(\|S\|)$ and mean square content $E(\|S\|^2) < \infty$ equals the lower bound at (3.20), provided $k \geq 2$ and the set of all such shapes is nonempty. To see this, consider a shape $S = S(\delta)$ which is simply a very thin shell of thickness $\delta X(\delta)$ and radius $\delta^{-1/(k-1)}$ for small δ, where $X(\delta)$ is a positive random variable. If $X(\delta) \to X$ as $\delta \to 0$ then the k-dimensional content of the shell is asymptotic to $X s_k$ as $\delta \to 0$, where s_k equals the surface content of a unit k-dimensional sphere. Therefore by choosing $X(\delta)$ suitably, with $X(\delta) \to Y s_k^{-1}$ in mean square (i.e., in L^2) as $\delta \to 0$, we may ensure that the shell has mean content $E(Y)$ and mean square content $E(Y^2)$ (both given in advance) for all small δ. If $\epsilon > 0$,

$$\int_{\mathbb{R}^k} (\exp[\rho E\{\|(\mathbf{x} + S) \cap S\|\}] - 1)\, d\mathbf{x}$$

$$\leq \rho \int_{\mathbb{R}^k} E\{\|(\mathbf{x} + S) \cap S\|\} \exp[\rho E\{\|(\mathbf{x} + S) \cap S\|\}]\, d\mathbf{x}$$

$$\leq \exp[\rho \sup_{|\mathbf{x}| > \epsilon} E\{\|(\mathbf{x} + S) \cap S\|\}] \rho \int_{\mathbb{R}^k} E\{\|(\mathbf{x} + S) \cap S\|\}\, d\mathbf{x}$$

$$\quad + \rho \int_{|\mathbf{x}| \leq \epsilon} E\{\|(\mathbf{x} + S) \cap S\|\} \exp[\rho E\{\|(\mathbf{x} + S) \cap S\|\}]\, d\mathbf{x}.$$

For fixed $\epsilon > 0$,

$$\sup_{|\mathbf{x}| > \epsilon} E\{\|(\mathbf{x} + S) \cap S\|\} \to 0$$

as $\delta \to 0$, and

$$\int_{|x| \leq \epsilon} E\{\|(\mathbf{x}+S) \cap S\|\} \exp[\rho E\{\|(\mathbf{x}+S) \cap S\|\}]\, dx$$
$$\leq E(\|S\|) \exp\{\rho E(\|S\|)\} v_k \epsilon^k,$$

where v_k equals the content of a unit k-dimensional sphere. By (3.19),

$$\int_{\mathbb{R}^k} E\{\|(\mathbf{x}+S) \cap S\|\}\, dx = E(\|S\|^2).$$

Combining these estimates we conclude that

$$\int_{\mathbb{R}^k} (\exp[\rho E\{\|(\mathbf{x}+S) \cap S\|\}] - 1)\, dx \to \rho E(\|S\|^2)$$

as $\delta \to 0$, and so by the definition (3.14), $\sigma^2(S)$ converges to the lower bound at (3.20).

If S is a sphere of fixed radius l then

$$\sigma^2(S) = (2l)^k \rho\|\mathcal{R}\| s_k \exp(-2\rho l^k v_k) \int_0^1 x^{k-1}\left[\exp\left\{\rho l^k B(x)\right\} - 1\right] dx,$$

where

$$B(x) \equiv \begin{cases} 2\pi^{(k-1)/2}\{\Gamma(\tfrac{1}{2}+\tfrac{1}{2}k)\}^{-1}\int_x^1 (1-y^2)^{(k-1)/2}\, dy & \text{if } 0 \leq x \leq 1 \\ 0 & \text{if } x > 1 \end{cases}$$

denotes the content of the lens of intersection of two unit k-dimensional spheres centered $2x$ apart.

Result (3.12) in Theorem 3.4 is essentially a "weak law of large numbers" for vacancy within a large region. We complement it with a central limit theorem. Suppose the region $\mathcal{R}$ is Riemann measurable and $0 < \|\mathcal{R}\| < \infty$.

Theorem 3.5. *Assume $E(\|S\|^2) < \infty$, and let V be as in Theorem 3.4. If $\delta \to 0$ as $\lambda \to \infty$ in such a manner that $\delta^k \lambda \to \rho$, where $0 \leq \rho < \infty$, then*

$$\lambda^{1/2}\{V - E(V)\} \to N(0, \sigma^2)$$

in distribution, where σ^2 is defined at (3.14).

It is instructive to rephrase Theorems 3.4 and 3.5 for a Boolean model within an expanding region. Fix $\mathcal{R}$, and let V_l denote vacancy within the region $\mathcal{R}_l \equiv l\mathcal{R}$ ($\mathcal{R}$ rescaled by the factor l in each dimension) arising from a Boolean model C_0 in which the driving Poisson process has *fixed* intensity ρ and each shape is distributed as S. Then as l increases, vacancy increases roughly in proportion to the content of $\mathcal{R}_l$:

$$V_l / \|\mathcal{R}_l\| \to \exp\{-\rho E(\|S\|)\} \qquad (3.21)$$

in probability and in L^1. Variance of vacancy increases in proportion to the content of $\mathcal{R}_l$ if and only if shape content has finite mean square, and if that condition holds then $\mathrm{var}(V_l) \sim \sigma_1^2 \|\mathcal{R}_l\|$ and

$$\{V_l - E(V_l)\} / \|\mathcal{R}_l\|^{1/2} \to N(0, \sigma_1^2) \qquad (3.22)$$

in distribution, where

$$\sigma_1^2 \equiv \exp\{-2\rho E(\|S\|)\} \int_{\mathbb{R}^k} (\exp[\rho E\{\|(\mathbf{x}+S) \cap S\|\}] - 1)\, d\mathbf{x}.$$

Results (3.21) and (3.22) can be employed to approximate the distribution of vacancy within large regions. On occasion it is useful to strengthen (3.21) to a strong "law of large numbers." That problem can be viewed in the context of ergodic theory. Vacancy within a large region may be closely approximated by a sum of vacancies within smaller, equal-sized regions, and those component vacancies form a stationary process when indexed by k-vectors of integers. An ergodic theorem (Appendix II) applies to such sums, and thus the strong law for vacancy emerges, as we now show.

As in Section 3.1, let C be the Boolean model in which shapes are distributed as S and the driving Poisson process has intensity λ. Assume $\mathcal{R}$ is a Riemann measurable region with $0 < \|\mathcal{R}\| < \infty$, and put $\mathcal{R}_l \equiv l\mathcal{R}$. Let V_l denote the vacancy within $\mathcal{R}_l$ arising from C. We permit l to increase, *keeping all other parameters (including λ) fixed.*

Theorem 3.6. *As $l \to \infty$,*

$$V_l / \|\mathcal{R}_l\| \to \exp\{-\lambda E(\|S\|)\}$$

almost surely.

We conclude this section with proofs of Theorems 3.4–3.6. Modified versions of the proof of Theorem 3.6 may be used to establish strong laws for many statistics defined on Boolean models.

Proof of Theorem 3.4. Result (3.11) in Theorem 3.4 is immediate from (3.5). The proof of each other result in Theorem 3.4 rests on formula (3.6) for the variance of vacancy. Note that in the present case, S must be replaced by δS in that formula.

Result (3.12) will follow if we prove that

$$E|V - E(V)|^2 \to 0.$$

In view of (3.6) and the inequality $e^x - 1 \le xe^x$, valid for all $x \ge 0$, we have

$$E|V - E(V)|^2 = \text{var}(V) \le \int\!\!\int_{\mathcal{R}^2} \lambda E\left\{ \|(\mathbf{x}_1 - \mathbf{x}_2 + \delta S) \cap \delta S\| \right\} d\mathbf{x}_1 \, d\mathbf{x}_2$$

$$= \delta^k \lambda \int\!\!\int_{\mathcal{R}^2} E\left[\left\| \left\{ \delta^{-1}(\mathbf{x}_1 - \mathbf{x}_2) + S \right\} \cap S \right\| \right] d\mathbf{x}_1 \, d\mathbf{x}_2.$$

The integrand in this expression does not exceed $E(\|S\|)$, and so the result $\text{var}(V) \to 0$ will follow by dominated convergence if we show that for almost all $(\mathbf{x}_1, \mathbf{x}_2) \in \mathcal{R}^2$,

$$E\left[\left\| \left\{ \delta^{-1}(\mathbf{x}_1 - \mathbf{x}_2) + S \right\} \cap S \right\| \right] \to 0.$$

Let $\mathbf{x} = \mathbf{x}_1 - \mathbf{x}_2$, and suppose $|\mathbf{x}| = 2d > 0$. Then

$$(\delta^{-1}\mathbf{x} + S) \cap S \subseteq \left\{ \delta^{-1}\mathbf{x} + \mathbf{y} : |\mathbf{y}| > \delta^{-1}d, \mathbf{y} \in S \right\}$$

$$\cup \left\{ \delta^{-1}\mathbf{x} + \mathbf{y} : |\mathbf{y}| \le \delta^{-1}d, \delta^{-1}\mathbf{x} + \mathbf{y} \in S \right\}$$

$$\subseteq \left\{ \delta^{-1}\mathbf{x} + \mathbf{y} : |\mathbf{y}| > \delta^{-1}d, \mathbf{y} \in S \right\}$$

$$\cup \left\{ \delta^{-1}\mathbf{x} + \mathbf{y} : |\delta^{-1}\mathbf{x} + \mathbf{y}| > \delta^{-1}2d - \delta^{-1}d, \delta^{-1}\mathbf{x} + \mathbf{y} \in S \right\}.$$

Consequently,

$$E\left\{ \|(\delta^{-1}\mathbf{x} + S) \cap S\| \right\} \le 2E\left[\left\| \left\{ \mathbf{y} : |\mathbf{y}| > \delta^{-1}d, \mathbf{y} \in S \right\} \right\| \right] \to 0$$

as $\delta \to 0$ as required.

We shall prove (3.14) last of all. That result implies that (3.13) holds if $E(\|S\|^2) < \infty$ (note Exercise 3.6). It then remains only to show that (3.13) implies $E(\|S\|^2) < \infty$. Now, formula (3.6) and the inequality $e^x - 1 \ge x$ yield

$$\exp\left\{ 2\delta^k \lambda E(\|S\|) \right\} \lambda \, \text{var}(V)$$

$$\ge \lambda^2 \int\!\!\int_{\mathcal{R}^2} E\{ \|(\mathbf{x}_1 - \mathbf{x}_2 + \delta S) \cap \delta S\| \} \, d\mathbf{x}_1 \, d\mathbf{x}_2. \quad (3.23)$$

Since $\mathcal{R}$ is Riemann measurable and $\|\mathcal{R}\| > 0$ then $\mathcal{R}$ contains a small sphere of radius $2\epsilon > 0$. We assume, to simplify notation, that the sphere is centered at the origin. Then

$$\int\int_{\mathcal{R}^2} E\{\|(\mathbf{x}_1 - \mathbf{x}_2 + \delta S) \cap \delta S\|\} \, d\mathbf{x}_1 \, d\mathbf{x}_2$$

$$\geq \delta^{3k} \int_{|\mathbf{x}_1| \leq 2\epsilon/\delta} d\mathbf{x}_1 \int_{|\mathbf{x}_2| \leq 2\epsilon/\delta} E\{\|(\mathbf{x}_1 - \mathbf{x}_2 + S) \cap S\|\} \, d\mathbf{x}_2$$

$$\geq \delta^{3k} \int_{|\mathbf{x}_1| \leq \epsilon/\delta} d\mathbf{x}_1 \int_{|\mathbf{x}_1 - \mathbf{x}_2| \leq \epsilon/\delta} E\{\|(\mathbf{x}_1 - \mathbf{x}_2 + S) \cap S\|\} \, d\mathbf{x}_2$$

$$= v_k \epsilon^k \delta^{2k} \int_{|\mathbf{x}| \leq \epsilon/\delta} E\{\|(\mathbf{x} + S) \cap S\|\} \, d\mathbf{x}. \qquad (3.24)$$

Let $T \equiv S \cap \{\mathbf{x} : |\mathbf{x}| \leq r\}$, where $r > 0$ is fixed. Then

$$\int_{|\mathbf{x}| \leq \epsilon/\delta} E\{\|(\mathbf{x} + S) \cap S\|\} \, d\mathbf{x} \geq \int_{|\mathbf{x}| \leq \epsilon/\delta} E\{\|(\mathbf{x} + T) \cap T\|\} \, d\mathbf{x}$$

$$= E\left\{ \int_{|\mathbf{x}| \leq \epsilon/\delta} d\mathbf{x} \int_{(\mathbf{x}+T) \cap T} d\mathbf{y} \right\}$$

$$= E\left[\int_T d\mathbf{y} \int_{(\mathbf{y}-T) \cap \{\mathbf{x}:|\mathbf{x}| \leq \epsilon/\delta\}} d\mathbf{x} \right]$$

$$= E\left[\int_T \|T \cap \{\mathbf{x} : |\mathbf{x} - \mathbf{y}| \leq \epsilon/\delta\}\| \, d\mathbf{y} \right].$$

If δ is so small that $\epsilon/\delta > 2r$, then $|\mathbf{x} - \mathbf{y}| \leq \epsilon/\delta$ for each $\mathbf{x}, \mathbf{y} \in T$. In that case,

$$\int_{|\mathbf{x}| \leq \epsilon/\delta} E\{\|(\mathbf{x} + S) \cap S\|\} \, d\mathbf{x} \geq E\left(\int_T \|T\| \, d\mathbf{y} \right) = E(\|T\|^2). \qquad (3.25)$$

Combining (3.23)–(3.25) we see that for each $r > 0$, and all sufficiently large λ,

$$\exp\left\{ 2\delta^k \lambda E(\|S\|) \right\} \lambda \, \text{var}(V) \geq v_k \epsilon^k (\delta^k \lambda)^2 E[\|S \cap \{\mathbf{x} : |\mathbf{x}| \leq r\}\|^2].$$

In consequence, $\lambda \, \text{var}(V)$ is bounded as $\lambda \to \infty$ only if

$$E[\|S \cap \{\mathbf{x} : |\mathbf{x}| \leq r\}\|^2]$$

is bounded uniformly in r. The latter condition entails $E(\|S\|^2) < \infty$.

Finally, we establish (3.14). Assume $E(\|S\|^2) < \infty$. In view of (3.6),

$$\exp\left\{2\delta^k \lambda E(\|S\|)\right\} \operatorname{var}(V)$$
$$= \delta^k \int_{\mathcal{R}} d\mathbf{x} \int_{\delta^{-1}(\mathbf{x}-\mathcal{R})} (\exp[\delta^k \lambda E\{\|(\mathbf{y}+S)\cap S\|\}] - 1)\, d\mathbf{y}.$$

Now, the function

$$f_\delta(\mathbf{x}) \equiv \int_{\delta^{-1}(\mathbf{x}-\mathcal{R})} (\exp[\delta^k \lambda E\{\|(\mathbf{y}+S)\cap S\|\}] - 1)\, d\mathbf{y}$$

is dominated by the constant

$$\int_{\mathbb{R}^k} (\exp[tE\{\|(\mathbf{y}+S)\cap S\|\}] - 1)\, d\mathbf{y}$$
$$\leq \exp\{tE(\|S\|)\} \int_{\mathbb{R}^k} tE\{\|(\mathbf{y}+S)\cap S\|\}\, d\mathbf{y}$$
$$= \exp\{tE(\|S\|)\}\, tE(\|S\|^2),$$

where $t \equiv \sup(\delta^k \lambda)$ [note (3.19)]. Since $\mathcal{R}$ is Riemann measurable and $\|\mathcal{R}\| > 0$ then for almost all $\mathbf{x} \in \mathcal{R}$, $\mathbf{x} - \mathcal{R}$ contains a sphere centered on the origin, in which case $\delta^{-1}(\mathbf{x} - \mathcal{R}) \to \mathbb{R}^k$ as $\delta \to 0$. Therefore

$$f_\delta(\mathbf{x}) \to c \equiv \int_{\mathbb{R}^k} (\exp[\rho E\{\|(\mathbf{y}+S)\cap S\|\}] - 1)\, d\mathbf{y}$$

as $\lambda \to \infty$, whence by dominated convergence,

$$\exp\left\{2\delta^k \lambda E(\|S\|)\right\} \lambda \operatorname{var}(V) \to \rho \int_{\mathcal{R}} c\, d\mathbf{x} = \rho c \|\mathcal{R}\|$$

as $\lambda \to \infty$. This proves (3.14). $\square$

Proof of Theorem 3.5. The proof is substantially simpler if, with probability 1, the set S is bounded:

$$P(|\mathbf{x}| \leq c \text{ for all } \mathbf{x} \in S) = 1, \tag{3.26}$$

for some $c > 0$. The first step in the proof consists of a truncation argument that permits us to impose this restriction.

Step (i). Given $c > 0$, define $T(c)$ to be the k-dimensional (closed) sphere centered at the origin, and put $S_i' \equiv S_i \cap T(c)$ and $S_i'' \equiv S_i \cap T(c)^\sim$. Define S' and S'' in an analogous manner. Thus, S_i' and S_i'' have the same distributions as S' and S'', respectively. We may decompose the basic Boolean model C, generated by shapes S_i, $i \geq 1$, and driven by a Poisson process $\{\xi_i, i \geq 1\}$, into two models C' and C'', where in C', S_i is replaced by S_i', and in C'', S_i is replaced by S_i''. Define

$$\chi_0(\mathbf{x}) \equiv \begin{cases} 1 & \text{if } \mathbf{x} \text{ is covered by } C \text{ but not by } C' \\ 0 & \text{otherwise,} \end{cases}$$

and put

$$V_0 \equiv \int_{\mathcal{R}} \chi_0(\mathbf{x})\, d\mathbf{x}. \tag{3.27}$$

Then

$$V' \equiv V + V_0 \tag{3.28}$$

is the amount of $\mathcal{R}$ not covered by the Boolean model C'. We shall prove that

$$\lim_{c \to \infty} \limsup_{\lambda \to \infty} \lambda \operatorname{var}(V_0) = 0. \tag{3.29}$$

Observe that

$$E\{\chi_0(\mathbf{x})\} = P(\mathbf{x} \text{ covered by } C \text{ but not by } C')$$
$$= P(\mathbf{x} \text{ not covered by } C') - P(\mathbf{x} \text{ not covered by } C),$$

while

$$E\{\chi_0(\mathbf{x}_1)\chi_0(\mathbf{x}_2)\} = P(\text{both } \mathbf{x}_1, \mathbf{x}_2 \text{ covered by } C;$$
$$\text{neither } \mathbf{x}_1 \text{ nor } \mathbf{x}_2 \text{ covered by } C')$$
$$= P(\text{neither } \mathbf{x}_1 \text{ nor } \mathbf{x}_2 \text{ covered by } C')$$
$$- P(\text{neither } \mathbf{x}_1 \text{ nor } \mathbf{x}_2 \text{ covered by } C', \text{ and}$$
$$\text{either } \mathbf{x}_1 \text{ or } \mathbf{x}_2 \text{ not covered by } C)$$
$$= P(\text{neither } \mathbf{x}_1 \text{ nor } \mathbf{x}_2 \text{ covered by } C')$$
$$+ P(\text{neither } \mathbf{x}_1 \text{ nor } \mathbf{x}_2 \text{ covered by } C)$$
$$- P(\mathbf{x}_1 \text{ not covered by } C, \mathbf{x}_2 \text{ not covered by } C')$$
$$- P(\mathbf{x}_2 \text{ not covered by } C, \mathbf{x}_1 \text{ not covered by } C').$$

Therefore

$$\text{cov}\{\chi_0(\mathbf{x}_1), \chi_0(\mathbf{x}_2)\}$$

$$= \left\{ P(\text{ neither } \mathbf{x}_1 \text{ nor } \mathbf{x}_2 \text{ covered by } C) \right.$$

$$\left. - \prod_{i=1}^{2} P(\mathbf{x}_i \text{ not covered by } C) \right\}$$

$$+ \left\{ P(\text{neither } \mathbf{x}_1 \text{ nor } \mathbf{x}_2 \text{ covered by } C') \right.$$

$$\left. - \prod_{i=1}^{2} P(\mathbf{x}_i \text{ not covered by } C') \right\}$$

$$- \{ P(\mathbf{x}_1 \text{ not covered by } C, \mathbf{x}_2 \text{ not covered by } C')$$

$$- P(\mathbf{x}_1 \text{ not covered by } C) P(\mathbf{x}_2 \text{ not covered by } C') \}$$

$$- \{ P(\mathbf{x}_2 \text{ not covered by } C, \mathbf{x}_1 \text{ not covered by } C')$$

$$- P(\mathbf{x}_2 \text{ not covered by } C) P(\mathbf{x}_1 \text{ not covered by } C') \}.$$

$$(3.30)$$

Now,

$$P(\mathbf{x}_1 \text{ not covered by } C, \mathbf{x}_2 \text{ not covered by } C')$$

$$= P(\text{for all } i \geq 1, \mathbf{x}_1 \notin \boldsymbol{\xi}_i + \delta S_i \text{ and } \mathbf{x}_2 \notin \boldsymbol{\xi}_i + \delta S_i')$$

$$= P(\text{for all } i \geq 1, \boldsymbol{\xi}_i \notin \mathbf{x}_1 - \delta S_i \text{ and } \boldsymbol{\xi}_i \notin \mathbf{x}_2 - \delta S_i')$$

$$= P\{\text{for all } i \geq 1, \boldsymbol{\xi}_i \notin (\mathbf{x}_2 - \mathbf{x}_1 + \delta S_i) \cup \delta S_i'\}$$

$$= \exp[-\lambda E\{\|(\mathbf{x}_2 - \mathbf{x}_1 + \delta S) \cup \delta S'\|\}]$$

$$= \exp[-\lambda E(\|\delta S\|) - \lambda E(\|\delta S'\|) + \lambda E\{\|(\mathbf{x}_2 - \mathbf{x}_1 + \delta S) \cap \delta S'\|\}],$$

the third equality following from symmetry and homogeneity of the Poisson process. [Compare the proof of (3.4).] By (3.4),

$$P(\mathbf{x}_1 \text{ not covered by } C) \, P(\mathbf{x}_2 \text{ not covered by } C')$$

$$= \exp\{-\lambda E(\|\delta S\|) - \lambda E(\|\delta S'\|)\} .$$

Thus,

$P(\mathbf{x}_1$ not covered by C, $\mathbf{x}_2$ not covered by $C')$

$\qquad - P(\mathbf{x}_1$ not covered by $C)\ P(\mathbf{x}_2$ not covered by $C')$

$\qquad = \exp\{-\lambda E(\|\delta S\|) - \lambda E(\|\delta S'\|)\}$

$\qquad\quad \times (\exp[\lambda E\{\|(\mathbf{x}_2 - \mathbf{x}_1 + \delta S) \cap \delta S'\|\}] - 1)$

$\qquad = \exp\{-2\lambda E(\|\delta S\|)\}\ (\exp[\lambda E\{\|(\mathbf{x}_2 - \mathbf{x}_1 + \delta S) \cap \delta S'\|\}] - 1)$

$\qquad\quad + [\exp\{\lambda E(\|\delta S''\|)\} - 1]\exp\{-2\lambda E(\|\delta S\|)\}$

$\qquad\quad \times (\exp[E\{\|(\mathbf{x}_2 - \mathbf{x}_1 + \delta S) \cap \delta S'\|\}] - 1)$

$\qquad\quad + \exp\{-2\lambda E(\|\delta S\|)\}\ (\exp[\lambda E\{\|(\mathbf{x}_2 - \mathbf{x}_1 + \delta S) \cap \delta S'\|\}]$

$$- \exp[\lambda E\{\|(\mathbf{x}_2 - \mathbf{x}_1 + \delta S) \cap \delta S\|\}]). \qquad (3.31)$$

The first term on the right-hand side of (3.31) equals

$$P(\text{neither } \mathbf{x}_1 \text{ nor } \mathbf{x}_2 \text{ covered by } C) - \prod_{i=1}^{2} P(\mathbf{x}_i \text{ not covered by } C).$$

We may now deduce from (3.31) and the inequality $e^x - 1 \le xe^x$, $x > 0$, that

$\Big| \{P(\mathbf{x}_1$ not covered by C, $\mathbf{x}_2$ not covered by $C')$

$\qquad - P(\mathbf{x}_1$ not covered by $C)P(\mathbf{x}_2$ not covered by $C')\}$

$\qquad - \Big\{ P(\text{neither } \mathbf{x}_1 \text{ nor } \mathbf{x}_2 \text{ covered by } C)$

$\qquad\quad - \prod_{i=1}^{2} P(\mathbf{x}_i \text{ not covered by } C) \Big\} \Big|$

$\qquad \le \lambda E(\|\delta S''\|)\lambda E\{\|(\mathbf{x}_2 - \mathbf{x}_1 + \delta S) \cap \delta S'\|\}$

$\qquad\quad + \lambda E\{\|(\mathbf{x}_2 - \mathbf{x}_1 + \delta S) \cap \delta S''\|\}\ .$

A similar argument shows that

$$\left| \{P(\mathbf{x}_2 \text{ not covered by } \mathcal{C}, \ \mathbf{x}_1 \text{ not covered by } \mathcal{C}') \right.$$

$$- P(\mathbf{x}_2 \text{ not covered by } \mathcal{C})P(\mathbf{x}_1 \text{ not covered by } \mathcal{C}')\}$$

$$- \left\{ P(\text{neither } \mathbf{x}_1 \text{ nor } \mathbf{x}_2 \text{ covered by } \mathcal{C}') \right.$$

$$\left. \left. - \prod_{i=1}^{2} P(\mathbf{x}_i \text{ not covered by } \mathcal{C}') \right\} \right|$$

$$\leq \lambda E(\|\delta S''\|)\lambda E\{\|(\mathbf{x}_1 - \mathbf{x}_2 + \delta S') \cap \delta S'\|\}$$
$$+ \lambda E\{\|(\mathbf{x}_1 - \mathbf{x}_2 + \delta S'') \cap \delta S'\|\} \ .$$

Substituting both these estimates into (3.30), we find that

$$|\text{cov}\{\chi_0(\mathbf{x}_1), \chi_0(\mathbf{x}_2)\}| \leq 2\delta^k \lambda^2 E(\|S''\|) E\{\|(\mathbf{x}_2 - \mathbf{x}_1 + \delta S) \cap \delta S\|\}$$
$$+ 2\lambda E\{\|(\mathbf{x}_2 - \mathbf{x}_1 + \delta S) \cap \delta S''\|\} \ . \qquad (3.32)$$

From (3.27) and (3.32) we see that

$$\text{var}(V_0) = \int\!\!\int_{\mathcal{R}^2} \text{cov}\{\chi_0(\mathbf{x}_1), \chi_0(\mathbf{x}_2)\} \, d\mathbf{x}_1 \, d\mathbf{x}_2$$

$$\leq \int_{\mathcal{R}} d\mathbf{x}_1 \int_{\mathbb{R}^k} |\text{cov}\{\chi_0(\mathbf{x}_1), \chi_0(\mathbf{x}_2)\}| \, d\mathbf{x}_2$$

$$\leq 2\delta^k \lambda^2 \|\mathcal{R}\| E(\|S''\|) \int_{\mathbb{R}^k} E\{\|(\mathbf{x} + \delta S) \cap \delta S)\|\} \, d\mathbf{x}$$

$$+ 2\lambda \|\mathcal{R}\| \int_{\mathbb{R}^k} E\{\|(\mathbf{x} + \delta S) \cap \delta S''\|\} \, d\mathbf{x}$$

$$= 2\delta^{3k}\lambda^2 \|\mathcal{R}\| E(\|S''\|) E(\|S\|^2) + 2\delta^{2k}\lambda \|\mathcal{R}\| E(\|S\|\|S''\|),$$

the last equality following via (3.19). Therefore

$$\limsup_{\lambda \to \infty} \lambda \, \text{var}(V_0) \leq 2\rho^3 \|\mathcal{R}\| E(\|S''\|) E(\|S\|^2)$$

$$+ 2\rho^2 \|\mathcal{R}\| E(\|S\|\|S''\|).$$

Since

$$E(\|S''\|) = E\{\|S \cap \mathcal{T}(c)\|\} \to 0$$

as $c \to \infty$, and likewise $E(\|S\|\|S''\|) \to 0$ as $c \to \infty$, then (3.29) is proved. This completes step (i) of the proof.

Define V' as in (3.28). If we show that

$$\lambda^{1/2}\{V' - E(V')\} \to N(0,\sigma_c^2) \tag{3.33}$$

in distribution as $\lambda \to \infty$, for some σ_c^2, and if

$$\sigma_c^2 \to \sigma^2 \tag{3.34}$$

as $c \to \infty$ [the quantity $\sigma^2 = \sigma^2(S)$ being defined at (3.14)], then it will follow from (3.29) that

$$\lambda^{1/2}\{V - E(V)\} \to N(0,\sigma^2) \tag{3.35}$$

in distribution as $\lambda \to \infty$. This would establish Theorem 3.5. We may prove (3.33) by deriving (3.35) under assumption (3.26). As for (3.34), that condition is easily checked by examining the formula for $\sigma^2(S)$ when S is truncated. Therefore we may complete our proof by establishing (3.35) under condition (3.26). This constitutes step (ii) below.

Step (ii). Let d be a large positive constant, and let c be as at (3.26). Divide all of $\mathbb{R}^k$ into a regular lattice of k-dimensional cubes of sidelength $cd\delta$, each cube separated from its nearest neighbors by a spacing strip of width $2c\delta$. Figure 3.4 illustrates the configuration when $k = 2$. The vacancy V within $\mathcal{R}$ may be written as

$$V = V_1 + V_2 + V_3,$$

where V_1 denotes vacancy within the union of those cubes wholly within $\mathcal{R}$ (call this region $\mathcal{A}_1$; it equals the cross-hatched area in Figure 3.4); V_2 denotes vacancy within all the rectangular boxes forming the spacings between cubes, and which are contained wholly within $\mathcal{R}$ (call this region $\mathcal{A}_2$; it is left blank in Figure 3.4); and V_3 equals the vacancy within the intersection of $\mathcal{R}$ with all those cubes or spacings which are not completely within or without $\mathcal{R}$ (call this region $\mathcal{A}_3$; it is shaded in Figure 3.4).

Since $\mathcal{R}$ is Riemann measurable then the content of $\mathcal{R}$, evaluated over increasingly fine dissections, is approximable arbitrarily closely by both inner and outer sums. In the case of the dissection described above, the difference between inner and outer sums is no greater than $\|\mathcal{A}_3\|$, and so

$$\|\mathcal{A}_3\| \to 0 \tag{3.36}$$

as $\lambda \to \infty$. Furthermore, the total content of all those spacings between cubes that intersect $\mathcal{R}$ and lie entirely within $\mathcal{R}$ is dominated

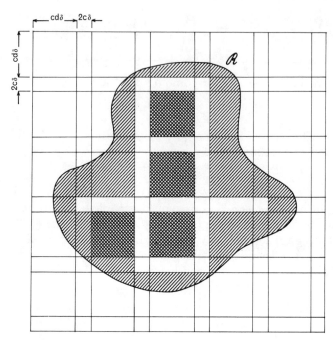

Figure 3.4. Definition of $\mathcal{A}_1$, $\mathcal{A}_2$, and $\mathcal{A}_3$. The "cubes" $\mathcal{D}_i$ are cross-hatched; their union is $\mathcal{A}_1$. The shaded region is $\mathcal{A}_3$, and the blank region within $\mathcal{R}$ is $\mathcal{A}_2$. [A definition of $\mathcal{D}_i$ is given following (3.41).]

by a constant multiple of $(1/cd\delta)c\delta = d^{-1}$. Therefore

$$\|\mathcal{A}_2\| \leq \text{const.}d^{-1}, \tag{3.37}$$

where the constant does not depend on d. We may deduce from (3.6) that the vacancy V_i within region $\mathcal{A}_i$ satisfies

$$\text{var}(V_i) \leq \int\int_{\mathcal{A}_i{}^2} \lambda E\{\|(\mathbf{x}_1 - \mathbf{x}_2 + \delta S) \cap \delta S\|\}\, d\mathbf{x}_1\, d\mathbf{x}_2$$
$$\leq \lambda \int_{\mathcal{A}_i} d\mathbf{x}_1 \int_{\mathrm{IR}^k} E\{\|(\mathbf{x}_1 - \mathbf{x}_2 + \delta S) \cap \delta S\|\}\, d\mathbf{x}_2$$
$$= \delta^{2k}\lambda\|\mathcal{A}_i\|E(\|S\|^2),$$

the last line following from (3.19). In view of (3.36) and (3.37), this implies

$$\lim_{\lambda\to\infty} \lambda\,\text{var}(V_3) = 0 \tag{3.38}$$

and

$$\lim_{d\to\infty} \limsup_{\lambda\to\infty} \lambda\,\text{var}(V_2) = 0. \tag{3.39}$$

Consequently, our goal of the central limit theorem (3.35) will be achieved if we prove that

$$\{V_1 - E(V_1)\}/(\mathrm{var}\, V_1)^{1/2} \to N(0,1) \qquad (3.40)$$

in distribution, and

$$\lim_{d\to\infty} \limsup_{\lambda\to\infty} |\lambda\,\mathrm{var}(V_1) - \sigma^2| = 0. \qquad (3.41)$$

Let $n = n(\lambda)$ denote the number of small cubes of side length $cd\delta$ that make up region $\mathcal{A}_1$, and let $\mathcal{D}_i$ denote the ith of these cubes, for $1 \le i \le n$. Write U_i for the contribution to V_1 from $\mathcal{D}_i$; then $V_1 = \sum_i U_i$. Since each random shape is contained within a sphere of radius $c\delta$, and the cubes $\mathcal{D}_i$ are distant at least $2c\delta$ apart, then no random shape can intersect more than one cube. [Note that we are assuming condition (3.26).] Therefore the variables U_i are independently distributed. They are obviously identically distributed, given λ. Hence

$$\mathrm{var}(V_1) = \sum_i \mathrm{var}(U_i) = n\,\mathrm{var}(U_1)$$

$$= n \exp\{-2\lambda E(\|\delta S\|)\}$$

$$\times \int\!\!\int_{\mathcal{D}_1{}^2} (\exp[\lambda E\{\|(\mathbf{x}_1 - \mathbf{x}_2 + \delta S)\cap \delta S\|\}] - 1)\, d\mathbf{x}_1\, d\mathbf{x}_2$$

$$\sim n\delta^{2k} \exp\{-2\rho E(\|S\|)\}$$

$$\times \int\!\!\int_{\mathcal{D}^2} (\exp[\rho E\{\|(\mathbf{x}_1 - \mathbf{x}_2 + S)\cap S\|\}] - 1)\, d\mathbf{x}_1\, d\mathbf{x}_2,$$

where $\mathcal{D}$ is any k-dimensional cube of side length cd with the same orientation as $\mathcal{D}_1$. Also,

$$E|U_i - E(U_i)|^3 \le \|\mathcal{D}_i\|\,\mathrm{var}(U_i) = (cd\delta)^k\,\mathrm{var}(U_i),$$

and so, since $n = O(\lambda)$ as $\lambda \to \infty$,

$$\left\{\sum_i E|U_i - E(U_i)|^3\right\}\Big/\left\{\sum_i \mathrm{var}(U_i)\right\}^{3/2} \le (cd\delta)^k\Big/\left\{\sum_i \mathrm{var}(U_i)\right\}^{1/2}$$

$$= O(\lambda^{-1}\lambda^{1/2}) \to 0$$

as $\lambda \to \infty$. The result (3.40) follows from this estimate and Lyapounov's central limit theorem (Appendix II). To prove (3.41), notice that it suffices to show

$$\lim_{d\to\infty} \limsup_{\lambda\to\infty} \left|\frac{\mathrm{var}(V) - \mathrm{var}(V_1)}{\mathrm{var}(V)}\right| = 0. \qquad (3.42)$$

Now,

$$
\begin{aligned}
|\operatorname{var}(V) - \operatorname{var}(V_1)| \\
= |E[\{V - V_1 - E(V - V_1)\}\{V + V_1 - E(V + V_1)\}]| \\
= |E[\{V_2 + V_3 - E(V_2 + V_3)\} \\
\times \{2V - V_2 - V_3 - E(2V - V_2 - V_3)\}]| \\
\leq 4\{\operatorname{var}(V_2) + \operatorname{var}(V_3)\}^{1/2} \{\operatorname{var}(V) + \operatorname{var}(V_2) + \operatorname{var}(V_3)\}^{1/2},
\end{aligned}
$$

$$(3.43)$$

using the Cauchy-Schwarz inequality. Result (3.42) follows from (3.14), (3.38), (3.39), and (3.43). $\quad\square$

Proof of Theorem 3.6. The case $E(\|S\|) = \infty$ follows from Theorem 3.1, and so we assume $E(\|S\|) < \infty$.

Since $\mathcal{R}$ is Riemann measurable then for any $\epsilon > 0$ there exists a finite sequence $\{\mathcal{D}_i\}$ of bounded, disjoint k-dimensional cubes such that the symmetric difference

$$
\left(\bigcup_i \mathcal{D}_i\right) \triangle \mathcal{R} = \left\{\left(\bigcup_i \mathcal{D}_i\right) \cap \tilde{\mathcal{R}}\right\} \cup \left\{\left(\bigcup_i \mathcal{D}_i\right)^{\sim} \cap \mathcal{R}\right\}
$$

has content less than ϵ. The vacancy within $l(\bigcup_i \mathcal{D}_i) = \bigcup_i(l\mathcal{D}_i)$ differs from the vacancy within $l\mathcal{R}$ by no more than $l^k \epsilon$, and so it suffices to prove that for each such sequence $\{\mathcal{D}_i\}$,

$$
V\left(l\bigcup_i \mathcal{D}_i\right) \Big/ \left\|l\bigcup_i \mathcal{D}_i\right\| \to \exp\{-\lambda E(\|S\|)\} \qquad (3.44)
$$

almost surely. Since the cubes $\mathcal{D}_i$ are disjoint then the vacancy within $l\bigcup_i \mathcal{D}_i$ equals the sum of the vacancies within sets $l\mathcal{D}_i$, and so (3.44) will follow if we show that for any single cube $\mathcal{D}$,

$$
V(l\mathcal{D})/\|l\mathcal{D}\| \to \exp\{-\lambda E(\|S\|)\} \qquad (3.45)
$$

almost surely. For ease of notation we take $\mathcal{D} \equiv [0,1]^k$.

Given a vector $\mathbf{i} = (i_1,\ldots,i_k) \in \mathbb{Z}^k$, let $V_{\mathbf{i}}$ denote the vacancy within the unit cube

$$
\mathcal{D}_{\mathbf{i}} \equiv \prod_{j=1}^k [i_j, i_j + 1].
$$

Result (3.45) is equivalent to

$$
n^{-k} \sum_{0 \leq \mathbf{i} \leq n-1} V_{\mathbf{i}} \to \exp\{-\lambda E(\|S\|)\} \qquad (3.46)
$$

almost surely as $n \to \infty$, where $\mathbf{0} \equiv (0,\ldots,0)$, $\mathbf{n} \equiv (n,\ldots,n)$, $\mathbf{1} \equiv (1,\ldots,1)$, and $\mathbf{0} \leq (i_1,\ldots,i_k) \leq \mathbf{n-1}$ means $0 \leq i_j \leq n-1$ for each j.

Let $c > 0$ be large and fixed, let $T(c)$ denote the k-dimensional (closed) sphere centered at the origin and of radius c, and put $S_i' \equiv S_i \cap T(c)$ and $S_i'' \equiv S_i \cap T(c)^\sim$. Write V_i' and V_i'' for vacancies within $\mathcal{D}_i$ resulting from the coverage processes obtained by replacing each shape S_j by S_j' and S_j'', respectively. Then

$$V_i' - (1 - V_i'') \leq V_i \leq V_i'. \qquad (3.47)$$

Since the shapes S_j' are of radius no more than c, then vacancies within regions separated by at least $2c$ units are stochastically independent. Therefore we may write

$$\sum_{0 \leq i \leq n-1} V_i' = \sum_{i \in \mathcal{I}_1(n)} V_i' + \ldots + \sum_{i \in \mathcal{I}_m(n)} V_i',$$

where the variables $\{V_i', i \in \mathcal{I}_j(n)\}$ are stochastically independent for each j and n, the sequence of sets $\{\mathcal{I}_j(n), n \geq 1\}$ is increasing for each j, the number $\#\mathcal{I}_j(n)$ of elements in each $\mathcal{I}_j(n)$ diverges to $+\infty$, $m \geq 1$ depends only on c (not on n), and $\bigcup_j \mathcal{I}_j(n) = \{i : 0 \leq i \leq n-1\}$. By the strong law of large numbers (Appendix II),

$$\{\#\mathcal{I}_j(n)\}^{-1} \sum_{i \in \mathcal{I}_j(n)} V_i' \to E(V_0')$$

almost surely as $n \to \infty$, for $1 \leq j \leq m$. Adding over j we conclude that

$$n^{-k} \sum_{0 \leq i \leq n-1} V_i' \to E(V_0') = \exp[-\lambda E\{\|S \cap T(c)\|\}] \qquad (3.48)$$

almost surely.

The sequence $\{V_i''\}$ is stationary and uniformly bounded, and so by the ergodic theorem (Appendix II),

$$n^{-k} \sum_{0 \leq i \leq n-1} V_i'' \to U,$$

say, almost surely and in L^p for each $p \geq 1$. Convergence in L^1 entails

$$E(U) = E(V_0'') = \exp[-\lambda E\{\|S \cap T(c)^\sim\|\}],$$

and so

$$n^{-k} \sum_{0 \leq i \leq n-1} (1 - V_i'') \to 1 - U \qquad (3.49)$$

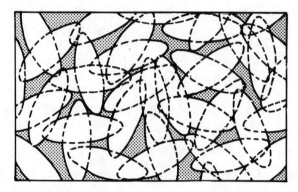

Figure 3.5. A high-intensity Boolean model of randomly-oriented, lens-shaped sets in $k = 2$ dimensions. Uncovered region is shaded.

almost surely, where for each $\epsilon > 0$,

$$P(1 - U > \epsilon) \leq \epsilon^{-1}(1 - \exp[-\lambda E\{\|S \cap \mathcal{T}(c)^{\sim}\|\}])$$
$$\leq \epsilon^{-1} \lambda E\{\|S \cap \mathcal{T}(c)^{\sim}\|\}.$$

The right-hand side here can be made arbitrarily small by choosing c sufficiently large, and similarly the right-hand side of (3.48) may be made arbitrarily close to $\exp\{-\lambda E(\|S\|)\}$ by choosing c large. The desired result (3.46) now follows from (3.47), (3.48), and (3.49). $\square$

3.5. SIZE AND STRUCTURE OF UNCOVERED REGIONS IN A HIGH-INTENSITY BOOLEAN MODEL

In some applications of Boolean model theory, the intensity of random shapes per unit content of space can be very high. For example, simple physical assumptions suggest that a very large fraction of the night sky should be covered by clusters of galaxies. A Boolean model such as this looks markedly different from the "moderate-intensity" Boolean models that we have studied so far. Compare Figures 3.1 and 3.5. Uncovered "chinks" of space are quite rare in high-intensity Boolean models, and those that do exist are very small.

When a connected, uncovered region is magnified to such an extent that it assumes a "reasonable" size, the surfaces of the shapes forming the region's boundary appear to be approximately flat (or

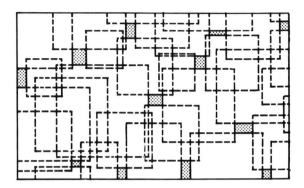

Figure 3.6. A high-intensity Boolean model of random squares, with axes aligned, in $k = 2$ dimensions. Uncovered region is shaded.

linear, in the case of $k = 2$ dimensions). Uncovered regions tend to be bounded by a relatively small number of random sets, and it is unlikely that vertices of sets form part of the boundary. For these reasons, uncovered regions are shaped approximately like convex k-dimensional "polyhedra"—polygons in two dimensions, polyhedra in higher dimensions.

It is clear that the shape of these uncovered chinks will depend on the orientation of random sets in the Boolean model. If these sets are isotropic then uncovered regions will be irregularly shaped, as in Figure 3.5. But if the shapes tend to be aligned in similar ways—for example, if they are all rectangular prisms of various sizes, aligned in the same direction—then chinks will tend to be highly regular in shape, as in Figure 3.6. We give most emphasis to the case of isotropy, treating fixed orientation only by example.

The simplest way to approach vacancy in a high-intensity Boolean model is often to observe that the vacancy within a region $\mathcal{R}$ is just the sum of the contents of those small, uncovered chinks of space that are within $\mathcal{R}$. Therefore a study of size and structure of these chinks is basic to an understanding of vacancy. Section 3.6 will apply the results in the present section on chink shape, to obtain approximations to the distribution of vacancy and to the probability that a region is completely covered when the Boolean model is of high intensity.

We begin with definitions and notation needed later in this section. First of all, we introduce the concept of "convergence in distribution" for random sets. Remember that in Section 3.1 we stated that our random sets would be either random closed sets or random open sets. During this and the next section we insist that the random shapes generating a Boolean model be *open*. The assumption of openness means that each uncovered chink is guaranteed *closed*. The weak limits of these chinks, after suitable rescaling, will be bounded with probability 1. Therefore they will also be compact. This observation simplifies the theory of weak convergence.

As in Section 3.1, let $\mathcal{K}'$ be the class of nonempty compact subsets of $\mathbb{R}^k$, and let ρ be the Hausdorff metric on $\mathcal{K}'$ [defined in (3.2)]. If $\{Z_\lambda, \lambda > 0\}$ and Z are random elements of $\mathcal{K}'$ (i.e., random compact sets), we say that $Z_\lambda \to Z$ (i.e., Z_λ converges to Z) in distribution as $\lambda \to \infty$, if for all functions $g : \mathcal{K}' \to \mathbb{R}$ that are bounded and uniformly continuous (with respect to the metric ρ), we have

$$E\{g(Z_\lambda)\} \to E\{g(Z)\} \qquad (3.50)$$

as $\lambda \to \infty$.

This definition synchronizes with the general definition of weak convergence in a metric space [see, for example, Billingsley (1968, pp. 11–12)]. To generalize it a little, suppose $\{Z'_\lambda, \lambda > 0\}$ and Z are random sets such that

$$P(Z'_\lambda \in \mathcal{K}') \to 1 \quad \text{as} \quad \lambda \to \infty \quad \text{and} \quad P(Z \in \mathcal{K}') = 1. \qquad (3.51)$$

Define $Z_\lambda = Z'_\lambda$ if $Z'_\lambda \in \mathcal{K}'$, and $Z_\lambda = \{0\}$ otherwise. We say that $Z'_\lambda \to Z$ in distribution if $Z_\lambda \to Z$ in distribution, or equivalently, if (3.50) holds for all bounded, uniformly continuous functions g.

It is often the case that the *structure* of a set is of greatest importance, not the set's location. In that situation, little is to be gained by distinguishing between a set and its translated images. This motivates the concept of *essential convergence in distribution*. If $\{Z'_\lambda\}$ and Z are random sets satisfying (3.51), we say that $Z'_\lambda \to Z$ essentially in distribution if (3.50) holds for all bounded, uniformly continuous, *translation invariant* functions g.

Next we need the notion of a *Poisson field* of $(k-1)$-dimensional planes in $\mathbb{R}^k$. Let $\xi_1 < \xi_2 < \dots$ be points of a stationary Poisson process $\mathcal{Q}$ on $[0, \infty)$ with intensity μ, and let $\Theta_1, \Theta_2, \dots$ be unit vectors distributed independently and uniformly over the unit sphere surface

$$\Omega \equiv \left\{ \mathbf{x} \in \mathbb{R}^k : |\mathbf{x}| = 1 \right\},$$

independent also of $\mathcal{Q}$. Construct the $(k-1)$-dimensional plane Π_i whose normal to the origin is of length ξ_i and has inclination Θ_i. The set $\{\Pi_1, \Pi_2, \dots\}$ is called a Poisson field of $(k-1)$-dimensional

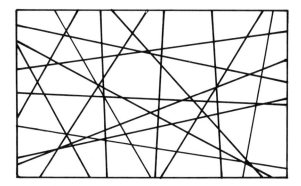

Figure 3.7. Poisson field of lines in the plane.

planes in $\mathbb{R}^k$ with intensity μ.* The field is completely homogeneous and isotropic—that is, strictly stationary and distributionally invariant under rotations. When $k = 2$ we have a Poisson field of lines in the plane (see Figure 3.7). Miles (1971) makes a thorough study of the general case, and Solomon (1978, Chapter 2) gives a detailed treatment of the case $k = 2$.

The closure $\overline{S}$ of $S \subseteq \mathbb{R}^k$ is the set of all limit points of sequences from S, the interior S° of S is the union of all open sets contained in S, and the surface or boundary of S is defined by

$$\partial S \equiv \overline{S} \cap (S^\circ)^\sim.$$

The k-dimensional content of S is given by $\|S\| = \|S\|_k$, and the $(k - 1)$-dimensional surface content of ∂S by

$$\|\partial S\|_{k-1} \equiv \lim_{\epsilon \to 0} (2\epsilon)^{-1} \|(\partial S)^\epsilon\|,$$

provided the limit exists. [Define $(\partial S)^\epsilon$ as in (3.1).] If the random set S is a sphere in $\mathbb{R}^k$ with (random) radius R, then $E(\|S\|_k) = v_k E(R^k)$ and $E(\|\partial S\|_{k-1}) = s_k E(R^{k-1})$.

We are now in a position to describe properties of size and structure of vacant "chinks" in high-intensity Boolean models. Consider a model in which the driving Poisson process has intensity λ and all shapes are distributed as the isotropic random set S. Assume

---

*The definition of a Poisson field of planes with intensity μ is not completely universal. Some authors take the variables ξ_i, $-\infty < i < \infty$ to be points of a Poisson process of intensity μ on the *whole line* $(-\infty, \infty)$, and take the orientations Θ_i to be uniformly distributed over a hemisphere. In our notation, such a Poisson field would have intensity 2μ.

that $0 < E(\|S\|) < \infty$, $0 < E(\|\partial S\|_{k-1}) < \infty$, and the surface of S is reasonably smooth. Let $\mathcal{V}_\lambda$ be an arbitrary connected, uncovered region in the Boolean model. As λ increases, $\mathcal{V}_\lambda$ tends to shrink in size, and we have to scale it up by an amount λ in each dimension to keep its content approximately constant. The following theorem describes asymptotic properties of the rescaled set,

$$\lambda \mathcal{V}_\lambda \equiv \{\lambda \mathbf{x} : \mathbf{x} \in \mathcal{V}_\lambda\}.$$

Theorem 3.7. *As $\lambda \to \infty$, the random set $\lambda\mathcal{V}_\lambda$ converges essentially in distribution to $\mathcal{V}^0$, where $\mathcal{V}^0$ is an arbitrary convex polygonal cell formed by a Poisson field of $(k-1)$-dimensional planes in $\mathbb{R}^k$ with intensity $\mu \equiv E(\|\partial S\|_{k-1})$.*

At the end of this section we give a proof of Theorem 3.7 in the case where S is a sphere of random radius. A proof for the case where S is a random polyhedron appears in Hall (1985e), and many other cases (e.g., cylinders, doughnuts, dumbells etc.) may be treated similarly. The crucial assumptions are that (i) S be isotropic, and (ii) the surface of S be reasonably smooth. It is not necessary to suppose that S be connected.

It is easy to simulate and depict the limiting set $\mathcal{V}^0$ in Theorem 3.7 (see, for example, Figure 3.7). However, only in the case $k = 1$ is the distribution of $\mathcal{V}^0$ available with any degree of explicitness. There, we know from Chapter 2 that if the set S is a line segment of random length then the spacing $\mathcal{V}_\lambda$ between two successive clumps of segments is exponentially distributed with mean λ^{-1}. Therefore $\lambda\mathcal{V}_\lambda$ has precisely an exponential distribution with unit mean, for each $\lambda > 0$. When $k = 1$, a Poisson field of $(k-1)$-dimensional "planes" with intensity μ is just a Poisson point process on the line with intensity $\frac{1}{2}\mu$, and so the random variable $\mathcal{V}^0$ is exponentially distributed with mean $2\mu^{-1}$. Since S is a line segment,

$$\mu = E(\|\partial S\|_{k-1}) = \text{number of endpoints of } S = 2.$$

Therefore Theorem 3.5 correctly predicts that when $\lambda \to \infty$ (indeed, for each $\lambda > 0$), $\lambda\mathcal{V}_\lambda$ has an exponential distribution with unit mean.

A little more generally, but still in the case $k = 1$, suppose each realization of the random shape S is a union of disjoint random line segments, and the mean number of segments in the union equals n. Then

$$\mu = E(\|\partial S\|_{k-1}) = \text{mean number of endpoints of } S = 2n,$$

and so $\lambda \mathcal{V}_\lambda$ has an exponential limit distribution with mean $2(2n)^{-1}$ $= n^{-1}$. Note that in no circumstance does the limit distribution of $\mathcal{V}^0$ depend on the mean content of the set S; it depends only on mean surface content.

Even though the exact distribution of $\mathcal{V}^0$ is not known, moments of quantities related to $\mathcal{V}^0$ may be found by indirect methods. In Appendix IV we show that

$$E(\|\mathcal{V}^0\|) = \left\{ \pi^{1/2} \Gamma(k+1) \Big/ \Gamma\left(\frac{k+1}{2}\right) \right\} \left\{ 2\Gamma\left(\frac{k+1}{2}\right) \Big/ \mu\Gamma\left(\frac{k}{2}\right) \right\}^k$$
(3.52)

and

$$E(\|\mathcal{V}^0\|^2) = \left\{ 2\pi^{1/2} \Gamma(k)\Gamma(k+1) \Big/ \Gamma\left(\frac{k}{2}\right)\Gamma\left(\frac{k+1}{2}\right)\Gamma\left(\frac{k+1}{2}\right) \right\}$$
$$\times \left\{ 2\pi^{1/2}\Gamma\left(\frac{k+1}{2}\right) \Big/ \mu\Gamma\left(\frac{k}{2}\right) \right\}^{2k},$$
(3.53)

for $k \geq 1$. Further information on the distribution of $\mathcal{V}^0$ is available in two and three dimensions. For example, let R denote the radius of the largest sphere contained within $\mathcal{V}^0$. Then R is exponentially distributed with mean μ^{-1}. Skewness of the distribution of $\|\mathcal{V}^0\|$ can be calculated from the third moment,

$$E(\|\mathcal{V}^0\|^3) = \begin{cases} 2^8 \pi^7 / (7\mu^6) & \text{if } k = 2 \\ 2^{24} \cdot 21\pi / \mu^9 & \text{if } k = 3 . \end{cases}$$
(3.54)

See Miles (1964a, 1964b, 1972).

We treat next a particular case in which random sets have fixed orientation. Take S to be a k-dimensional cube of random side length but with fixed orientation, and consider the Boolean model in which all sets are distributed as S. Uncovered "chinks" of space tend to be random rectangular prisms, and a random process of prisms may be defined as follows. Let $\mathcal{P}_i$, $1 \leq i \leq k$, be independent stationary Poisson processes on the real line, with common intensity ν. Place the points of $\mathcal{P}_i$ along the ith coordinate axis in $\mathbb{R}^k$ for each $1 \leq i \leq k$, and through each Poisson point on an axis draw a plane orthogonal to that axis. We call the resulting sequence of planes an *orthogonal Poisson system* with intensity ν. Let $\mathcal{V}_\lambda$ be an arbitrary connected, uncovered region in the Boolean model.

Theorem 3.8. *Let S be a k-dimensional cube of fixed orientation and random side length L satisfying $0 < E(L^k) < \infty$. As $\lambda \to \infty$, the random set $\lambda\mathcal{V}_\lambda$ converges essentially in distribution to $\mathcal{V}^1$, where $\mathcal{V}^1$ is an arbitrary rectangular prism formed by an orthogonal Poisson system of $(k-1)$-dimensional planes in $\mathbb{R}^k$, with intensity $\nu \equiv E(L^{k-1})$ and axes aligned with the axes of S.*

Figure 3.8 illustrates the Boolean model of aligned random "cubes" (squares) in the case $k = 2$.

It is easy to show that for an orthogonal Poisson system of planes with intensity ν,

$$E(\|\mathcal{V}^1\|) = \nu^{-k} \quad \text{and} \quad E(\|\mathcal{V}^1\|^2) = 2^k \nu^{-2k}. \tag{3.55}$$

We conclude this section with proofs of Theorem 3.7 (in the case of random spheres) and Theorem 3.8.

Proof of Theorem 3.7 for random radius spheres. Without loss of generality, S is centered at the origin. Let R denote its radius. Then $E(R^k) = v_k^{-1} E(\|S\|) < \infty$. We study the distribution of the vacant region containing the origin, given that the origin is uncovered.

First we derive the distribution of

$$Y_i \equiv \text{distance from origin to that sphere surface } i\text{th}$$
$$\text{closest to origin,}$$

for $i \geq 1$. See Figure 3.8 for a picture in the case $k = 2$.

The conditional distribution of $(Y_1, \ldots, Y_m)$, given that the origin is not covered, has joint density

$$f_1(y_1, \ldots, y_m \mid \mathbf{0} \text{ not covered}) = p^{-1} f_2(y_1, \ldots, y_m),$$

where

$$p \equiv P(\mathbf{0} \text{ not covered}) = \exp\{-\lambda E(\|S\|)\}$$

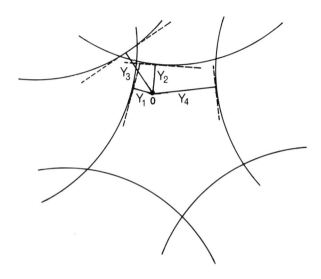

Figure 3.8. Distance, Y_i, from origin to ith nearest sphere surface.

[see (3.4)] and

$$f_2(y_1,\ldots,y_m)\,dy_1\ldots dy_m$$

$$= \left\{ \prod_{i=1}^{m} P(\text{nearest portion of some sphere is distant}\right.$$

$$\left. \text{between } y_i \text{ and } y_i + dy_i \text{ from } \mathbf{0}) \right\}$$

$$\times P(\text{no parts of any spheres are distant} \le y_m \text{ from } \mathbf{0})$$

$$= \left[\prod_{i=1}^{m} \left\{ \lambda s_k E(y_i + R)^{k-1}\,dy_i \right\} \right] \exp\left\{ -\lambda v_k E(y_m + R)^k \right\}$$

$$= \lambda^m s_k^m \left\{ \prod_{i=1}^{m} E(y_i + R)^{k-1} \right\} \exp\left\{ -\lambda v_k E(y_m + R)^k \right\} dy_1 \ldots dy_m.$$

Therefore

$$f_1(y_1,\ldots,y_m \mid \mathbf{0} \text{ not covered}) = \lambda^m s_k^m \left\{ \prod_{i=1}^{m} E(y_i + R)^{k-1} \right\}$$

$$\times \exp\left[-\lambda v_k \left\{ E(y_m + R)^k - E(R^k) \right\} \right], \qquad (3.56)$$

for $0 < y_1 < \cdots < y_m < \infty.$

Next we rescale each dimension by the factor λ, setting $Z_i = \lambda Y_i$ and $z_i = \lambda y_i$ for $1 \leq i \leq m$. Observe that as $\lambda \to \infty$,

$$
\begin{aligned}
s_k E(y_i + R)^{k-1} &= s_k E(\lambda^{-1} z_i + R)^{k-1} \\
&= s_k E(R^{k-1}) + O(\lambda^{-1}) \\
&= E(\|\partial S\|_{k-1}) + O(\lambda^{-1}),
\end{aligned}
$$

uniformly in bounded z_i, and that

$$
\begin{aligned}
\lambda v_k \{ E(y_m + R)^k - E(R^k) \} &= z_m k v_k E(R^{k-1}) + O(\lambda^{-1}) \\
&= z_m E(\|\partial S\|_{k-1}) + O(\lambda^{-1}),
\end{aligned}
$$

uniformly in bounded z_m, since $k v_k = s_k$. Let $\mu \equiv E(\|\partial S\|_{k-1})$. Substituting these estimates into (3.56) we deduce that the conditional distribution of $(Z_i, \ldots, Z_m)$, given that the origin is not covered, has joint density

$$
f_3(z_1, \ldots, z_m \mid \mathbf{0} \text{ not covered}),
$$

which satisfies

$$
f_3(z_1, \ldots, z_m \mid \mathbf{0} \text{ not covered}) = \mu^m \exp(-\mu z_m) + O(\lambda^{-1}) \quad (3.57)
$$

as $\lambda \to \infty$, uniformly in $0 < z_1 < \cdots < z_m < C$ for any $C < 0$.

Let $W_1, \ldots, W_m$ be independent exponential variables with means μ^{-1}, and define $\xi_i \equiv \sum_{j=1}^{i} W_j$ for $1 \leq j \leq m$. Then $(\xi_1, \ldots, \xi_m)$ has density

$$
\mu^m \exp(-\mu x_m), \qquad 0 < x_1 < \cdots < x_m < \infty.
$$

Comparing this density with the right-hand side of (3.57), we see that the distribution of $(Z_1, \ldots, Z_m)$ converges to that of $(\xi_1, \ldots, \xi_m)$ as $\lambda \to \infty$. The joint distribution of $(\xi_1, \ldots, \xi_m)$ is that of the first m points in a Poisson process of intensity μ on the positive half-line.

Let $\mathcal{T}(r)$ be the k-dimensional closed sphere of radius r centered at the origin, and let Θ_i denote a unit vector in the direction of the normal to the origin from the ith nearest sphere surface. Each Θ_i is distributed over Ω, the surface of the k-dimensional unit sphere centered at the origin. It follows by symmetry that for each $\lambda > 0$, the variables $\Theta_1, \ldots, \Theta_M$ are stochastically independent, independent also of $(Z_1, \ldots, Z_m)$, and distributed uniformly over Ω. If $R(i)$ denotes the radius of the sphere ith closest to the origin, conditional on the origin being uncovered, then $\lambda R(i) \to \infty$ as in probability as $\lambda \to \infty$. These considerations, and the conclusion of the previous paragraph, lead us to the following result for our rescaled Boolean model. Let $T_1, \ldots, T_m$ be the m spheres from the model whose surfaces are closest to the origin. For each $r > 0$, the random compact set

$$T(r) \cap \left(\bigcup_{i=1}^{m} T_i \right)^{\sim},$$

conditional on the origin being uncovered, converges in distribution to the random compact set formed by the intersection with $T(r)$ of m $(k-1)$-dimensional planes whose orientations are independent and uniformly distributed over Ω, and whose perpendicular distances from the origin are distributed as the first m points of a Poisson process on the positive half-line with intensity μ. The positions and orientations of these planes are the same as those of the m planes closest to the origin in a Poisson field of $(k-1)$-dimensional planes with intensity μ.

On an intuitive level this completes the proof of Theorem 3.7. However, we should check the formal definition of essential convergence in distribution. To do this, let Z'_λ have the distribution of the connected, uncovered set containing the origin, given that the origin is uncovered, and let Z be the convex cell containing the origin and formed by our Poisson field of lines. The information garnered so far establishes property (3.51). (Recall that the spheres are assumed to be *open*.) Let $Z_\lambda = Z'_\lambda$ if Z'_λ is compact, and $Z_\lambda = \{0\}$ otherwise. We shall prove (3.50) for bounded, uniformly continuous g.

Let $\mathcal{E}_m$ denote the event that the connected, uncovered set containing the origin is completely determined by the vector $\mathbf{v}_m$ whose elements are the lengths and orientations of the normals from the origin to the m nearest sphere surfaces, and the radii of these spheres. Let $\mathcal{F}_m$ be the event that in a Poisson field of $(k-1)$-dimensional planes with intensity μ, the cell containing the origin is completely determined by the m planes closest to the origin. We have already established that as $\lambda \to \infty$, $\mathbf{v}_m$ converges in distribution to its counterpart for a random field, in which the radius components are all equal to $+\infty$. Therefore as $\lambda \to \infty$,

$$P(\mathcal{E}_m \mid \mathbf{0} \text{ not covered}) \to P(\mathcal{F}_m),$$

and for any bounded and continuous g,

$$E\{g(Z_\lambda) I(\mathcal{E}_m)\} \to E\{g(Z) I(\mathcal{F}_m)\}.$$

By choosing m so large that $P(\mathcal{F}_m) > 1 - \epsilon$, we see that

$$\limsup_{\lambda \to \infty} |E\{g(Z_\lambda)\} - E\{g(Z)\}| \leq 2\epsilon \sup |g|,$$

for all $\epsilon > 0$. Therefore $E\{g(Z_\lambda)\} \to E\{g(Z)\}$, as had to be shown.

We have established that the connected, uncovered set containing the origin, given that the origin is uncovered, converges in distribution to the convex cell containing the origin and formed by a Poisson field of planes having intensity μ. From this it follows that an arbitrary uncovered region converges essentially in distribution to an arbitrary convex cell formed by a Poisson field of planes. (Of course, the origin cell has properties slightly different from those of an "arbitrary" cell, due to a k-dimensional version of length-biased sampling. Likewise for the uncovered set containing the origin, given that the origin is uncovered.) $\Box$

Proof of Theorem 3.8. We may take the axes of our coordinate system in $\mathbb{R}^k$ to be perpendicular to faces of the random cubes. Rescale each of the dimensions of $\mathbb{R}^k$ by the factor λ. Our original Boolean model $\mathcal{C}$ is now transformed to a model $\mathcal{C}^*$, in which the underlying Poisson process has intensity $\lambda^{-(k-1)}$ and each shape has the distribution of λS. By using techniques similar to those in the proof of Theorem 3.7 we may prove that the probability that a corner of a random cube is part of the boundary of the uncovered set containing the origin, conditional on the origin being uncovered, converges to 0 as $\lambda \to \infty$. Therefore, with probability tending to 1 as $\lambda \to \infty$, the uncovered set is a rectangular prism whose faces are perpendicular to the axes.

Conditional on the event that the origin is uncovered, let X_i (respectively, Y_i) denote the distance along the ith coordinate axis in the positive (negative) direction to the first cube surface. We shall derive the probability that X_i takes a value between x_i and $x_i + dx_i$, and that Y_i takes a value between y_i and $y_i + dy_i$, for $1 \leq i \leq k$.

The probability that some cube intersects the ith coordinate axis between x_i and $x_i + dx_i$ [respectively, $-(y_i + dy_i)$ and $-y_i$], and has its center on the opposite side of x_i $(-y_i)$ to the origin, equals

$$\lambda^{-(k-1)} E(\lambda L)^{k-1} dx_i = E(L^{k-1}) dx_i$$

$[E(L^{k-1}) dy_i]$. The probability that no cubes protrude into a given rectangular prism containing the origin and having side lengths $(x_1 + y_1), \ldots, (x_k + y_k)$, conditional on the origin being uncovered, is

$$\exp\left[-\lambda^{-(k-1)} E\left\{\prod_{i=1}^{k}(x_i + y_i + \lambda L)\right\}\right] \Big/ \exp\left\{-\lambda^{-(k-1)} E(\lambda L)^k\right\}$$

$$= \exp\left\{-E(L^{k-1}) \sum_{i=1}^{k}(x_i + y_i) + O(\lambda^{-1})\right\}.$$

Therefore

$$P(x_i < X_i \le x_i + dx_i \text{ and } y_i < Y_i \le y_i + dy_i \text{ for } 1 \le i \le k)$$

$$= \left[\prod_{i=1}^{k} \left\{ E(L^{k-1}) \, dx_i E(L^{k-1}) \, dy_i \right\} \right]$$

$$\times \exp\left\{ -E(L^{k-1}) \sum_{i=1}^{k} (x_i + y_i) + O(\lambda^{-1}) \right\}$$

$$= \left[\left\{ E(L^{k-1}) \right\}^{2k} \exp\left\{ -E(L^{k-1}) \sum_{i=1}^{k} (x_i + y_i) \right\} + O(\lambda^{-1}) \right]$$

$$\times dx_1 \ldots dx_k \, dy_1 \ldots dy_k.$$

In consequence, the components of the vector $(X_1, \ldots, X_k, Y_1, \ldots, Y_k)$ are asymptotically independent and identically distributed, each having the exponential distribution with mean $1/E(L^{k-1})$. This is the same distribution as the sequence $(\xi_{11}, \xi_{21}, \ldots, \xi_{k1}, -\xi_{1,-1}, -\xi_{2,-1}, \ldots, -\xi_{k,-1})$, where $\{\xi_{ij}, -\infty < j < \infty \text{ and } j \ne 0\}$, $1 \le i \le k$, are independent stationary Poisson processes on the real line, each with the same intensity $E(L^{k-1})$, notated so that

$$\ldots < \xi_{i,-2} < \xi_{i,-1} < 0 < \xi_{i1} < \xi_{i2} < \ldots .$$

Thus, the random set defining the uncovered region surrounding the origin, given that the origin is uncovered, has asymptotically the same distribution as the origin cell in an orthogonal Poisson system of $(k-1)$-dimensional planes with intensity $E(L^{k-1})$. $\square$

3.6. APPROXIMATE DISTRIBUTION OF VACANCY IN A CRITICAL HIGH-INTENSITY BOOLEAN MODEL

Our aim in this section is to establish an approximation to the distribution of vacancy within a region $\mathcal{R}$ in the case where the probability that $\mathcal{R}$ is completely covered is close to neither 0 nor 1. A result of this type permits us to approximate the chance that $\mathcal{R}$ is completely covered, for the probability of complete coverage is often just the size of the atom at 0 in the distribution of vacancy (see Theorem 3.3 and the remarks preceding that result). We say that a high-intensity Boolean model in which the probability of coverage is neither close to 0 nor close to 1, is *critical.* As we shall show, criticality can be expressed as a concise condition involving Poisson intensity and mean shape content [see condition (3.62) below].

To introduce criticality, consider the case of a one-dimensional Boolean model $\mathcal{C}$ in which the driving Poisson process has intensity λ and shapes are line segments distributed as δS, where $S \equiv [0, L]$. If $\alpha \equiv E(L) < \infty$ and

$$(\log x) \int_x^\infty P(L > y)\, dy \to 0$$

as $x \to \infty$, and if $\delta = \delta(\lambda)$ and λ vary together in such a manner that

$$\alpha\delta\lambda = \log(\lambda/u) + o(1) \tag{3.58}$$

as $\lambda \to \infty$, where u is a fixed positive number, then Theorem 2.4 (Chapter 2, Section 2.4) tells us that the length C of an arbitrary clump is asymptotically exponentially distributed with mean u^{-1}. Clump lengths and spacing lengths are all independently distributed, spacings between clumps are precisely exponential with mean λ^{-1}, and of course spacings shrink to 0 as $\lambda \to \infty$. Therefore, as explained in Section 2.5, the coverage process $\mathcal{C}$ resembles a Poisson process $\mathcal{Q}$ of intensity u in which tiny independent and exponentially distributed spacings are centered at points of $\mathcal{Q}$, and the intervals between successive spacings represent clumps. The number of points of $\mathcal{Q}$ within an interval $[0, t]$ is Poisson-distributed with mean ut, and total vacancy $V = V([0, t])$ within $[0, t]$ equals the sum of the lengths of those spacings or parts of spacings that lie within $[0, t]$. The chance that a spacing overlaps the point 0 (or t) equals $e^{-\alpha\delta\lambda}$, and so converges to 0 as $\lambda \to \infty$ [use (3.58)]. Therefore as $\lambda \to \infty$,

$$\lambda V \to \sum_{i=1}^N Z_i \tag{3.59}$$

in distribution, where $N, Z_1, Z_2, \ldots$ are independent random variables, N is Poisson-distributed with mean ut, and each Z_i is exponential with unit mean.

The random variable on the right-hand side of (3.59) has an atom at the origin, of size $P(N = 0) = e^{-ut}$, and a continuous component on $(0, \infty)$ with the distribution of $\sum_{1 \leq i \leq N} Z_i$ conditional on $N \geq 1$. The Laplace-Stieltjes transform of this continuous component is given by

$$E\left\{\exp\left(-s\sum_{i=1}^{N}Z_i\right)\;\middle|\;N\geq 1\right\}=E\left\{(s+1)^{-N}\mid N\geq 1\right\}$$

$$=(1-e^{-ut})^{-1}\sum_{n=1}^{\infty}(ut)^n(s+1)^{-n}e^{-ut}/n!$$

$$=(e^{ut}-1)^{-1}[\exp\{ut/(s+1)\}-1].$$

The Laplace-Stieltjes transform of the entire limiting distribution of V equals

$$e^{-ut}+(1-e^{-ut})(e^{ut}-1)^{-1}[\exp\{ut/(s+1)\}-1]$$
$$=\exp\{-ust/(s+1)\}. \qquad (3.60)$$

This type of distribution has been termed the "chi-square distribution with zero degrees of freedom" by Siegel (1979a). To see why, let the random variable X have a noncentral chi-square distribution with n degrees of freedom and noncentrality parameter 2μ. Then the distribution of X has Laplace-Stieltjes transform

$$E(e^{-sX})=(1+2s)^{-(n/2)}\exp\{-2\mu s/(2s+1)\}.$$

Taking $n=0$ and $\mu=ut$, we see that $\frac{1}{2}X$ has the Laplace-Stieltjes transform given in (3.60).

These results are summarized in the theorem below.

Theorem 3.9. *Let V denote the vacancy within $[0,t]$ arising from a Boolean model on the real line in which the driving Poisson process has intensity λ and sets are segments whose lengths have distribution function $G(\cdot/\delta)$. Suppose λ and $\delta=\delta(\lambda)$ are connected by formula (3.58), where u is a fixed positive constant and*

$$\alpha\equiv\int_0^{\infty}x\,dG(x)<\infty.$$

Assume the distribution function G does not have an atom at the origin, and satisfies

$$(\log x)\int_x^{\infty}\{1-G(y)\}\,dy\to 0 \qquad (3.61)$$

as $x\to\infty$. Then $\lambda V\to W$ in distribution as $\lambda\to\infty$, where W has Laplace-Stieltjes transform

$$E(e^{-sW})=\exp\{-ust/(s+1)\}.$$

One corollary of Theorem 3.9 is that

$$P([0,t] \text{ completely covered}) = P(V = 0) \to P(N = 0) = e^{-ut}$$

as $\lambda \to \infty$.

In the case of $k = 1$ dimension the condition of criticality is given by (3.58), relating mean shape content to Poisson intensity. In $k \geq 1$ dimensions that condition becomes

$$\alpha \delta^k \lambda = \log(\lambda/u) + (k - 1) \log \log \lambda + o(1) \qquad (3.62)$$

as $\lambda \to \infty$, assuming that shapes in the Boolean model are distributed as δS with $E(\|S\|) = \alpha$, and that the driving Poisson process has intensity λ. Recall from the foregoing discussion that we must assume more than just $E(\|S\|) < \infty$. In particular the condition imposed in Theorem 3.9 that G has no atom at the origin becomes $P(\|S\| > 0) = 1$, and condition (3.61) is replaced by

$$(\log r) E[\|S\| I\{S \cap \mathcal{T}(r)^{\sim} \neq \phi\}] \to 0 \quad \text{as} \quad r \to \infty,$$

where $\mathcal{T}(r)$ is the k-dimensional closed sphere of radius r centered at the origin. Finally, since our limit theorem for vacancy hinges on the description in Theorem 3.7 of uncovered "chinks" of space, we assume that the random shape S is isotropic and smooth—examples include spheres and isotropic polygons, cylinders etc. It is not necessary to assume S is connected.

Let $\mathcal{R}$ be a Riemann-measurable subset of $\mathbb{R}^k$. The total vacancy $V = V(\mathcal{R})$ within $\mathcal{R}$ equals the sum of vacancies $V_1, V_2, \ldots$ of connected uncovered chinks of space contained within $\mathcal{R}$, plus the content within $\mathcal{R}$ of uncovered regions that intersect the boundary of $\mathcal{R}$. It may be proved that

(i) $V_1, V_2, \ldots$ are asymptotically independent, and $(\delta^{k-1}\lambda)^k V_i \to \|\mathcal{V}^0\|$ in distribution as $\lambda \to \infty$, where $\|\mathcal{V}^0\|$ denotes the content of an arbitrary convex cell formed from a Poisson field of $(k - 1)$-dimensional planes in $\mathbb{R}^k$ with intensity $\mu \equiv E(\|\partial S\|_{k-1})$;

(ii) the number of uncovered chinks of space within $\mathcal{R}$ is asymptotically Poisson-distributed and independent of the contents $V_1, V_2, \ldots$ of the chinks;

(iii) the chance that some uncovered chink straddles the boundary of $\mathcal{R}$ converges to 0 as $\lambda \to \infty$.

[The second part of property (i) follows from Theorem 3.7, on changing scale in the obvious way.]

We may write

$$\lambda(\log\lambda)^{k-1}V = \alpha^{k-1}\sum_{i=1}^{M}Y_i + R, \qquad (3.63)$$

where R is the rescaled content within $\mathcal{R}$ of chinks of uncovered space straddling the boundary, M is the total number of chinks wholly contained within $\mathcal{R}$, and

$$Y_i \equiv \alpha^{-(k-1)}\lambda(\log\lambda)^{k-1}V_i.$$

Condition (3.62) declares that

$$(\delta^{k-1}\lambda)^k = (\delta^k\lambda)^{k-1}\lambda \sim \alpha^{-(k-1)}\lambda(\log\lambda)^{k-1}, \qquad (3.64)$$

and so by property (i) above, the distribution of each Y_i converges to that of $\|\mathcal{V}^0\|$ as $\lambda \to \infty$. By property (iii), $P(R=0) \to 1$, and by property (ii), M is asymptotically Poisson-distributed. To get an idea of the mean ν of the limit distribution of M, notice that the left-hand side of (3.63) has mean

$$\lambda(\log\lambda)^{k-1}E(V) = \lambda(\log\lambda)^{k-1}\|\mathcal{R}\|\exp(-\alpha\delta^k\lambda) \qquad \text{[see (3.5)]}$$
$$\sim \lambda(\log\lambda)^{k-1}\|\mathcal{R}\|\exp\left\{-\log(\lambda/u) - (k-1)\log\log\lambda\right\}$$
$$= u\|\mathcal{R}\|,$$

using (3.62), while the right-hand side of (3.63) has mean asymptotic to

$$\alpha^{k-1}\nu E(Y_i) \sim \alpha^{k-1}\nu E(\|\mathcal{V}^0\|).$$

Therefore $\nu = u\|\mathcal{R}\|/\{\alpha^{k-1}E(\|\mathcal{V}^0\|)\}$. A formula for $E(\|\mathcal{V}^0\|)$ is given at (3.52).

Thus, we obtain the following limit theorem: As $\lambda \to \infty$,

$$\lambda(\log\lambda)^{k-1}V \to \alpha^{k-1}\sum_{i=1}^{N}Z_i$$

in distribution, where $N, Z_1, Z_2, \ldots$ are independent random variables, N is Poisson-distributed with mean

$$\nu \equiv u\|\mathcal{R}\|/\{\alpha^{k-1}E(\|\mathcal{V}^0\|)\}$$
$$= u\|\mathcal{R}\|_k\left\{\Gamma\left(\frac{k+1}{2}\right)\bigg/\pi^{1/2}\Gamma(k+1)\right\}\left\{\Gamma\left(\frac{k}{2}\right)\bigg/2\Gamma\left(\frac{k+1}{2}\right)\right\}^k$$
$$\times \{E(\|\partial S\|_{k-1})\}^k \{E(\|S\|_k)\}^{-(k-1)}, \qquad (3.65)$$

and the Z_i's have the distribution of $\|\mathcal{V}^0\|$.

A formal proof in the case of random-radius spheres and random polygons is given in Hall (1985e). First, second, and third moments of the limit distribution may be found using formulas (3.52)–(3.54) for the first three moments of $\|\mathcal{V}^0\|$. In particular,

$$\text{var}\left(\alpha^{k-1}\sum_{i=1}^{N}Z_i\right) = \alpha^{2(k-1)}\nu\,\text{var}(\|\mathcal{V}^0\|). \qquad (3.66)$$

This result may be used to establish an approximation to the probability that a given region is completely covered, as follows. Observe from (3.62) that

$$u = \lambda(\log\lambda)^{k-1}\exp\left\{-\lambda E(\|\delta S\|_k)\right\} + o(1). \qquad (3.67)$$

Writing

$$t_k \equiv \left\{\Gamma\left(\frac{k+1}{2}\right)\Big/\pi^{1/2}\Gamma(k+1)\right\}\left\{\Gamma\left(\frac{k}{2}\right)\Big/2\Gamma\left(\frac{k+1}{2}\right)\right\}^k,$$

we see from (3.62), (3.64), (3.65), and (3.67) that

$$\nu = ut_k\|\mathcal{R}\|_k\left\{E(\|\partial\delta S\|_{k-1})\right\}^k\left\{E(\|\delta S\|_k)\right\}^{-(k-1)}$$
$$\sim \lambda^k t_k\|\mathcal{R}\|_k\left\{E(\|\partial\delta S\|_{k-1})\right\}^k\exp\left\{-\lambda E(\|\delta S\|_k)\right\}.$$

Therefore in a Boolean model where all shapes are distributed as S (a smooth-surfaced, isotropic open set), and the driving Poisson process has intensity λ, the probability of complete coverage is given by

$$P(V = 0) \simeq \exp[-\lambda^k t_k\|\mathcal{R}\|_k\left\{E(\|\partial S\|_{k-1})\right\}^k\exp\left\{-\lambda E(\|S\|_k)\right\}]. \qquad (3.68)$$

It is significant that this approximation depends on the distribution of S only through the mean k-dimensional content and mean $(k-1)$-dimensional surface content of S. Other characteristics of S play a secondary role in determining the probability of complete coverage, at least when S is a small set and Poisson intensity is large.

Let us return briefly to the case $k = 1$, and take S to be a disjoint union of intervals. Note that $t_1 = \frac{1}{2}$. Let $l \equiv E(\|S\|_1)$ denote the mean total length of all intervals, and let n be the mean number of disjoint intervals, assumed finite. Then $E(\|\partial S\|_0) = 2n$, and so by (3.68),

$$P(V = 0) \simeq \exp(-\lambda\|\mathcal{R}\|ne^{-\lambda l}).$$

This formula was established earlier in Theorem 2.5 (Chapter 2, Section 2.5) in the special case where $n = 1$ and $\mathcal{R}$ was an interval of length t.

A key assumption in the work above is that the shape S be isotropic. Versions of the result hold in other situations. We consider here only the case of k-dimensional cubes with random side length and fixed alignment.

Let V denote vacancy within the Riemann-measurable set $\mathcal{R} \subseteq \mathbb{R}^k$, with $0 < \|\mathcal{R}\| < \infty$. Let S be a k-dimensional cube with random side length L and fixed orientation, and consider the Boolean model $\mathcal{C}$ in which all shapes are distributed as δS and the driving Poisson process has intensity λ. Put $\alpha \equiv E(L^k)$, and let $\mathcal{V}^1$ be the rectangular prism defined in Theorem 3.8.

Theorem 3.10. *Assume $P(L > 0) = 1$,*

$$(\log r)E\left\{L^k I(L > r)\right\} \to 0$$

as $r \to \infty$, and $\delta = \delta(\lambda) \to 0$ as $\lambda \to \infty$ in such a manner that

$$E(L^k)\delta^k\lambda = \log(\lambda/u) + (k-1)\log\log\lambda + o(1), \qquad (3.69)$$

where u is a fixed positive constant. Then

$$\lambda(\log\lambda)^{k-1}V \to \alpha^{k-1}\sum_{i=1}^{N} Z_i$$

in distribution as $\lambda \to \infty$, where $N, Z_1, Z_2, \ldots$ are independent random variables, N has a Poisson distribution with mean

$$\nu \equiv u\|\mathcal{R}\|/\{\alpha^{k-1}E(\|\mathcal{V}^1\|)\}$$
$$= u\|\mathcal{R}\|\left\{E(L^{k-1})\right\}^k \left\{E(L^k)\right\}^{-(k-1)},$$

and the Z_i's are distributed as $\|\mathcal{V}^1\|$.

The criticality condition here is (3.69), and is identical to (3.62). The proof of Theorem 3.10 is set as Exercise 3.10.

It follows from the definition of $\mathcal{V}^1$ (see Theorem 3.8) that $\|\mathcal{V}^1\|$ is distributed as $\{E(L^{k-1})\}^{-k} \prod_{1 \leq i \leq k} X_i$, where $X_1, \ldots, X_k$ are independent exponential variables with unit mean.

We conclude this section by setting our work on criticality into the context of general limit theory for vacancy. We have been examining vacancy within a fixed region $\mathcal{R}$ arising from a Boolean model in which shapes are distributed as δS and the driving Poisson process has intensity λ. The scale parameter $\delta = \delta(\lambda)$ was permitted to decrease as λ increased, and the exact rate of increase was governed by criticality conditions (3.62) and (3.69). Earlier, in Section 3.4, we used a similar model to study behavior of vacancy within an expanding region $\mathcal{R}_l$, for a Boolean model of fixed intensity. The key difference between the two models is the rate at which δ decreases with increasing λ. In Section 3.4 we assumed that δ and λ varied in such a manner that $\delta^k \lambda \to \rho$, where ρ was the intensity of the Boolean model observed within $\mathcal{R}_l$ (see the discussion between Theorems 3.5 and 3.6). A slightly slower rate of decrease of δ is dictated by the criticality conditions; in fact, they require $\delta^k \lambda (\log \lambda)^{-1} \to$ const. as $\lambda \to \infty$.

Of course, there is a vast range of other possible configurations for δ and λ. The model in Section 3.4 might be termed a "moderate-intensity" Boolean model, and a model where $\delta = \delta(\lambda) \to 0$ such that $\delta^k \lambda \to 0$, a "low-intensity" or "sparse" model. Limit theorems may be derived in all cases, and each case can be thought of as modeling a particular type of real-world phenomenon. The fact that λ is increasing (and $\mathcal{R}$ fixed) means that on rescaling, we are in effect modeling the coverage pattern derived from a fixed-intensity Boolean model within an expanding region. Varying δ adjusts shape size relative to the distance between "centers" of random sets in the pattern. A low-intensity or sparse Boolean model is one in which relatively few sets intersect, so that most clumps of sets contain just one set (the so-called "singleton clumps" or "isolated sets"; see Chapter 4, Section 4.4).

As a general rule, vacancy in a subcritical Boolean model—that is, in a model where $\delta \to 0$ more slowly than is allowed by criticality conditions such as (3.62) and (3.69)—is asymptotically normally distributed, and

$$\{V - E(V)\} / \{\text{var}(V)\}^{1/2} \to N(0,1)$$

in distribution. The method of proof of results such as this is very close to that employed to establish Theorem 3.5 in Section 3.4, and so we do not go into details here. Of course, regularity conditions are required on the distribution of the shape S—even Theorem 3.5 does not hold if $E(\|S\|^2) = \infty$. Furthermore, asymptotic properties of var(V) can be quite different from those outlined in Theorem 3.4; see Exercises 3.11 and 3.12.

The coverage pattern has a unique character in the case of a *very* low intensity Boolean model. We next give a heuristic argument explaining why. Suppose the driving Poisson process is $\mathcal{P} \equiv \{\xi_i\}$, and that shapes δS_i are "centered" at points of $\mathcal{P}$, so that vacancy within $\mathcal{R}$ equals the total content of $\mathcal{R}$ not covered by any of the random sets $\xi_i + \delta S_i$. If $\delta = \delta(\lambda) \to 0$ sufficiently quickly as $\lambda \to \infty$, then with high probability, no two random sets centered within $\mathcal{R}$ will intersect with one another or with the boundary of $\mathcal{R}$. In this case a very good approximation to vacancy is given by

$$\|\mathcal{R}\| - \sum_{i \in \mathcal{I}} \delta^k \|S_i\|,$$

where $\mathcal{I}$ is the set of indices i such that $\xi_i \in \mathcal{R}$. Therefore

$$V - E(V) \simeq -\left\{ \sum_{i \in \mathcal{I}} \delta^k \|S_i\| - E(N)\delta^k E(\|S\|) \right\},$$

where N denotes the number of indices in $\mathcal{I}$. Of course, N has a Poisson distribution with mean $\lambda \|\mathcal{R}\|$, and N is independent of the variables $\|S_i\|$. Therefore as $\lambda \to \infty$,

$$\mathrm{var}(V) \sim E\left\{ \mathrm{var}\left(\sum_{i \in \mathcal{I}} \delta^k \|S_i\| \, \Big| \, N \right) \right\} + \mathrm{var}\left\{ E\left(\sum_{i \in \mathcal{I}} \delta^k \|S_i\| \, \Big| \, N \right) \right\}$$

$$= \delta^{2k} \lambda \|\mathcal{R}\| E(\|S\|^2),$$

and

$$\{V - E(V)\} / (\delta^{2k}\lambda)^{1/2} \to N(0, \|\mathcal{R}\| E(\|S\|^2))$$

in distribution. (Note the central limit theorem in Appendix II for the sum of a random number of random variables.)

3.7. BOUNDS FOR PROBABILITY OF COVERAGE

The work in the previous section on critical high-intensity Boolean models supplied us with an approximation to the chance that a given set $\mathcal{R}$ is completely covered. However, in theoretical work we often need more explicit estimates of coverage probability, such as upper or lower bounds. Upper bounds may be obtained by simple moment inequalities—see formula (3.7), for example. In some circumstances, strict lower bounds of similar form may be constructed.

This section presents an example to illustrate the possibilities and the techniques.

Assume $k = 2$. Suppose the shape S is a disk of fixed radius r, and $\mathcal{R}$ is a unit square. Put $a \equiv \pi r^2$ (the content of S), and let V denote vacancy within $\mathcal{R}$ resulting from a Boolean model in which shapes are distributed as S and Poisson intensity equals λ.

Theorem 3.11. *For each $\lambda \geq 1$ and $0 < r \leq \frac{1}{2}$,*

$$0.05 \min \left\{ 1, (1 + a\lambda^2)e^{-a\lambda} \right\} < P(V > 0) < 3 \min \left\{ 1, (1 + a\lambda^2)e^{-a\lambda} \right\}.$$

Proof.

(i) *Upper bound.* Observe that

$$P(V > 0) = p_1 + p_2 + p_3,$$

where

$$
\begin{aligned}
p_1 &\equiv P(\text{no disks centered within } \mathcal{R}) \\
&= e^{-\lambda} \\
&= e^{-\lambda(1-a) - a\lambda} \\
&\leq e^{-a\lambda},
\end{aligned}
$$

$$
\begin{aligned}
p_2 &\equiv P(\text{at least one disk centered within } \mathcal{R}, \text{ but none} \\
&\qquad \text{of these disks intersects any other disk}) \\
&\leq P(\geq 1 \text{ disk centered within } \mathcal{R}) \\
&\qquad \times P(\text{a given disk intersects no other disks}) \\
&= (1 - e^{-\lambda}) \exp \left\{ -\lambda \pi (2r)^2 \right\} \\
&\leq e^{-a\lambda},
\end{aligned}
$$

and

$$
\begin{aligned}
p_3 &\equiv P(\mathcal{R} \text{ not covered, at least one disk centered within } \mathcal{R}, \\
&\qquad \text{and at least one of these disks intersects another disk}).
\end{aligned}
$$

Therefore

$$P(V > 0) \leq 2e^{-a\lambda} + p_3. \tag{3.70}$$

Our next task is to derive an upper bound to p_3.

Define a *crossing* to be any point of intersection of the boundaries of two disks. A crossing is said to be *uncovered* if it is not an interior point of a third disk. If $\mathcal{R}$ is not covered, if one or more disks are centered within $\mathcal{R}$, and if at least one of those disks intersects another disk, then at least one of the disks has two or more uncovered crossings on its boundary. Therefore

$$p_3 \le P(M \ge 2) \le E(M)/2,$$

where M denotes the total number of uncovered crossings on disks centered within $\mathcal{R}$. Now, the expected number of disks whose centers lie between $r + x$ and $r + x + dx$ from the center of some given disk $\mathcal{T}$ of radius r, equals $\lambda 2\pi (r + x)\,dx$; the expected number of crossings of the boundary of $\mathcal{T}$ equals

$$2 \int_0^r \lambda 2\pi (r + x)\,dx = 6\lambda \pi r^2 = 6a\lambda;$$

the probability that a given crossing is uncovered equals $e^{-a\lambda}$; and the expected number of disks centered within $\mathcal{R}$ equals λ. Therefore

$$E(M) = \lambda \cdot 6a\lambda \cdot e^{-a\lambda},$$

whence

$$p_3 \le 3a\lambda^2 e^{-a\lambda}.$$

Substituting into (3.70) we see that

$$P(V > 0) < 3(1 + a\lambda^2)e^{-a\lambda}.$$

The desired upper bound follows from this result and the trivial estimate, $P(V > 0) < 3$.

(ii) *Lower bound.* We base the lower bound on the inequality,

$$P(V > 0) = 1 - P(V = 0) \ge (EV)^2 / E(V^2), \tag{3.71}$$

which follows from (3.7).

Let

$$\chi(\mathbf{x}) = \begin{cases} 1 & \text{if } \mathbf{x} \text{ is not covered} \\ 0 & \text{otherwise.} \end{cases}$$

Then

$$E(V^2) = E\left\{ \int\!\!\int_{\mathcal{R}^2} \chi(\mathbf{x}_1)\chi(\mathbf{x}_2)\,d\mathbf{x}_1\,d\mathbf{x}_2 \right\}$$

$$= \int\!\!\int_{\mathcal{R}^2 \cap \{|\mathbf{x}_1 - \mathbf{x}_2| \le 2r\}} E\{\chi(\mathbf{x}_1)\chi(\mathbf{x}_2)\}\,d\mathbf{x}_1\,d\mathbf{x}_2$$

$$+ \int\!\!\int_{\mathcal{R}^2 \cap \{|\mathbf{x}_1 - \mathbf{x}_2| > 2r\}} E\{\chi(\mathbf{x}_1)\chi(\mathbf{x}_2)\}\,d\mathbf{x}_1\,d\mathbf{x}_2. \tag{3.72}$$

If $|\mathbf{x}_1 - \mathbf{x}_2| = x \leq 2r$, then

$$E\{\chi(\mathbf{x}_1)\chi(\mathbf{x}_2)\} = \exp[-\lambda\{2a - r^2 B(x/2r)\}],$$

where

$$B(u) \equiv 4\int_u^1 (1-y^2)^{1/2}\,dy = \pi - 4\int_0^u (1-y^2)^{1/2}\,dy$$

denotes the area of the lens of intersection of two unit disks centered $2u$ apart. Now,

$$\int_0^u (1-y^2)^{1/2}\,dy = (u/2)\left\{u^{-1}\arcsin u + (1-u^2)^{1/2}\right\}$$
$$\geq (u/2)\arcsin 1$$
$$= (\pi/4)u,$$

since $u^{-1}\arcsin u + (1-u^2)^{1/2}$ is decreasing on $(0,1)$. Therefore

$$I_1 \equiv \iint_{\mathcal{R}^2 \cap \{|\mathbf{x}_1 - \mathbf{x}_2| \leq 2r\}} E\{\chi(\mathbf{x}_1)\chi(\mathbf{x}_2)\}\,d\mathbf{x}_1\,d\mathbf{x}_2$$
$$\leq e^{-a\lambda}\int_{\mathcal{R}} d\mathbf{x}_1 \int_0^{2r} 2\pi x \exp\left\{-4\lambda r^2(\pi/4)(x/2r)\right\}\,dx$$
$$= 8ae^{-a\lambda}\int_0^1 ye^{-a\lambda y}\,dy.$$

If $a\lambda \leq 1$ we bound the right-hand side by

$$8ae^{-a\lambda}\int_0^1 y\,dy = 4ae^{-a\lambda} \leq \pi e^{-a\lambda},$$

while if $a\lambda > 1$ we bound it by

$$8ae^{-a\lambda}\int_0^\infty ye^{-a\lambda y}\,dy = 8a^{-1}\lambda^{-2}e^{-a\lambda}.$$

Furthermore, since $E\{\chi(\mathbf{x})\} = e^{-a\lambda}$ for all $\mathbf{x}$, and $\chi(\mathbf{x}_1)$ and $\chi(\mathbf{x}_2)$ are independent for $|\mathbf{x}_1 - \mathbf{x}_2| > 2r$,

$$I_2 \equiv \int_{\mathcal{R}^2 \cap \{|\mathbf{x}_1 - \mathbf{x}_2| > 2r\}} E\{\chi(\mathbf{x}_1)\chi(\mathbf{x}_2)\}\,d\mathbf{x}_1\,d\mathbf{x}_2$$
$$\leq e^{-2a\lambda}.$$

Combining these estimates with (3.72) we see that

$$(EV)^2/E(V^2) = e^{-2a\lambda}/(I_1 + I_2)$$
$$> \begin{cases} e^{-2a\lambda}/(\pi e^{-a\lambda} + e^{-2a\lambda}) & \text{if } a\lambda \leq 1 \\ e^{-2a\lambda}/(8a^{-1}\lambda^{-2}e^{-a\lambda} + e^{-2a\lambda}) & \text{if } a\lambda > 1. \end{cases}$$

(3.73)

In the case $a\lambda \leq 1$, it follows from (3.71) and (3.73) that

$$P(V > 0) > (\pi e + 1)^{-1} > 0.10 > (0.05)\min\left\{1, (1 + a\lambda^2)e^{-a\lambda}\right\}.$$

When $a\lambda > 1$ we have $a\lambda^2 > 1$, and so

$$P(V > 0) > a\lambda^2 e^{-a\lambda}/(8 + a\lambda^2 e^{-a\lambda})$$
$$\geq (1 + a\lambda^2)e^{-a\lambda}\Big/2\left\{8 + (1 + a\lambda^2)e^{-a\lambda}\right\}$$
$$\geq (1/18)\min\left\{1, (1 + a\lambda^2)e^{-a\lambda}\right\}$$
$$> (0.05)\min\left\{1, (1 + a\lambda^2)e^{-a\lambda}\right\}.$$

This completes our derivation of the lower bound. □

3.8. EXPECTED VACANCY IN THE PRESENCE OF CLUSTERING OR CLUMPING

Until now we have concentrated on properties of vacancy in a Boolean model. There, the driving point process is a stationary Poisson process. It is of interest to see how properties of vacancy alter when this process is changed to permit a degree of interaction between points. We shall treat two cases: that of a Poisson cluster process and that of a Cox process. Each leads to a coverage process possessing a degree of clustering or clumping not present in a Boolean model, and so expected vacancy for a given density of sets per unit content tends to increase. For the sake of simplicity we take the shapes to be fixed (that is, nonrandom) sets.

The tendency toward increased vacancy suggested by our examples is not universal. A coverage process driven by a "hard-core" stationary point process $\mathcal{P}$, such as a point process where points are located at vertices of a regular lattice placed randomly into

$\mathbb{R}^k$, usually results in lesser expected vacancy than occurs in a Boolean model with the same intensity of sets per unit content (see Exercise 3.13). As a rule, clustering or clumping point processes tend to give increased vacancy, while hard-core processes and other processes in which points "avoid" one another result in reduced vacancy.

Let $\mathcal{P} \equiv \{\xi_i, i \geq 1\}$ be a stationary point process in $\mathbb{R}^k$. We shall give explicit definitions of the versions of $\mathcal{P}$ needed here, but the reader will find a general definition of a stationary point process in the next chapter. Take $\mathcal{S} \subseteq \mathbb{R}^k$ to be a fixed set, and consider the coverage process $\mathcal{C} = \{\xi_i + \mathcal{S}, i \geq 1\}$. Let V be the vacancy within a region $\mathcal{R}$ resulting from $\mathcal{C}$. Using the argument leading to (3.5) we conclude that

$$E(V) = \|\mathcal{R}\| P(\mathbf{0} \text{ not covered}) = \|\mathcal{R}\| P(\text{for all } i, \xi_i \notin -\mathcal{S}). \quad (3.74)$$

Case (i). $\mathcal{P}$ Is a Poisson Cluster Point Process

Here the points of $\mathcal{P}$ are the "progeny" of "parent points." The parent points $\{\boldsymbol{\eta}_i, i \geq 1\}$ form a stationary Poisson process $\mathcal{P}_0$ in $\mathbb{R}^k$ with intensity λ_0. Each parent produces progeny, represented by points in space, in an independent and identically distributed way. The number N_i of progeny born to a parent point $\boldsymbol{\eta}_i$ has the distribution of N, which does not depend on i. Suppose

$$P(N = n) = p_n, \qquad n \geq 0. \quad (3.75)$$

The jth child of $\boldsymbol{\eta}_i$ is the point $\boldsymbol{\eta}_i + \boldsymbol{\eta}_{ij}$, for $1 \leq j \leq N_i$. Conditional on N_i and $\boldsymbol{\eta}_i$, and on the locations of all progeny of all parents other than the ith, the vectors $\boldsymbol{\eta}_{ij}$ $(1 \leq j \leq N_i)$ are independent and identically distributed with density h defined on $\mathbb{R}^k$. The points $\{\boldsymbol{\eta}_i + \boldsymbol{\eta}_{ij}, i \geq 1 \text{ and } 1 \leq j \leq N_i\}$ comprise a *Poisson cluster point process* $\mathcal{P} \equiv \{\xi_i, i \geq 1\}$.

Conditional on $\boldsymbol{\eta}_i = \mathbf{x}$ and $N_i = n$, the chance that none of the points $\boldsymbol{\eta}_i + \boldsymbol{\eta}_{ij}$, $1 \leq j \leq N_i$, lies within $-\mathcal{S}$ equals $p_1(\mathbf{x})^n$, where

$$p_1(\mathbf{x}) \equiv 1 - \int_{-\mathbf{x}-\mathcal{S}} h(\mathbf{y})\,dy = 1 - \int_{\mathcal{S}} h(-\mathbf{x}-\mathbf{y})\,dy.$$

Conditional only on $\boldsymbol{\eta}_i = \mathbf{x}$, the chance that none of the progeny of $\boldsymbol{\eta}_i$ lies within $-\mathcal{S}$ equals

$$p_2(\mathbf{x}) \equiv E\left\{ p_1(\mathbf{x})^N \right\}.$$

Since the points $\boldsymbol{\eta}_i$ comprise a stationary Poisson process $\mathcal{P}_0$ with intensity λ_0, then the number M of points $\boldsymbol{\eta}_i$ which have at least one child lying within $-\mathcal{S}$ must be Poisson-distributed with mean

$$\nu \equiv \int_{\mathbb{R}^k} \lambda_0 \{1 - p_2(\mathbf{x})\} \, d\mathbf{x}.$$

(This is intuitively clear, but a formal proof may proceed using methodology of marked point processes discussed in Chapter 4, Section 4.1.) The probability that no random sets in the coverage process $\{\boldsymbol{\xi}_i + \mathcal{S}, i \geq 1\}$ cover the origin equals the chance that M takes the value 0. That is, it equals $e^{-\nu}$. We may now deduce from (3.74) that

$$E(V) = \|\mathcal{R}\| e^{-\nu}, \qquad (3.76)$$

where

$$\nu = \lambda_0 \sum_{n=1}^{\infty} p_n \int_{\mathbb{R}^k} \left[1 - \left\{ 1 - \int_S h(\mathbf{x} - \mathbf{y}) \, d\mathbf{y} \right\}^n \right] d\mathbf{x} \qquad (3.77)$$

and p_n is as in (3.75).

Two special cases are worth treating separately.

(a) *Partial splitting.* Here each parent has one or two children. Thus $p_n = 0$ for $n \geq 3$, $p_1 = p$ and $p_2 = 1 - p$, where $0 \leq p \leq 1$. Therefore

$$\lambda_0^{-1} \nu = p \int_{\mathbb{R}^k} d\mathbf{x} \int_S h(\mathbf{x} - \mathbf{y}) \, d\mathbf{y}$$

$$+ (1 - p) \int_{\mathbb{R}^k} \left[2 \int_S h(\mathbf{x} - \mathbf{y}) \, d\mathbf{y} - \left\{ \int_S h(\mathbf{x} - \mathbf{y}) \, d\mathbf{y} \right\}^2 \right] d\mathbf{x}$$

$$= (2 - p)\|\mathcal{S}\| - (1 - p) \int_{\mathbb{R}^k} \left\{ \int_S h(\mathbf{x} - \mathbf{y}) \, d\mathbf{y} \right\}^2 d\mathbf{x}.$$

No splitting occurs if $p = 1$, in which case we recover a Boolean model driven by a Poisson process with intensity λ_0 and having shapes fixed at $\mathcal{S}$.

(b) *Poisson descendents.* Here N has a Poisson distribution with mean μ; that is,

$$p_n = \mu^n e^{-\mu} / n!, \qquad n \geq 0.$$

Then

$$\nu = \lambda_0 \int_{\mathbb{R}^k} \left[1 - \exp\left\{ -\mu \int_S h(\mathbf{x} - \mathbf{y}) \, d\mathbf{y} \right\} \right] d\mathbf{x}.$$

We return now to the general formulas (3.76) and (3.77). Noting that

$$1 - (1 - x)^n \leq nx$$

if $0 \leq x \leq 1$, we see that

$$\lambda_0^{-1} \nu \leq \sum_{n=1}^{\infty} p_n n \int_{\mathbb{R}^k} dx \int_S h(\mathbf{x} - \mathbf{y}) \, d\mathbf{y} = \mu \|\mathcal{S}\|,$$

where $\mu \equiv E(N)$. Consequently,

$$E(V) = \|\mathcal{R}\| e^{-\nu} \geq \|\mathcal{R}\| \exp(-\lambda_0 \mu \|\mathcal{S}\|). \qquad (3.78)$$

Now, the total density of random sets in $\mathcal{C}$ per unit content of $\mathbb{R}^k$ equals $\lambda_0 \mu$. This is also the "average" intensity of the driving point process $\mathcal{P}$. Therefore the right-hand side of (3.78) equals expected vacancy for a Boolean model having the same type of set and the same density of sets as the coverage process $\mathcal{C}$. Thus, (3.78) confirms that clustering has the effect of increasing expected vacancy per unit content.

Case (ii). $\mathcal{P}$ is a Cox Point Process

If points in a point process are regarded as individuals in a population then their density in space may be thought of as an environmental factor. The density can vary from place to place, depending on the suitability of the environment for "survival" of points. A Cox point process permits the introduction of stochastic environmental heterogeneity into a model for the distribution of centers of sets. It is doubly stochastic in nature, in the sense that the stochastic variability of an ordinary Poisson point process is supplemented by permitting the intensity function to be random.

Let $\Lambda(\mathbf{x})$, $\mathbf{x} \in \mathbb{R}^k$, be a nonnegative real-valued random field (a stochastic process indexed in $\mathbb{R}^k$). Conditional on $\Lambda(\mathbf{x}) = \lambda(\mathbf{x})$ for $\mathbf{x} \in \mathbb{R}^k$, let $\mathcal{P}(\lambda)$ be an inhomogeneous Poisson process with intensity function λ. Then $\mathcal{P} \equiv \mathcal{P}(\Lambda)$ is a *Cox point process*. To ensure stationarity of $\mathcal{P}$ we ask that Λ be stationary; that is, for all $\mathbf{x}$ the distributional properties of the translated stochastic process $\Lambda_\mathbf{x}$ defined by $\Lambda_\mathbf{x}(\mathbf{y}) \equiv \Lambda(\mathbf{x} + \mathbf{y})$ do not depend on $\mathbf{x}$.

Conditional on $\Lambda \equiv \lambda$ (a given function), the chance that no points of $\mathcal{P}$ lie within $-\mathcal{S}$ equals

$$\exp\left\{ -\int_S \lambda(\mathbf{x}) \, d\mathbf{x} \right\}.$$

Therefore the unconditional probability that no points of $\mathcal{P}$ lie within $-\mathcal{S}$ is

$$E\left[\exp\left\{-\int_{\mathcal{S}} \Lambda(\mathbf{x})\,dx\right\}\right].$$

It now follows from (3.74) that expected vacancy within a region $\mathcal{R}$ equals

$$E(V) = \|\mathcal{R}\| E\left[\exp\left\{-\int_{-\mathcal{S}} \Lambda(\mathbf{x})\,dx\right\}\right].$$

Since the stochastic process Λ is stationary then $\lambda_0 \equiv E\{\Lambda(\mathbf{x})\}$ is a constant, not depending on $\mathbf{x}$. We may now deduce via Jensen's inequality (e.g., Billingsley 1979, p. 240; Chung 1974, p. 47) that

$$E(V) \geq \|\mathcal{R}\| \exp\left[-E\left\{\int_{-\mathcal{S}} \Lambda(\mathbf{x})\,dx\right\}\right] = \|\mathcal{R}\| \exp(-\lambda_0 \|\mathcal{S}\|).$$

The right-hand side equals expected vacancy for a Boolean model having the same type of set and same density of sets as the coverage process $\mathcal{C}$. Thus it emerges that the clustering and clumping associated with a Cox process distribution of set centers has the effect of increasing expected vacancy per unit content.

EXERCISES

3.1 Let $\mathcal{C}$ be a Boolean model in $\mathbb{R}^k$. For fixed $n \geq 2$, let $\mathcal{D}_{ni}$ denote the unit cube

$$(0,1)^{k-1} \times (ni, ni+1), \qquad i \geq 1.$$

Let $V(\mathcal{D}_{ni})$ be the vacancy within $\mathcal{D}_{ni}$. Prove that for each fixed $m \geq 1$, the variables $V(\mathcal{D}_{ni})$, $1 \leq i \leq m$, are asymptotically independent as $n \to \infty$. [Hint: Assume shapes generating $\mathcal{C}$ are distributed as S. First eliminate the cases $E(\|S\|) = 0$ and $E(\|S\|) = \infty$. Then approximate to each $V(\mathcal{D}_{ni})$ by vacancy from a Boolean model where shapes, having the distribution of S' say, are essentially bounded; that is, $P(\operatorname{rad} S' \leq r) = 1$ for some $r < \infty$.]

3.2 Let $\mathcal{C}$ be a Boolean model in $\mathbb{R}^k$ where the driving Poisson process $\mathcal{P} = \{\xi_i, i \geq 1\}$ has intensity λ and each shape S_i is distributed as S. Put

$$\mathcal{U} \equiv \bigcup_{i=1}^{\infty} (\xi_i + S_i).$$

(i) Prove that if the probability of $\tilde{\mathcal{U}}$ being at most countably infinite (i.e., of containing no more than a countable infinity of elements) is positive when $\lambda = \lambda_0 > 0$, then the probability of $\tilde{\mathcal{U}}$ being empty equals 1 when $\lambda = 2\lambda_0$.

(ii) Show that if $P(\tilde{\mathcal{U}} \neq \phi) < 1$ for all $\lambda > 0$, then

$$P(\tilde{\mathcal{U}} \text{ uncountably infinite}) = 1$$

for all $\lambda > 0$.

3.3 Take $k = 1$, and let S denote the open version of the set treated in Theorem 3.2. Then the Boolean model pattern is generated by many intervals of lengths l_i, $i \geq 1$. Assume $\sum_{1 \leq i < \infty} l_i = \infty$. Show that with probability 1, each endpoint of every interval is an interior point of another interval, in fact, of an infinite number of other intervals. Therefore if the real line is incompletely covered by these intervals, it is not because intervals "abut," missing out the points at which they abut.

3.4 Consider the Boolean model in $\mathbb{R}^k$ ($k \geq 1$) where the driving Poisson process has intensity $\lambda > 0$ and each shape is distributed as S.

(i) Take S to be an open sphere of random radius. Prove that if $\mathcal{R}$ is an open subset of $\mathbb{R}^k$ then the event

$$\{V(\mathcal{R}) = 0; \ \mathcal{R} \text{ not completely covered}\}$$

has probability 0. [The case where S is closed and $E(\|S\|) < \infty$ follows from Theorem 3.3.]

(ii) Show that if S contains a sphere S', and $E(\|S'\|) = \infty$, then with probability 1 all of $\mathbb{R}^k$ is completely covered.

3.5 Consider a Boolean model in which shapes are distributed as S and Poisson intensity is λ. Prove that if $\mathbb{R}^k$ is completely covered with probability 1 for all $\lambda > 0$, then the probability that each point in $\mathbb{R}^k$ is covered by an infinite number of random sets equals 1 for all $\lambda > 0$.

3.6 Assume $0 < \rho < \infty$, $0 < \|\mathcal{R}\| < \infty$ and $E(\|S\|) < \infty$. Prove that the asymptotic variance $\sigma^2(S)$ defined at (3.14) is finite if and only if $E(\|S\|^2) < \infty$.

3.7 Let $\sigma^2(S)$ denote the asymptotic variance defined at (3.14), and take $S \equiv S_0 \cup pS_1$, where S_0 is a sphere of unit radius centered at the origin and S_1 is a very thin concentric shell of fixed content $v > 0$ and distant $t \geq 0$ from the surface of S_0. Prove that for small v, the value of t that minimizes $\sigma^2(S)$ is strictly positive. This means that a Boolean model of interpenetrating spheres can be made "more stable," in the sense of reducing the large-region variance of vacancy, by taking some of the mass of each sphere away from its surface and putting it in a thin concentric shell. [Hint: Derive an expansion for

$$\tau^2(S) \equiv \int_{\mathbb{R}^k} [\exp\{\rho \|(\mathbf{x} + S) \cap S\|\} - 1]\, d\mathbf{x},$$

of the form

$$\tau^2(S) = \tau^2(S_0) + vc_1(t) + o(v)$$

as $v \to 0$, where $c_1(t) > 0$. Hence deduce that

$$\sigma^2(S) = \sigma^2(S_0) + vc_2(t) + o(v)$$

for a function $c_2(t)$, and show that the minimum over $t \geq 0$ of $c_2(t)$ occurs at a point $t_0 > 0$.]

3.8 Let $\mathcal{V}^1$ be an arbitrary rectangular prism formed by an orthogonal Poisson system of $(k-1)$-dimensional planes with intensity ν (see Theorem 3.8). Prove that

$$E(\|\mathcal{V}^1\|) = \nu^{-k} \quad \text{and} \quad E(\|\mathcal{V}^1\|^2) = 2^k \nu^{-2k}.$$

3.9 State and prove a version of Theorem 3.8 for the case where S is a k-variate prism of fixed orientation, whose side lengths $L_1, \ldots, L_k$ have an arbitrary k-variate distribution satisfying $0 < E(L_1 \ldots L_k) < \infty$.

3.10 (i) Prove Theorem 3.10.

(ii) Use Theorem 3.10 to obtain an approximation for the probability that a given region $\mathcal{R}$ is completely covered by a Boolean model in which shapes are distributed as k-dimensional cubes with axes aligned.

3.11 Let Ω denote the set of unit vectors in $\mathbb{R}^k$. Assume that the random shape S has the following properties:

$$E(\|S\|^2) < \infty;$$

there exists a positive function $f(\theta)$, bounded away from 0, such that

$$E\{\|(x\theta+S)\cap S\|\} = \alpha - xf(\theta) + o(x)$$

uniformly in $\theta \in \Omega$ as $x \downarrow 0$, where $\alpha \equiv E(\|S\|)$. Consider a Boolean model in which all shapes are distributed as δS, and Poisson intensity λ diverges. Prove that if $\delta = \delta(\lambda) \to 0$ as $\lambda \to \infty$, in such a manner that $\delta^k\lambda \to \infty$, then the vacancy V within a fixed, Riemann-measurable region $\mathcal{R}$ with $0 < \|\mathcal{R}\| < \infty$, satisfies

$$\text{var}(V) \sim \lambda^{-1}(\delta^k\lambda)^{1-k}\exp(-\alpha\delta^k\lambda)\|\mathcal{R}\|\tau^2 \qquad (3.\text{A})$$

as $\lambda \to \infty$, where

$$\tau^2 \equiv \Gamma(k)\int_\Omega \{f(\theta)\}^{-k}\,d\theta.$$

[Hint: Show that

$$\text{var}(V) \le \delta^k\exp(-2\alpha\delta^k\lambda)\|\mathcal{R}\|$$
$$\times \int_{\mathbb{R}^k}(\exp[\delta^k\lambda E\{\|(x+S)\cap S\|\}]-1)\,dx,$$

and if $\mathcal{R} \equiv \bigcup_{1 \le i \le n}\mathcal{D}_i$ is a disjoint union of k-dimensional cubes $\mathcal{D}_i$ with respective side lengths $2t_i$,

$$\text{var}(V) \ge \delta^k\exp(-2\alpha\delta^k\lambda)(1-\epsilon)^k\|\mathcal{R}\|$$
$$\times \int_{|x|\le\delta^{-1}\epsilon t}(\exp[\delta^k\lambda E\{\|(x+S)\cap S\|\}]-1)\,dx$$

for any $0 < \epsilon < 1$, where $t \equiv \min t_i$. Then prove that if $0 < \gamma_1 < \infty$, $0 < \gamma_2 \le \infty$, $\gamma_1 \to \infty$ and γ_2 is bounded away from 0,

$$\int_{|x|<\gamma_2}(\exp[\gamma_1 E\{\|(x+S)\cap S\|\}]-1)\,dx \sim \gamma_1^{-k}\exp(\alpha\gamma_1)\tau^2.]$$

(ii) The asymptotic formula (3.A) for variance of vacancy depends on shape orientation only through $\tau^2 = \tau^2(S)$. Prove that if the shape S^* is a uniformly, randomly oriented (i.e., isotropic) version of S, then $\tau^2(S^*) \le \tau^2(S)$, so that random orientation has the effect of reducing (or at least, not increasing) asymptotic variance.

3.12 Consider a Boolean model in which all shapes are distributed as δS, where $E(\|S\|^2) < \infty$ and Poisson intensity λ diverges. Prove that if $\delta^k \lambda \to 0$ as $\lambda \to \infty$ then the vacancy V within a fixed, Riemann-measurable region $\mathcal{R}$ with $0 < \|\mathcal{R}\| < \infty$ satisfies

$$\mathrm{var}(V) \sim \delta^{2k} \lambda \|\mathcal{R}\| E(\|S\|^2).$$

[The argument sketched at the end of Section 3.6 does not encompass this general case.]

3.13 Give an example to illustrate the claim made in Section 3.8 that coverage processes driven by "hard-core" stationary point processes can result in reduced expected vacancy per unit content, relative to that for a Boolean model of the same intensity. [Hint: Use formula (3.74), and consider the point process in which points are sited at vertices of a cubic lattice placed into $\mathbb{R}^k$ with random orientation.]

3.14 Consider a Boolean model $\{\xi_i + S_i, i \geq 1\}$, in which shapes are distributed as S and the driving Poisson process has intensity λ. Given a Borel-measurable set $S \subseteq \mathbb{R}^k$, define $S - S \equiv \{\mathbf{x} - \mathbf{y} : \mathbf{x} \in S, \mathbf{y} \in S\}$. By adapting the argument leading to (3.4), or otherwise, show that

$$P\left\{ S \cap \bigcup_{i=1}^{\infty} (\xi_i + S_i) \neq \emptyset \right\} = 1 - \exp(-\lambda E\|S - S\|).$$

[The left-hand side, a function of S, is called the *capacity function* or *capacity functional*. See Ohser (1980).]

3.15 Show that if $S \subseteq \mathbb{R}^k$ is Lebesgue measurable and $0 < \|S\| < \infty$ then

$$\|(\mathbf{x} + S) \cap \tilde{S}\| > 0$$

whenever $\mathbf{x} \neq \mathbf{0}$.

NOTES

Section 3.1
Theories of random sets are developed and explored by Choquet (1955), Matheron (1967, 1975), and Kendall (1974). See also Fortet and Kambouzia (1975), Ripley (1976), Cressie (1978), Davy (1978), Norburg (1984), Stoyan, Kendall, and Mecke (1987, Chapter 6), and the references therein. Ripley (1981, Section 9.1) surveys key components of these theories, and motivates definitions by considering

practical problems arising from digitizing pictures of sets. Matheron (1975, p. 27ff) discusses random closed sets, (p. 48) random open sets, and (p. 14) the Hausdorff metric. He and other authors define a Boolean model as a generalized Poisson process on a class of sets. While this definition is technically attractive, it is a little too abstract for some purposes. Hadwiger and Giger (1968) show how the Boolean model arises from basic assumptions of uniformity and independence.

Miles and Serra (1978), a collection of papers on "Geometrical Probability and Biological Structures," is an excellent source for discussion of random set theory from a nontechnical viewpoint; see, for example, Serra (1978).

Sections 3.2 and 3.3
The development in Section 3.2 of formulas for moments of vacancy follows closely early work of Robbins (1944, 1945), Bronowski and Neyman (1945), Garwood (1947), Santaló (1947), Solomon (1953), Takács (1958), and Ailam (1966) on moment properties connected with random sets. See also Siegel (1978a), Greenberg (1980), Pitts (1981), and Stam (1984). Domb (1947) has given formulas for Laplace transforms of moments of vacancy within an interval $[0, t]$, interpreted as functions of t, in the case of a Boolean model of fixed-length line segments in $k = 1$ dimension. That paper also derives atoms of the distribution of vacancy in the case of fixed-length segments, using Laplace transforms rather than the techniques in Section 3.3.

Section 3.4
Moran (1973a), inspired by thermodynamic applications and by conjectures in Widom and Rowlinson (1970) and Melnyk and Rowlinson (1971), proves a precursor of the central limit theorem in Section 3.4. His result is for fixed-radius spheres distributed in $k = 3$ dimensions. See also Hammersley, Lewis, and Rowlinson (1975). An important generalization of the context of our work is the case where the point process describing centers of random sets is simply stationary, rather than stationary and Poisson. Baddeley (1980a) and Mase (1982) establish central limit theorems for vacancy in that case. Their assumptions include mixing conditions on the set $\bigcup_i (\xi_i + S_i)$ and moment conditions on the radii of random shapes. Among other results, Baddeley and Mase illustrate the use of ergodic theory to prove strong laws, of the type in Section 3.4. See also Cowan (1978).

Sections 3.5 and 3.6
The description of uncovered "chinks" of space in Sections 3.5 and 3.6 has its genesis in a virology problem considered by Moran and Fazekas de St. Groth (1962). These authors suggested that uncovered chinks should be distributed like convex cells formed by a Poisson field of planes, in the case where the random sets are circular caps distributed on a three-dimensional sphere. This postulate suggests an approximation to the probability that the sphere is completely covered, analogous to that given by formula (3.68) in Section 3.6. See Hall (1985e) for a rigorous account in the case of random sets distributed in Euclidean space. Closely related problems are treated by Wendel (1962), Silverman (1964), Gilbert (1965), Miles (1969), Flatto and Newman (1977), Siegel (1979a, 1979b), and Janson (1983b). Moran and Fazekas de St. Groth (1962) check the performance of their approximation via a highly ingenious physical (as opposed to numerical) simulation. See Stoyan (1986) for a discussion of the relationship between a "typical" cell and the cell containing the origin.

Section 3.7
The bounds for probability of complete coverage established in Section 3.7, based on moments of vacancy and on the concept of uncovered crossings, are inspired by work of Gilbert (1965) on the coverage of a sphere by circular caps. This approach has many further applications.

Section 3.8
The discussion in Section 3.8 of vacancy in coverage processes driven by non-Poisson point processes is easily generalized to other cases, and to properties of vacancy other than expected value. The reader is referred to Cox and Isham (1980), Ripley (1977, 1981), and Diggle (1983) for excellent accounts of the rich variety of available point processes. Technical descriptions are available in Matthes, Kerstan, and Mecke (1978), Mecke and Stoyan (1980a), and Daley and Vere-Jones (1988). Hanisch (1981) was one of the first to discuss coverage processes in great generality. Exercise 3.14 gives the Boolean model version of Hanisch's formula for the capacity function.

CHAPTER FOUR
Counting and Clumping

4.1. INTRODUCTION

This chapter is devoted to problems of counting and clumping in a coverage process. For the most part we assume that the process is a Boolean model in $\mathbb{R}^k$ (see Chapter 3, Section 3.1 for a definition), but on occasion we look at more general coverage processes in Euclidean space, where the driving point process is assumed only to be stationary, not necessarily to be Poisson. Section 4.1 reviews definitions and notation, and introduces marked point process and "Poisson-ness" arguments used later in the chapter. Section 4.2 treats problems connected with the number of random sets which intersect a given fixed set. Attention is confined to the cases $k = 1$, 2, and 3, and problems of edge effects are discussed. Section 4.3 introduces the notion of curvature in $k = 2$ dimensions, and shows how measurements of curvature can be used to estimate quantities such as the expected number of clumps minus voids per unit area. Tangent count estimators of intensity are studied in the context of curvature. Section 4.4 examines the mean and variance of the number of isolated sets, or clumps of order one, centered within a given region. The remainder of the chapter treats problems of higher-order clumping. Section 4.5 studies clumps of arbitrary (finite) order in sparse k-dimensional Boolean models, while Section 4.6 provides an upper bound to the expected total number of clumps per unit area in a two-dimensional Boolean model, and discusses formulas for the expected total number of clumps per unit area. Both Sections 4.5 and 4.6 discuss formulas for the expected order of an arbitrary clump. Section 4.7 treats the phenomenon of infinite clumping (percolation) in a k-dimensional Boolean model.

As this summary indicates, problems of inference are discussed in the present chapter when they relate directly to properties of counting and clumping. Chapter 5 will develop other facets of statistical inference.

We turn now to notation used frequently in this chapter. A *convex set* S (a subset of $\mathbb{R}^k$) is a set with the property that for each $\mathbf{x}, \mathbf{y} \in S$, the straight line segment $t\mathbf{x} + (1-t)\mathbf{y}$ $(0 < t < 1)$ linking $\mathbf{x}$ to $\mathbf{y}$ lies entirely within S. Examples when $k = 2$ include disks, ellipses, triangles, rectangles, hexagons, etc. Recall that a random set (defined in Chapter 3, Section 3.1) is assumed to be always closed or always open. Therefore a random convex set is always closed or always open.

If $S_1, S_2 \subseteq \mathbb{R}^k$, we write "$S_1$ intersects S_2" to mean that $S_1 \cap S_2 \neq \emptyset$. Some authors describe this as "S_1 hits S_2" or "$S_1 \uparrow S_2$." In our notation the upward-pointing arrow refers to the limit, S, of an increasing sequence of sets S_n:

$$S_n \uparrow S \equiv \bigcup_{n=1}^{\infty} S_n.$$

The *content* $\|S\| = \|S\|_k$ of a set $S \subseteq \mathbb{R}^k$ is defined to be the Lebesgue measure of S. Let ∂S denote the surface or boundary of S. The *surface content* of S is given by

$$\|\partial S\|_{k-1} \equiv \lim_{\epsilon \to 0} (2\epsilon)^{-1} \|(\partial S)^\epsilon\|_k,$$

assuming this limit exists, where $(\partial S)^\epsilon$ is defined by (3.1). Let

$$\mathcal{T}(r) \equiv \left\{ \mathbf{x} \in \mathbb{R}^k : |\mathbf{x}| \leq r \right\}$$

be the closed sphere of radius r centered at the origin. The set $\Omega = \Omega_k$ is defined to be the surface or boundary of $\mathcal{T}(1)$. Equivalently, it is the set of all unit vectors in $\mathbb{R}^k$. An *isotropic*, or *randomly and uniformly oriented*, random set was defined in Chapter 3, Section 3.1.

Given a nonempty set $S \subseteq \mathbb{R}^k$, put

$$\bar{v}(S) \equiv \inf \left\{ \|\mathcal{T}(r)\| : r \geq 0 \text{ and } S \subseteq \mathbf{x} + \mathcal{T}(r) \text{ for some } \mathbf{x} \in \mathbb{R}^k \right\},$$

the content of the smallest sphere containing S; and

$$\underline{v}(S) \equiv \sup \left\{ ||T(r)|| : r \geq 0 \text{ and } \mathbf{x} + T(r) \subseteq S \text{ for some } \mathbf{x} \in \mathbb{R}^k \right\},$$

the content of the largest sphere contained in S. The radius $\text{rad}(S)$ of S equals the radius of the smallest sphere containing S. If we define

$$v_k \equiv ||T(1)||_k = \pi^{k/2}/\Gamma\left(1 + \tfrac{1}{2}k\right)$$

and

$$s_k \equiv ||\partial T(1)||_{k-1} = 2\pi^{k/2}/\Gamma(k/2),$$

then $s_k = kv_k$ and

$$\text{rad}(S) = \left\{ v_k^{-1}\bar{v}(S) \right\}^{1/k}.$$

If S is a random set then $\text{rad}(S)$ is an extended real-valued random variable, taking values on $[0, \infty]$. We say that $\text{rad}(S)$ is *essentially bounded* if for some $0 < r < \infty$,

$$P\{\text{rad}(S) \leq r\} = 1.$$

A point process $\mathcal{P} \equiv \{\xi_i, i \geq 1\}$ in $\mathbb{R}^k$ is said to be *stationary* if for each Borel set $S \subseteq \mathbb{R}^k$ and each $\mathbf{x} \in \mathbb{R}^k$, the distribution of the number of indices i such that $\xi_i \in S$ is the same as the distribution of the number of indices such that $\xi_i \in \mathbf{x} + S$. In this case there exists $\lambda \geq 0$ such that the expected number of points of $\mathcal{P}$ in S equals $\lambda||S||$, for all Borel sets S. (Notes at the end of the chapter give references.) We always assume that $0 < \lambda < \infty$, and call λ the *intensity* of $\mathcal{P}$. We further suppose that $\mathcal{P}$ is *simple*; that is, $\xi_i = \xi_j$ if and only if $i = j$. We say that $\mathcal{P}$ is *ergodic*, or metrically transitive, if every translation-invariant event associated with $\mathcal{P}$ has measure 0 or 1 [see Doob (1953, p. 457)].

Let $\{S_i, i \geq 1\}$ be a sequence of independent and identically distributed random sets (which we call "shapes"), independent of the stationary point process $\mathcal{P}$. We call the collection of sets $\mathcal{C} \equiv \{\xi_i + S_i, i \geq 1\}$ a *coverage process*, and call ξ_i the *center* of the random set $\xi_i + S_i$. The set $\xi_i + S_i$ is *centered within* the region $\mathcal{R}$ if and only if $\xi_i \in \mathcal{R}$. Throughout this chapter we reserve the name "coverage process" for a sequence of interpenetrating sets constructed in this specific way. We say that $\mathcal{C}$ is *driven* by $\mathcal{P}$ and *generated* by the shapes S_i. In our notation, the random shapes generating a coverage process are never empty. If $\mathcal{P}$ happens to be a stationary Poisson process then $\mathcal{C}$ is a Boolean model. In that case, if we replace the sequence of shapes $\{S_i, i \geq 1\}$ by $\{\eta_i + S_i, i \geq 1\}$, still independent and identically distributed and independent of $\mathcal{P}$, we obtain a new Boolean model $\mathcal{C}' \equiv \{(\xi_i + \eta_i) + S_i, i \geq 1\}$ with the same properties as $\mathcal{C}$ (see Chapter 1, Section 1.7, and Chapter 3, Section 3.1).

Two very simple arguments are surprisingly useful for establishing counting and clumping results. The first is based on the notion of *marked point processes* (Matthes 1963). Suppose independent and identically distributed shapes $\{S_i, i \geq 1\}$ are centered at points of a stationary point process $\mathcal{P} = \{\xi_i, i \geq 1\}$ (independent of the shapes), forming a coverage process $\mathcal{C} = \{\xi_i + S_i, i \geq 1\}$. The points of $\mathcal{P}$ can be thought of as being indexed or *marked* by the types of shapes associated with them. This is most easily seen when the shapes have a "discrete" distribution with "atoms" at distinct, fixed shapes $S_1, S_2, \ldots$, that is, when shapes are distributed as S where

$$P(S = S_i) = p_i, \quad i \geq 1, \quad \text{and} \quad \sum_{i=1}^{\infty} p_i = 1. \tag{4.1}$$

In this case the points in $\mathcal{P}$ may be divided among processes $\mathcal{P}_1$, $\mathcal{P}_2, \ldots$, where $\mathcal{P}_i$ contains precisely those points at which the shape S_i is centered. If $\mathcal{P}$ is stationary with intensity λ then $\mathcal{P}_i$ is stationary with intensity $\lambda_i = \lambda p_i$. Thus, we can consider $\mathcal{C}$ to arise by superimposing a sequence of coverage processes $\mathcal{C}_i \equiv \{\xi + S_i : \xi \in \mathcal{P}_i\}$, the shapes generating $\mathcal{C}_i$ all being identical to S_i.

The processes $\mathcal{C}_i$ are not necessarily stochastically independent of one another, although they would be independent if the original point process $\mathcal{P}$ were Poisson. However, independence is not required for many purposes, such as computing expectations: the mean of a sum of random variables equals the sum of the means, with or without independence. For example, suppose we wish to compute the expected value of the number N of times that random sets from a coverage process $\mathcal{C}$ intersect a fixed set S_0. If the point process driving $\mathcal{C}$ is stationary with intensity λ and if all shapes are identical to the *fixed* set S, then $E(N)$ is just $\lambda \|\mathcal{A}(S)\|$, where

$$\mathcal{A}(S) \equiv \{x : (x + S) \cap S_0 \neq \emptyset\}.$$

In the case where shapes have the discrete distribution defined by (4.1), $E(N)$ equals the sum of the expected numbers of times that sets from the individual coverage processes $\mathcal{C}_i$ intersect S_0:

$$E(N) = \sum_{i=1}^{\infty} \lambda_i \|\mathcal{A}(S_i)\| = \lambda \sum_{i=1}^{\infty} p_i \|\mathcal{A}(S_i)\|,$$

that is,
$$E(N) = \lambda E\{\|\mathcal{A}(S)\|\}. \tag{4.2}$$

In this example the marks on the points in the point process may be taken to be the integers $1,2,\dots$. A general marked point process in $\mathbb{R}^k$ has its marks in a complete separable metric space, which we call a Polish space. A marked point process may be thought of as a point process taking values in the product space $\mathbb{R}^k \times \mathcal{M}$, where $\mathcal{M}$ is the Polish space. If the random shapes with which we are working take values in such a space, then the Campbell theorem for marked point processes guarantees formula (4.2). More generally still, if the random shapes S_i are distributed as S which takes values in a Polish space $\mathcal{M}$, and if f is a nonnegative, measurable function on $\mathbb{R}^k \times \mathcal{M}$, then the Campbell theorem declares that

$$E\left\{ \sum_i f(\xi_i, S_i) \right\} = \lambda E\left\{ \int_{\mathbb{R}^k} f(\mathbf{x}, S)\, d\mathbf{x} \right\}$$

$$= \lambda \int_{\mathbb{R}^k} E\{f(\mathbf{x}, S)\}\, d\mathbf{x} \tag{4.3}$$

(see Stoyan, Kendall, and Mecke 1987, pp. 99–100; and Daley and Vere-Jones 1988, Section 12.1). The Campbell theorem is really a special form of Fubini's theorem on interchange of order of integration. We should stress that formula (4.3) is stated under the assumption that $S_1, S_2, \dots$ are independently distributed as S and are independent of $\mathcal{P} \equiv \{\xi_i, i \geq 1\}$.

The class $\mathcal{K}'$ of nonempty compact subsets of $\mathbb{R}^k$, with the Hausdorff metric ρ defined in Section 3.1, is a complete separable metric space (Federer 1969, Section 2.10.21). Therefore we may apply marked point process techniques to coverage processes in which marks are shapes and shapes are random compact sets. We may also apply them when shapes are random bounded open sets, noting the "equivalence class" argument in Section 3.1. [We may think of the Hausdorff metric ρ as a metric on the set of equivalence classes $\mathcal{E}(K)$, $K \in \mathcal{K}'$, where the nonempty set $A \subseteq \mathbb{R}^k$ is in $\mathcal{E}(K)$ if and only if $\rho(A, K) = 0$. Note that if $A, B \subseteq \mathbb{R}^k$ then $\rho(A, B) = \rho(\overline{A}, \overline{B}) = \rho(A, \overline{B}) = \rho(\overline{A}, B)$, where $\overline{A}$ and $\overline{B}$ denote the closures of A and B respectively.] There are various ways of treating more general cases, such as coverage processes in which shapes are random closed sets, not necessarily bounded. One approach is to use a marked point process argument in limiting form (see Exercise 4.1).

There is no reason why the mark at the point ξ_i must be the shape S_i. It might be a quite different characteristic, dependent on other sets in the coverage process; for example, it might equal the number of other sets that intersect $\xi_i + S_i$. Let M_i be the mark at the point ξ_i, taking values in a Polish space $\mathcal{M}$. Assume that the marked point process $\{(\xi_i, M_i), i \geq 1\}$ is stationary, by which we mean that for each Borel set $B \subseteq \mathbb{R}^k \times \mathcal{M}$ and each $\mathbf{x} \in \mathbb{R}^k$, the distribution of the number of indices such that $(\xi_i, M_i) \in B$ is the same as the distribution of the number of indices such that $(\xi_i + \mathbf{x}, M_i) \in B$. (This implies stationarity of the ordinary point process $\mathcal{P} \equiv \{\xi_i, i \geq 1\}$.) Let f be a nonnegative, measurable function on $\mathbb{R}^k \times \mathcal{M}$. The analogue of formula (4.3) in this context is

$$E\left\{\sum_i f(\xi_i, M_i)\right\} = \lambda E\left\{\int_{\mathbb{R}^k} f(\mathbf{x}, M)\, d\mathbf{x}\right\}$$

$$= \lambda \int_{\mathbb{R}^k} E\{f(\mathbf{x}, M)\}\, d\mathbf{x}, \qquad (4.4)$$

where M is a random mark having the so-called "mark distribution." The distribution of M is closely connected with Palm distribution theory. Suffice it to say here that if the point process $\mathcal{P}$ is ergodic as well as stationary, for example, if $\mathcal{P}$ is a Poisson process, then M has the distribution of the mark of an "arbitrary," or "typical," or "randomly chosen" point of $\mathcal{P}$. We shall have more to say about this phenomenon later in this section.

Another useful technique is the "Poisson-ness" argument, which we now describe. Suppose we can compute the mean of a random variable X, which might be the number of random sets in a stationary coverage process that intersect a given fixed set S_0. If the point process $\mathcal{P}$ driving the coverage process is Poisson then it is often the case that X is Poisson-distributed, and so knowing $E(X)$ gives us the entire distribution of X. Poisson-ness is usually intuitively clear from the fact that a coverage process driven by a Poisson point process consists of sets placed "at random" into $\mathbb{R}^k$. In more detail, suppose that with probability 1 the shapes generating a Boolean model have radius no more than r, and put

$$S_0^r \equiv \left\{\mathbf{x} \in \mathbb{R}^k : \text{ for some } \mathbf{y} \in S_0, |\mathbf{x} - \mathbf{y}| < r\right\}.$$

The number N of points of $\mathcal{P}$ within S_0^r is Poisson-distributed, and only shapes centered at those points can intersect S_0. Conditional on N, the events that individual sets intersect S_0 are independent and have the same probability. The sum of a Poisson number of independent, identically distributed indicator random variables is Poisson-distributed, and so the number of random sets intersecting

S_0 must be Poisson-distributed. The condition that the shapes have bounded radius may be removed by using a truncation and limiting (as $r \to \infty$) argument, as in the proof of Theorem 3.5.

The Poisson-ness argument may be thought of as an application of Rényi's theorem for thinned point processes. A thinned point process is obtained by systematically deleting points from an initial process, and Rényi's theorem states that stochastically independent, position-dependent thinning of a Poisson process produces another Poisson process.

We now work through an example in detail to illustrate use of the marked point process and Poisson-ness arguments. We compute the distribution of the number N of sets from a k-dimensional Boolean model that intersect a given Borel set S_0. Suppose initially that the point process $\mathcal{P}$ driving the coverage process $\mathcal{C} \equiv \{\xi_i + S_i\}$ is stationary with intensity λ, and that the shapes S_i are distributed as S, which is a random compact set or a random bounded open set. Then the expected number of sets from $\mathcal{C}$ that intersect S_0 equals

$$E(N) = E\left\{\sum_i f(\xi_i, S_i)\right\},$$

where $f(\mathbf{x}, S) = 1$ if $(\mathbf{x} + S) \cap S_0 \neq \emptyset$, 0 otherwise. By the Campbell theorem [formula (4.3) above],

$$\begin{aligned}
E(N) &= \lambda \int_{\mathbb{R}^k} P\{(\mathbf{x} + S) \cap S_0 \neq \emptyset\}\, d\mathbf{x} \\
&= \lambda E\left[\int_{\mathbb{R}^k} I\{(\mathbf{x} + S) \cap S_0 \neq \emptyset\}\, d\mathbf{x}\right] \\
&= \lambda E[\|\{\mathbf{x} : (\mathbf{x} + S) \cap S_0 \neq \emptyset\}\|] \\
&\equiv \mu,
\end{aligned} \qquad (4.5)$$

say. If $\mathcal{P}$ is Poisson then N has a Poisson distribution, and so

$$P(N = n) = \mu^n e^{-\mu}/n!, \quad \text{all } n \geq 0. \qquad (4.6)$$

If we take $S_0 \equiv \{\mathbf{x}\}$ (a singleton), and note that in this case $\mu = \lambda E(\|\mathbf{x} - S\|) = \lambda E(\|S\|)$, we see that the chance that no sets cover $\mathbf{x}$ equals

$$P(N = 0) = \exp\{-\lambda E(\|S\|)\}. \qquad (4.7)$$

This formula was proved in Section 3.2 using different methods.

Very often in the theory of coverage processes we require the notion of an "arbitrary" or "typical" set. For example, we may wish to calculate the expected number of sets in the process that intersect an arbitrary set in the process. Having selected the arbitrary set we usually wish to translate the origin to its center and view the coverage process from that position. That is, we wish to study the coverage pattern conditional on a point from the driving point process being at the origin. To do this rigorously we require the concept of *Palm distribution* (see bibiographic notes for references). Palm distribution theory is particularly simple when the driving point process is Poisson, and so we assume that the coverage process is a Boolean model whenever we consider an "arbitrary set."

The Palm distribution of a stationary Poisson process $\mathcal{P}$ in $\mathbb{R}^k$ "conditional on a point being sited at $\mathbf{x}$," is just the distribution of the process $\mathcal{P} \cup \{\mathbf{x}\}$. Indeed, this property characterizes the Poisson process. It is one aspect of the so-called lack-of-memory property of Poisson processes: once away from the point $\mathbf{x}$, the process "forgets" that it had a point there and behaves like an ordinary (i.e., unconditioned) Poisson process (see Chapter 1, Section 1.7). Random shapes, or marks, may now be assigned in the usual way. For example, if $S, S_1, S_2, \ldots$ are independent and identically distributed random shapes in $\mathbb{R}^k$, and $\mathcal{P} \equiv \{\xi_i, i \geq 1\}$ is a stationary Poisson process independent of the shapes, then the distribution of the Boolean model $\mathcal{C} \equiv \{\xi_i + S_i, i \geq 1\}$, conditional on one of the shapes being centered at $\mathbf{x}$, is that of the collection of sets $\mathcal{C} \cup \{\mathbf{x} + S\}$. Properties of an arbitrary or typical set in $\mathcal{C}$ relative to the other sets in $\mathcal{C}$ are (defined to be) those of $\mathbf{x} + S$ relative to random sets from $\mathcal{C}$.

Palm distribution theory allows us to be more explicit about the distribution of the random mark M appearing in formula (4.4). Assume that the coverage process $\mathcal{C} \equiv \{\xi_i + S_i, i \geq 1\}$ is a Boolean model, as described in the previous paragraph. Let $\mathcal{C}_i \equiv \{\xi_j + S_j : j \geq 1, j \neq i\}$ represent the sets in $\mathcal{C}$ excluding $\xi_i + S_i$. Suppose each point ξ_i of $\mathcal{P}$ is assigned a mark M_i which is a measurable function m of ξ_i, S_i, and $\mathcal{C}_i$:

$$M_i \equiv m(\xi_i, S_i, \mathcal{C}_i).$$

For example, if S is a subset of $\mathbb{R}^k$ and $\mathcal{Y}$ a collection of subsets of $\mathbb{R}^k$, then

$$m(\mathbf{x}, S, \mathcal{Y}) \equiv \sum_{Y \in \mathcal{Y}} I\{(\mathbf{x} + S) \cap Y \neq \emptyset\}$$

equals the number of sets in $\mathcal{Y}$ that intersect $\mathbf{x} + S$. The random mark M in (4.4) may be taken equal to

$$M \equiv m(0, S, \mathcal{C}). \tag{4.8}$$

4.2. NUMBER OF SETS THAT INTERSECT A GIVEN SET

Throughout this section we consider the coverage process $\mathcal{C}$ in which the driving stationary point process $\mathcal{P}$ has intensity λ and shapes are independently distributed as the random set S, independent also of $\mathcal{P}$. It will be common for us to suppose that $\mathcal{P}$ is Poisson, so that $\mathcal{C}$ is a Boolean model, but this assumption will be clearly indicated each time it is needed. We confine attention to $k = 1, 2$, and 3 dimensions.

Assume first that $k = 2$, and that the random shape S is isotropic and convex with probability 1. Let S_0 be any fixed convex subset of $\mathbb{R}^2$ (see Chapter 3, Section 3.1 for a definition of isotropy). We calculate

$$\nu(S_0) \equiv \text{mean number of random sets that intersect } S_0.$$

Suppose first that S is a uniform orientation of a nonrandom convex set S. There is no loss of generality in insisting that the centers of S and S_0, which we could take to be the centroids, are both sited at the origin (see Section 4.1). Consider the set $\mathcal{A}$ of all points $\mathbf{x} \in \mathbb{R}^2$ such that $\mathbf{x} + S$ intersects S_0. We wish to determine the area of $\mathcal{A}$, $\|\mathcal{A}\|_2$. Now, $\mathcal{A}$ is itself a convex set whose boundary equals the locus of all points $\mathbf{x}$ such that $\mathbf{x} + S$ just touches S_0. Let P denote a point whose vector coordinate $\mathbf{x}$ is such that $\mathbf{x} + S$ just touches S_0 at some point A. Refer to Figure 4.1 for notation. Slide $\mathbf{x} + S$ an infinitesimal amount along the boundary of S_0, while maintaining the orientation of S, so that the point of contact moves from A to B. This motion carries A into A' and P into P'. The resulting contribution to the area of $\mathcal{A}$ is the area defined by $OAPP'B$. If we were to integrate this area element, we would obtain $\|\mathcal{A}\|$.

Referring all the while to Figure 4.1, we see that

$$\text{area} \, (OAPP'B) = \text{area} \, (OAB) + \text{area} \, (APP'B). \qquad (4.9)$$

Since the orientation of S has been maintained, AP is parallel to $A'P'$. The lengths AB and AA' are infinitesimal, and so the region $APP'A'$ is really a parallelogram. Let Q divide PP' in the ratio $|AB| : |BA'|$ (refer to Figure 4.2). Then

$$\text{area} \, (APP'B) = \text{area} \, (APP'A') - \text{area} \, (P'A'B)$$
$$= \text{area} \, (APQB) + \text{area} \, (QA'B)$$
$$= \text{area} \, (APQB) + \text{area} \, (P'A'B).$$

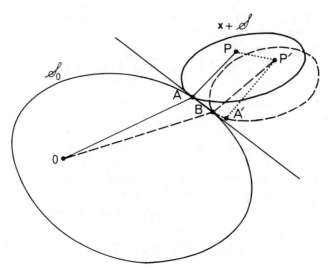

Figure 4.1. Translate of S just touching S_0.

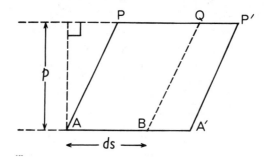

Figure 4.2. Detail of parallelogram $APP'A'$.

Let p denote the perpendicular distance from P to the tangent at A, and ds the distance from A to B. If the inward-pointing vector along which p is measured makes an angle ϕ relative to some fixed vector $\mathbf{v}$ marked on S, we call $p = p(\phi)$ the *support function* of S, relative to the center and fixed vector $\mathbf{v}$. Now,

$$\text{area}\,(APQB) = p\,ds,$$

and so

$$\text{area}\,(APP'B) = p\,ds + \text{area}\,(P'A'B).$$

Substituting into (4.9), we obtain

$$\text{area}\,(OAPP'B) = \text{area}\,(OAB) + \text{area}\,(P'A'B) + p\,ds. \qquad (4.10)$$

The infinitesimal area (OAB) is an element of the area of S_0, while area $(PA'B)$ is an element of the area of S. Therefore if we integrate (4.10), we obtain

$$\|A\| = \|S_0\| + \|S\| + \int_{\partial S_0} p\,ds, \qquad (4.11)$$

where the last-written integral is taken around the perimeter of S_0. Here, if $p = p(\phi)$ then ds denotes an element of the boundary of ∂S_0 at the point of contact of the tangent whose inward-pointing normal is in direction ϕ.

Throughout this argument we have assumed S to have fixed orientation. Now choose an origin O and vector $\mathbf{u}$ fixed in the plane, with $\mathbf{u}$ in the same direction as the vector $\mathbf{v}$ attached to S. Rotate S about O until $\mathbf{v}$ attains orientation θ $(0 \le \theta < 2\pi)$ relative to $\mathbf{u}$. Let $S(\theta)$ denote the resulting image of S. Under this motion, A transforms to $A(\theta)$, say, and $p(\phi)$ to $p(\phi + \theta)$. Integrating over all possible orientations θ, and noting that S is obtained from S by orienting S uniformly on $(0, 2\pi)$, we obtain a formula for the mean area $\mu(S_0, S)$ of the region into which centers of shapes intersecting S_0 must fall:

$$\mu(S_0, S) \equiv (2\pi)^{-1} \int_0^{2\pi} \|A(\theta)\|\, d\theta$$

$$= \|S_0\| + \|S\| + (2\pi)^{-1} \int_0^{2\pi} d\theta \int_{\partial S_0} p\{\phi(s) + \theta\}\, ds. \qquad (4.12)$$

Now,

$$\int_0^{2\pi} d\theta \int_{\partial S_0} p\{\phi(s) + \theta\}\, ds = \int_{\partial S_0} ds \int_0^{2\pi} p\{\phi(s) + \theta\}\, d\theta$$

$$= \|\partial S_0\| \int_0^{2\pi} p(\theta)\, d\theta.$$

The quantity $p(\theta)\,d\theta$ is an infinitesimal element of the perimeter of S, and so

$$\int_0^{2\pi} p(\theta)\, d\theta = \|\partial S\|_1.$$

Substituting these estimates into (4.12) we see that

$$\mu(S_0, S) = \|S_0\|_2 + \|S\|_2 + (2\pi)^{-1} \|\partial S_0\|_1 \|\partial S\|_1. \qquad (4.13)$$

Next we use the marked point process argument to extend this result to the general case, where the shape S is any isotropic random convex set. We may write $S = S'_\Theta$, where S' is a convex set, Θ is uniformly distributed on $(0, 2\pi)$ and independent of S', and S'_Θ denotes S' rotated to have orientation Θ. Take $f(\mathbf{x}, S) \equiv 1$ if $(\mathbf{x} + S) \cap S_0 \neq \phi$, $f(\mathbf{x}, S) = 0$ otherwise. Apply formula (4.3) (the Campbell theorem) with this f, to deduce that if independent shapes distributed as S are centered at points of a stationary point process with intensity λ then the expected number of random sets intersecting S_0 equals

$$\begin{aligned}
\nu(S_0) &= \lambda E\left\{ \int_{\mathbb{R}^k} f(\mathbf{x}, S)\, d\mathbf{x} \right\} \\
&= \lambda E\left[\| \{ \mathbf{x} : (\mathbf{x} + S'_\Theta) \cap S_0 \neq \phi \} \| \right] \\
&= \lambda E\left(E\left[\| \{ \mathbf{x} : (\mathbf{x} + S'_\Theta) \cap S_0 \neq \phi \} \| \mid S' \right] \right) \\
&= \lambda E\left\{ \mu(S_0, S') \right\} \\
&= \lambda \left\{ \|S_0\|_2 + E(\|S\|_2) + (2\pi)^{-1} \|\partial S_0\|_1 E(\|\partial S\|_1) \right\},
\end{aligned}$$

the last line following from (4.13). If the driving point process is Poisson then the number of random sets intersecting S_0 must be Poisson-distributed (see Section 4.1).

We have proved the following theorem. Note particularly the assumptions of convexity and isotropy.

Theorem 4.1. *Take $k = 2$. Consider the Boolean model in which each shape is distributed as the isotropic convex random set S, and the driving Poisson process has intensity λ. Let S_0 be a fixed convex subset of $\mathbb{R}^2$. The number of sets in the Boolean model that intersect S_0 is Poisson-distributed with mean*

$$\nu(S_0) = \lambda \left\{ \|S_0\|_2 + E(\|S\|_2) + (2\pi)^{-1} \|\partial S_0\|_1 E(\|\partial S\|_1) \right\}, \quad (4.14)$$

assumed finite.

The proof shows that formula (4.14) for the mean holds under the assumption that the driving point process is stationary; Poissonness is not needed there.

Without assuming that S_0 or S is convex, but assuming that the point process is stationary, the number of sets from the coverage process which intersect S_0 has mean

$$\lambda E[\| \{ \mathbf{x} : (\mathbf{x} + S) \cap S_0 \neq \emptyset \} \|].$$

If the point process is stationary then the number is Poisson-distributed with this mean. These results were proved in Section 4.1; see (4.5) and (4.6).

As a special case of Theorem 4.1, if we take $\mathcal{S}_0$ to be the singleton $\{\mathbf{x}\}$ (a degenerate convex set) for which $\|\mathcal{S}_0\| = \|\partial\mathcal{S}_0\|_1 = 0$, we see that the number N of sets in the Boolean model that cover $\mathbf{x}$ is Poisson-distributed with mean

$$E(N) = \lambda E(\|S\|_2) \tag{4.15}$$

—a formula that we derived in Section 4.1, but without the assumption that S be convex. If S is an isotropic line segment with mean length l then $E(\|S\|_2) = 0$ and $E(\|\partial S\|_1) = 2l$, and so the expected number of segments that intersect a fixed convex set $\mathcal{S}_0$ is

$$\nu(\mathcal{S}_0) = \lambda(\|\mathcal{S}_0\|_2 + \pi^{-1}l\|\partial\mathcal{S}_0\|_1).$$

If S is a disk with random radius R, and we take $\alpha \equiv E(\|S\|_2) = \pi E(R^2)$, then

$$\nu(\mathcal{S}_0) = \lambda\{\|\mathcal{S}_0\|_2 + \alpha + \|\partial\mathcal{S}_0\|_1 E(R)\}.$$

In this case if $\mathcal{S}_0$ is a disk of radius r then the probability that $\mathcal{S}_0$ is not intersected by any of the random disks in the Boolean model equals

$$e^{-\nu(\mathcal{S}_0)} = e^{-\alpha\lambda}\exp\left[-\lambda\pi\left\{r^2 + 2rE(R)\right\}\right].$$

From this it follows that if random disk radius R has probability density function f then the density of radius of isolated disks is proportional to

$$f(r)\exp\left[-\lambda\pi\left\{r^2 + 2rE(R)\right\}\right], \quad r > 0.$$

(An isolated set from the Boolean model is one that intersects no other random sets.)

Returning to the case of a general isotropic convex shape S, define $\alpha \equiv E(\|S\|_2)$ and $\beta = E(\|\partial S\|_1)$ to be the mean area and perimeter of S, respectively. Then the probability that no shapes intersect $\mathcal{S}_0$ equals

$$e^{-\nu(\mathcal{S}_0)} = e^{-\alpha\lambda}\exp\left[-\lambda\left\{\|\mathcal{S}_0\|_2 + (2\pi)^{-1}\beta\|\partial\mathcal{S}_0\|_1\right\}\right].$$

If we replace S_0 by S in this formula, and take expectations, we obtain the probability that in a Boolean model generated by convex shapes distributed as S, an arbitrary set is isolated (i.e., part of a clump of order one):

$$e^{-\alpha\lambda}E\left[\exp\left\{-\lambda(\|S\|_2 + (2\pi)^{-1}\beta\|\partial S\|_1)\right\}\right]. \qquad (4.16)$$

The isolation property will be studied in greater detail in Section 4.4.

The proof we have given of Theorem 4.1 closely follows classical lines of an argument in integral geometry, although we have not used terminology from that field. If $\mu(S_0, S)$ is defined as in (4.7) then $2\pi\mu(S_0, S)$ is called the *kinematic measure* of the set of motions of the moving, rigid set S that intersect S_0. It is just the Lebesgue measure of the set of all triples (x_1, x_2, θ) such that, if S_θ denotes S rotated through an angle θ about any fixed center and relative to any fixed vector in $\mathbb{R}^2$, then $(x_1, x_2) + S_\theta$ has nonempty intersection with S_0. (This measure is independent of the choice of center and vector in $\mathbb{R}^2$.) The last term on the right-hand side of (4.11) equals twice the *mixed area of Minkowski* for the sets S and S_0, and can be expressed alternatively as

$$\int_0^{2\pi} \left\{p(\phi)p_0(\phi) - p'(\phi)p_0'(\phi)\right\} d\phi,$$

where p and p_0 are support functions of S and S_0, respectively. See Santaló (1976, p. 93ff) for a proof making explicit use of integral geometry. Of course, the advantage of those arguments is that they apply to sets more general than convex ones. They produce a general version of *Blaschke's fundamental formula* (Santaló 1976, p. 113ff), of which (4.13) is but a special case. However, the general formula does not supply the mean number of random sets intersecting a fixed set S_0, without the assumptions of convexity in Theorem 4.1. Sometimes it gives first moments of certain statistics, but not other properties. For example, it tells us that the total number of times m distinct points are covered by sets from the Boolean model has mean equal to $m\lambda E(\|S\|)$, but this number is not necessarily Poisson distributed. [Of course, the value of the mean $m\lambda E(\|S\|)$ also follows from adding (4.15) over all m points.]

Next we treat the case of convex sets in $k = 3$ dimensions. We first introduce the notion of the *mean support* (sometimes called *mean perpendicular*) of a convex set in $\mathbb{R}^3$. Let O denote any origin

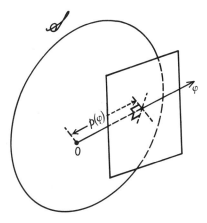

Figure 4.3. Definition of $p(\phi)$.

interior to S, and let Ω be the set of all unit vectors in $\mathbb{R}^3$. For any $\phi \in \Omega$, let $p(\phi)$ be the length of the perpendicular from O to the tangent plane to the surface of S whose outward-pointing normal is in direction ϕ (see Figure 4.3). The mean support of S is defined to be

$$\bar{p}(S) \equiv \|\Omega\|_2^{-1} \int_\Omega p(\phi)\,d\phi = (4\pi)^{-1} \int_\Omega p(\phi)d\phi.$$

Since $\bar{p}^*(\phi) \equiv \bar{p}(\phi) + \bar{p}(-\phi)$ (sometimes known as "thickness" or "Breite") equals the distance between two parallel tangent planes, then $\bar{p}^*(\phi)$ is independent of choice of O, and so

$$\bar{p}(S) = (8\pi)^{-1} \int_\Omega \bar{p}^*(\phi)\,d\phi$$

is also independent of choice of O.

Theorem 4.2. *Take $k = 3$. Consider the Boolean model in which each shape is distributed as the random isotropic convex set S, and the driving Poisson process has intensity λ. Let S_0 be a fixed convex subset of $\mathbb{R}^3$. The number of sets in the Boolean model that intersect S_0 is Poisson-distributed with mean*

$$\nu(S_0) \equiv \lambda[\|S_0\|_3 + E(\|S\|_3) + \|\partial S_0\|_2 E\{\bar{p}(S)\} + E(\|\partial S\|_2)\bar{p}(S_0)],$$
$$(4.17)$$

assumed finite.

Proof. As in the proof of Theorem 4.1, we first calculate the volume of the set $\mathcal{A}$ of all points $\mathbf{x} \in \mathbb{R}^3$ such that $\mathbf{x} + S$ intersects S_0, where S is any given convex set. Choose interior centers for S and S_0. The argument used to prove (4.10) gives us here the analogous formula,

$$dV = dv_0 + dv + p_0(-\omega)\,ds + p(\omega)\,ds_0.$$

The quantities dV, dv_0, and dv are elements of the volumes of $\mathcal{A}$, S_0, and S, respectively; $p_0(-\omega)$ and $p(\omega)$ are lengths of perpendiculars from the respective centers of S_0 and S to tangent planes whose outward-pointing normals are in the directions $-\omega$ and ω; and ds_0 and ds are elements of surface areas of S_0 and S at the places where these tangent planes touch S_0 and S, respectively. Integrating, we obtain

$$\|\mathcal{A}\| = \|S_0\| + \|S\| + \int_{\partial S} p_0(-\omega)\,ds + \int_{\partial S_0} p(\omega)\,ds_0. \qquad (4.18)$$

So far we have assigned S a fixed orientation. If we now integrate (4.18) over all possible orientations into which S can be rotated, we obtain a formula for the mean volume $\mu(S_0, S)$ of the region within which centers of translated and randomly oriented images of S intersecting S_0 must fall. Supposing S to be rotated through an angle ϕ about an axis with inclination $\boldsymbol{\theta}$, the quantities $\mathcal{A}$, $p_0(-\omega)$, and $p(\omega)$ change to $\mathcal{A}(\boldsymbol{\theta}, \phi)$, $p_0(-\omega \mid \boldsymbol{\theta}, \phi)$, and $p(\omega \mid \boldsymbol{\theta}, \phi)$, respectively. Recall from Chapter 3, Section 3.1 that isotropic rotation has $\boldsymbol{\theta}$ uniformly distributed on Ω and ϕ distributed on the interval $(0, 2\pi)$ with density $\pi^{-1}\sin^2(\tfrac{1}{2}\phi)$, independently of $\boldsymbol{\theta}$. Therefore

$$\mu(S_0, S) = \|\Omega\|_2^{-1}\pi^{-1}\int_0^{2\pi}\sin^2(\tfrac{1}{2}\phi)\,d\phi\int_\Omega \|\mathcal{A}(\boldsymbol{\theta}, \phi)\|\,d\boldsymbol{\theta}$$

$$= \|S_0\| + \|S\|$$

$$+ \|\Omega\|_2^{-1}\pi^{-1}\int_0^{2\pi}\sin^2(\tfrac{1}{2}\phi)\,d\phi\int_\Omega d\boldsymbol{\theta}\int_{\partial S} p_0(-\omega \mid \boldsymbol{\theta}, \phi)\,ds$$

$$+ \|\Omega\|_2^{-1}\pi^{-1}\int_0^{2\pi}\sin^2(\tfrac{1}{2}\phi)\,d\phi\int_\Omega d\boldsymbol{\theta}\int_{\partial S_0} p(\omega \mid \boldsymbol{\theta}, \phi)\,ds_0.$$

$$(4.19)$$

Now,

$$
\int_0^{2\pi} \sin^2\left(\tfrac{1}{2}\phi\right) d\phi \int_\Omega d\boldsymbol{\theta} \int_{\partial S} p_0(-\omega \mid \boldsymbol{\theta}, \phi)\, ds
$$

$$
= \int_0^{2\pi} \sin^2\left(\tfrac{1}{2}\phi\right) d\phi \int_{\partial S} ds \int_\Omega p_0(-\omega \mid \boldsymbol{\theta}, \phi)\, d\boldsymbol{\theta}
$$

$$
= \int_0^{2\pi} \sin^2\left(\tfrac{1}{2}\phi\right) d\phi \int_{\partial S} ds \|\Omega\|_2 \overline{p}(S_0)
$$

$$
= \pi \|\partial S\|_2 \|\Omega\|_2 \overline{p}(S_0),
$$

with a similar formula for the second integral in (4.19). Therefore

$$
\mu(S_0, S) = \|S_0\| + \|S\| + \|\partial S\|_2 \overline{p}(S_0) + \|\partial S_0\| \overline{p}(S),
$$

which is the three-dimensional analogue of (4.13). In a Boolean model in which all shapes are distributed as random, uniform orientations of S, the mean number of sets intersecting S must be $\lambda \mu(S_0, S)$. Appealing to the marked point process argument of Section 4.1 we see that if shapes are distributed as the isotropic convex set S, then the mean number of sets intersecting S_0 equals

$$
\nu(S_0) = E\left\{\|S_0\| + \|S\| + \|\partial S\|_2 \overline{p}(S_0) + \|\partial S_0\|_2 \overline{p}(S)\right\},
$$

as stated in Theorem 4.2. The distribution of this number must be Poisson with this mean (see Section 4.1). $\quad\square$

As the proof shows, formula (4.17) for the mean number of intersections remains true under the assumption that the driving point process is stationary; Poisson-ness is not needed there. Also, (4.17) is no more than a special case of general formulas in integral geometry.

The case of a Boolean model in $k = 1$ dimension is very easy. There, a convex set is simply a line segment. If the fixed set S_0 and the random shape S are both convex then the number of random sets in a Boolean model that intersect S_0 must be Poisson-distributed with mean

$$
\nu(S_0) \equiv \lambda \left\{\|S_0\|_1 + E(\|S\|_1)\right\}.
$$

Results related to Theorems 4.1 and 4.2 may be used to correct particle counts for edge effects. We consider only the case $k = 2$, and begin by describing the problem. Suppose sets in a coverage process are observed within a convex region $\mathcal{R}$ (a subset of $\mathbb{R}^2$), with the aim of estimating the unknown point process intensity, λ.

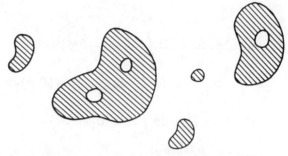

Figure 4.4. A shape comprised of five connected components, containing three voids.

Often $\mathcal{R}$ represents a microscope field (in which case it is usually circular) or a calibrated graticule (which may be square, rectangular, or circular). Those shapes that fall entirely within $\mathcal{R}$ are obviously "in." But how do we count shapes that overlap the boundary of $\mathcal{R}$? Some assistance is provided by the following theorem, which tells us the expected number of intersections of the boundary. Here we do not need to assume that S is convex, but we do suppose that S is isotropic. We allow S to be comprised of a number of disjoint connected components, and permit these components to contain voids (see Figure 4.4).

Theorem 4.3. *Take $k = 2$. Consider the coverage process in which each shape is distributed as the isotropic random set S, and the driving point process is stationary with intensity λ. Assume that the boundary of S is rectifiable, and that the total length of the boundary of S is finite with probability 1. Let $\mathcal{I}$ denote a fixed interval (line segment) in $\mathbb{R}^2$ of length t. The total number of intersections of $\mathcal{I}$ with parts of boundaries of sets from the Boolean model has mean*

$$2\pi^{-1}\lambda t E(\|\partial S\|_1).$$

In this theorem, intersections of parts of boundaries of sets with $\mathcal{I}$ are counted once for each intersection, not once for each set intersecting. Poisson-ness of the driving point process is not used in the proof below; only stationarity is required.

Proof. Let S be a fixed set, and assume initially that S is a uniformly distributed rotation of S. As in the proof of Theorem 4.1, we write this as $S = S_\Theta$ where Θ is uniformly distributed on $(0, 2\pi)$

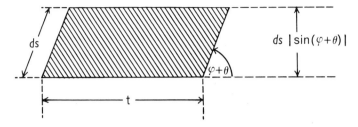

Figure 4.5. Parallelogram within which **x** must lie.

and S_θ denotes the image of S after rotation through θ about some center O and relative to some vector **u** fixed in the plane. We wish to calculate

$$\nu(\mathcal{I},S) \equiv \int_{\mathbb{R}^2} (\text{expected number of times } \mathbf{x} + S_\Theta \text{ intersects } \mathcal{I})\, d\mathbf{x}.$$

Let s denote the distance along the boundary ∂S of an arbitrary point Q on that boundary, measured from an origin P on ∂S, and suppose Q is carried to $Q(\theta)$ with coordinates $\mathbf{q}(\theta)$ by the motion of S to S_Θ. Define the function F by $F(s\,|\,\mathbf{x},\theta) = 1$ if $\mathbf{x} + \mathbf{q}(\theta)$ lies on $\mathcal{I}$, $F(s\,|\,\mathbf{x},\theta) = 0$ otherwise. Then

$$\nu(\mathcal{I},S) = (2\pi)^{-1} \int_0^{2\pi} \int_{\partial S} \int_{\mathbb{R}^2} d\mathbf{x} F(ds\,|\,\mathbf{x},\theta)\, d\theta.$$

If the element ds of ∂S is inclined at angle ϕ to the fixed vector **u** in the plane, then the inclination changes to $\phi + \theta$ after S is rotated to S_θ. If the image of this element after the motion of S to $\mathbf{x} + S_\Theta$ is to lie on $\mathcal{I}$, then $\mathbf{x}$ must fall within a parallelogram of area $t\,ds|\sin(\phi + \theta)|$ (see Figure 4.5). Therefore

$$\nu(\mathcal{I},S) = (2\pi)^{-1} \int_0^{2\pi} \int_{\partial S} t\,ds|\sin(\phi + \theta)|\, d\theta$$

$$= (2\pi)^{-1} t \int_{\partial S} ds \int_0^{2\pi} |\sin\theta|\, d\theta$$

$$= 2\pi^{-1} t \|\partial S\|_1. \tag{4.20}$$

To complete this proof we use the marked point process argument. Let $f(\mathbf{x},S)$ equal the number of times that the boundary of $\mathbf{x} + S$ intersects $\mathcal{I}$, and note that S has the distribution of S'_Θ where S' is a random set and Θ is independent of S' and uniformly distributed on $(0,2\pi)$. By formula (4.3), the expected number of times that sets in the coverage process intersect $\mathcal{I}$ equals

$$\lambda E\left\{\int_{\mathbb{R}^k} f(\mathbf{x}, S'_\ominus)\, d\mathbf{x}\right\} = \lambda E\left[\int_{\mathbb{R}^k} E\left\{f(\mathbf{x}, S'_\ominus) \mid S'\right\} d\mathbf{x}\right]$$
$$= \lambda E\left\{\nu(\mathcal{I}, S')\right\}$$
$$= \lambda E(2\pi^{-1}t\|\partial S'\|_1)$$
$$= \lambda 2\pi^{-1}t E(\|\partial S\|_1),$$

using (4.20) to obtain the second-last identity. $\quad\square$

To return to set counts within the convex observation region $\mathcal{R}$, let N_1 denote the total number of intersections of boundaries of sets from our stationary coverage process $\mathcal{C}$ with the boundary of $\mathcal{R}$. It follows from Theorem 4.3 that

$$E(N_1) = \int_{\partial\mathcal{R}} 2\pi^{-1}\lambda\, dt E(\|\partial S\|_1) = 2\pi^{-1}\lambda\|\partial\mathcal{R}\|_1 E(\|\partial S\|_1).$$

Assume that the shape S is convex with probability 1, and let N_2 equal the total number of sets from $\mathcal{C}$ which intersect $\mathcal{R}$. (Not all these sets necessarily lie entirely within $\mathcal{R}$.) By Theorem 4.1,

$$E(N_2) = \lambda\left\{\|\mathcal{R}\|_2 + E(\|S\|_2) + (2\pi)^{-1}\|\partial\mathcal{R}\|_1 E(\|\partial S\|_1)\right\},$$

and so

$$E\left(N_2 - \tfrac{1}{4}N_1\right) = \lambda\left\{\|\mathcal{R}\|_2 + E(\|S\|_2)\right\}.$$

Assuming $\|\mathcal{R}\|_2$ and $E(\|S\|_2)$ are both known,

$$\hat{\lambda} \equiv (N_1 - \tfrac{1}{4}N_2)/\left\{\|\mathcal{R}\|_2 + E(\|S\|_2)\right\} \tag{4.21}$$

is an unbiased estimator of intensity. We should stress that this statement is made under the assumption that shapes are convex and isotropic. The statement does not require the driving point process $\mathcal{P}$ to be Poisson; $\mathcal{P}$ need only be stationary with intensity λ. However, Poisson-ness is required in the discussion two paragraphs below.

The difference $N_2 - \tfrac{1}{4}N_1$ may be written in a slightly different form, which provides more information about the distribution of this quantity. Since both $\mathcal{R}$ and the random sets are convex, then with probability 1 any set which intersects $\mathcal{R}$ cuts the boundary of $\mathcal{R}$ in either two or four places. If the cuts occur in four places, then the contribution made by that particular set cancels from the

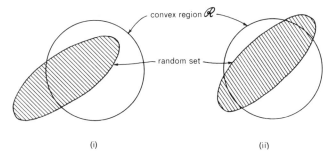

Figure 4.6. Sets that intersect boundary of $\mathcal{R}$. (i) Random set that protrudes into $\mathcal{R}$, but does not leave $\mathcal{R}$. (ii) Random set that both enters and leaves $\mathcal{R}$. Only sets of type (i) are counted in the total M_2. No sets of type (ii) are counted in formula (4.22).

formula $N_2 - \frac{1}{4}N_1$. If the cuts occur in just two places, then the contribution to $N_2 - \frac{1}{4}N_1$ equals $1 - \left(\frac{1}{4} \times 2\right) = \frac{1}{2}$. Therefore

$$N_2 - \tfrac{1}{4}N_1 = M_1 + \tfrac{1}{2}M_2, \qquad (4.22)$$

where M_1 equals the total number of sets contained entirely within the observation region $\mathcal{R}$, and M_2 equals the total number of sets that protrude into $\mathcal{R}$ from outside but do not both enter and leave $\mathcal{R}$ (see Figure 4.6). Therefore the estimator $\hat{\lambda}$ ignores sets that both enter and leave $\mathcal{R}$.

Assume that the coverage process $\mathcal{C}$ is a Boolean model. Then the random variables M_1 and M_2 appearing in (4.22) are independent and Poisson-distributed. As usual this result is easier to prove when $S \equiv \mathcal{S}$ is fixed, and so we begin with that case. Let $\mathcal{R}_1$ denote the set of points $\mathbf{x} \in \mathbb{R}^2$ for which $\mathbf{x} + S$ is wholly contained within $\mathcal{R}$, let $\mathcal{R}_2$ be the set of $\mathbf{x}$ such that $\mathbf{x} + S$ enters $\mathcal{R}$ from outside but does not leave $\mathcal{R}$ (that is, for which the boundaries of $\mathbf{x} + S$ and $\mathcal{R}$ meet in exactly 2 points). Then M_i equals the number of points from the Poisson process that lie within $\mathcal{R}_i$, for $i = 1$ and 2. Since $\mathcal{R}_1$ and $\mathcal{R}_2$ are, by definition, disjoint, then M_1 and M_2 must be independent and Poisson-distributed, with respective means $\lambda \|\mathcal{R}_1\|_2$ and $\lambda \|\mathcal{R}_2\|_2$. In the case of a general convex shape S, use the marked point process argument. Notice that since the driving point process is Poisson, then members of the sequence of coverage processes generated by the marked point process argument are stochastically independent of one another. Therefore the marked point process argument expresses each of M_1 and M_2 as sums of independent random variables.

Let $m_i \equiv E(M_i)$. Since $\hat{\lambda}$ defined at (4.21) is unbiased for λ, then by (4.22),

$$m_1 + \tfrac{1}{2}m_2 = E\left(N_1 - \tfrac{1}{4}N_2\right) = \lambda\left\{\|\mathcal{R}\|_2 + E(\|S\|_2)\right\},$$

whence

$$m_1 + \tfrac{1}{4}m_2 = \lambda\left\{\|\mathcal{R}\|_2 + E(\|S\|_2)\right\} - \tfrac{1}{4}m_2.$$

From these results and the Poisson-ness of N_1 and N_2 we may deduce that the estimator $\hat{\lambda}$ has variance

$$\begin{aligned}
\mathrm{var}\hat{\lambda} &= \left\{\mathrm{var}(M_1) + \tfrac{1}{4}\mathrm{var}(M_2)\right\} / \left\{\|\mathcal{R}\|_2 + E(\|S\|_2)\right\}^2 \\
&= (m_1 + \tfrac{1}{4}m_2) / \left\{\|\mathcal{R}\|_2 + E(\|S\|_2)\right\}^2 \\
&= \lambda\left\{\|\mathcal{R}\|_2 + E(\|S\|_2)\right\}^{-1} - m_2[4\left\{\|\mathcal{R}\|_2 + E(\|S\|_2)\right\}^2]^{-1}.
\end{aligned}$$

The smallest value m_2 can take is 0, which occurs when the shape S degenerates to a point. The largest value of m_2 arises when $m_1 = 0$, and is

$$m_2 = 2\lambda\left\{\|\mathcal{R}\|_2 + E(\|S\|_2)\right\},$$

when all shapes are too big to be contained wholly within $\mathcal{R}$. Combining the material in this paragraph we conclude that

$$\tfrac{1}{2}\lambda\left\{\|\mathcal{R}\|_2 + E(\|S\|_2)\right\}^{-1} \le \mathrm{var}(\hat{\lambda}) \le \lambda\left\{\|\mathcal{R}\|_2 + E(\|S\|_2)\right\}^{-1}. \quad (4.23)$$

If $M_1 + \tfrac{1}{2}M_2$ were Poisson-distributed then the variance of $\hat{\lambda}$ would be

$$\lambda\left\{\|\mathcal{R}\|_2 + E(\|S\|_2)\right\}^{-1}.$$

This observation, when compared with formula (4.23), suggests that a slightly conservative confidence interval for λ may be constructed by regarding

$$\hat{\lambda}\left\{\|\mathcal{R}\|_2 + E(\|S\|_2)\right\}$$

as though it were Poisson-distributed.

An important special case is that where the shape S is a line segment of random length (a degenerate convex set). There, $E(\|S\|_2) = 0$, and so $\hat{\lambda}$ reduces to

$$\hat{\lambda} = \left(M_1 + \tfrac{1}{2}M_2\right) / \|\mathcal{R}\|_2.$$

We can interpret this as a counting rule for random segments intersecting a convex region $\mathcal{R}$: count one for each segment wholly contained within $\mathcal{R}$, half for each segment that cuts the boundary of $\mathcal{R}$ just once, and zero for each segment that cuts the boundary twice. The total of all such counts, divided by the area of the region, is an unbiased estimator of point process intensity. This rule is often used in an industrial setting, for example where the segments model asbestos fibers and the region $\mathcal{R}$ denotes a portion of a filter sample.

We pause to combine our results into a theorem.

Theorem 4.4. *Take $k = 2$. Consider the Boolean model in which each shape is distributed as the isotropic convex random set S, with finite mean area and perimeter. Assume the observation region $\mathcal{R}$ is convex. Then the estimator*

$$\hat{\lambda} \equiv (N_2 - \tfrac{1}{4}N_1) / \{\|\mathcal{R}\|_2 + E(\|S\|_2)\}$$
$$= (M_1 + \tfrac{1}{2}M_2) / \{\|\mathcal{R}\|_2 + E(\|S\|_2)\}$$

is unbiased for λ, and has variance within the range $[\tfrac{1}{2}c, c]$ where

$$c \equiv \lambda / \{\|\mathcal{R}\|_2 + E(\|S\|_2)\}.$$

The variables M_1 and M_2 are independent and Poisson-distributed.

In practice the region $\mathcal{R}$ is very often circular, and then it is clear from considerations of symmetry that the assumption that S be isotropic is not required.

Of course, there are many other ways of estimating λ. For example, we might wish to avoid making assumptions about mean area $E(\|S\|_2)$ of random shapes by using vacancy to estimate $E(\|S\|_2)$. Recall from Chapter 3, formula (3.5) that for a Boolean model, the vacancy V within $\mathcal{R}$ has mean

$$E(V) = \|\mathcal{R}\| \exp\{-\lambda E(\|S\|_2)\}.$$

Therefore if we define

$$\hat{\rho} \equiv V / \|\mathcal{R}\|$$

to be the uncovered proportion of $\mathcal{R}$, then $-\lambda^{-1} \log \hat{\rho}$ estimates $E(\|S\|_2)$. We could estimate λ using this formula in place of $E(\|S\|_2)$ in the definition (4.21) of $\hat{\lambda}$, and define $\hat{\lambda}_1$ to be the solution of

$$\hat{\lambda}_1 = (N_1 - \tfrac{1}{4}N_2) / \left\{ \|\mathcal{R}\|_2 + (-\hat{\lambda}_1^{-1} \log \hat{\rho}) \right\};$$

that is,

$$\hat{\lambda}_1 = (N_1 - \tfrac{1}{4}N_2 + \log \hat{\rho}) / \|\mathcal{R}\|_2.$$

This estimator tends to slightly underestimate λ, since by Jensen's inequality (e.g., Billingsley 1979, p. 240; Chung 1974, p. 47) and concavity of the function $\log x$,

$$
\begin{aligned}
E(\hat{\lambda}_1) &\leq \{ E(N_1 - \tfrac{1}{4}N_2) + \log E(\hat{\rho}) \} / \|\mathcal{R}\|_2 \\
&= [\lambda \{ \|\mathcal{R}\|_2 + E(\|S\|_2) \} - \lambda E(\|S\|_2)] / \|\mathcal{R}\|_2 \\
&= \lambda .
\end{aligned}
$$

However, it may be proved that $\hat{\lambda}_1$ is consistent for λ as the observation region $\mathcal{R}$ increases.

Another simple procedure is to associate with each set in the Boolean model a specific point. This might be, for example, that point on the set furthest to the right, if such a point is uniquely defined. The set of associated points forms a stationary Poisson point process with intensity λ, and so the total number of associated points visible within the observation region $\mathcal{R}$ is Poisson-distributed with mean $\lambda\|\mathcal{R}\|$. Therefore the total number visible within $\mathcal{R}$, divided by the area of $\mathcal{R}$, is an unbiased estimator $\hat{\lambda}_2$ of λ with variance

$$
\mathrm{var}(\hat{\lambda}_2) = \lambda \|\mathcal{R}\|^{-1} . \tag{4.24}
$$

This technique is called the *associated point method*, and has the advantage that it requires no assumptions about convexity or isotropy of the random shapes, or about convexity of the observation region. Furthermore, $\hat{\lambda}_2$ is unbiased for λ under the assumption that the driving point process is stationary. On the other hand, the variance of $\hat{\lambda}_2$ is generally a little more than the variance of $\hat{\lambda}$; compare (4.23) and (4.24) in the Poisson case, and see Appendix A of Jensen and Sundberg (1986) for more general circumstances.

We should point out that practical application of these estimators requires the outline of each set in each clump within the observation field $\mathcal{R}$ to be visible. The estimates are of little use if clumps are completely "opaque," meaning that individual sets cannot be identified.

4.3. CURVATURE

In this section we introduce the notion of curvature and discuss its application to counting problems. *Throughout the section we confine attention to $k = 2$ dimensions.* Except for the last two paragraphs we assume that the driving point process is Poisson and stationary,

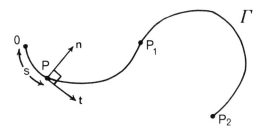

Figure 4.7. Tangent **t** and normal **n** to point P on curve Γ.

so that the coverage process is a Boolean model. We begin with a little elementary differential geometry.

The Cartesian coordinates of a general point $P = P(s)$ on a planar curve Γ may be expressed in parametric form as $\mathbf{x} = \mathbf{x}(s) = (x_1, x_2)$, where s equals the arc length measured from some fixed point O on the curve (see Figure 4.7). If the curve is smooth in a neighborhood of P then the derivative $\mathbf{t} \equiv d\mathbf{x}/ds$ exists and is a unit vector in the direction of the tangent at P. The vector $\mathbf{n} \equiv d\mathbf{t}/ds$ points in the direction of the normal at P, and has length

$$|\kappa| \equiv |\mathbf{n}| = |d\mathbf{t}/ds| = |d^2\mathbf{x}/ds^2|.$$

We call $|\kappa|$ the absolute value of the curvature of Γ at P.

If the tangent vector $\mathbf{t}$ makes an angle $\theta = \theta(s)$ $(-\pi < \theta \leq \pi)$ to the positive direction of the x_1-axis, then $\mathbf{t} = (\cos\theta, \sin\theta)$ and

$$\frac{d\mathbf{t}}{ds} = (-\sin\theta, \cos\theta)\frac{d\theta}{ds}.$$

Therefore an equivalent definition of absolute curvature at P is $|\kappa| \equiv |d\theta/ds|$, the rate of change of the inclination of the tangent at P.

We assign a sign to κ by convention, as follows. If we are traveling along Γ in such a direction that our motion is counterclockwise, we give κ a positive sign. Otherwise, we give it a negative sign. For example, if we traverse the curve in Figure 4.7 from left to right, then κ is positive between O and P_1 and negative between P_1 and P_2. It is 0 at P_1. This signed value of κ is called the *curvature* of Γ at $\mathbf{x}(s)$.

Suppose the curve Γ joining points P_1 and P_n is made up of $n - 1$ smooth arcs Γ_i, $1 \leq i \leq n - 1$, the ith joining P_i to P_{i+1} (refer to Figure 4.8). Let θ_i, $-\pi < \theta_i \leq \pi$, be the signed angle between the tangents at P_i, the sign being determined by the convention in the

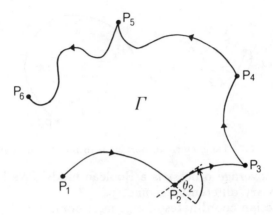

Figure 4.8. A curve comprised of five smooth arcs. Arrows indicate the direction of travel.

previous paragraph. We call θ_i the *turning angle* at the *junction* P_i. The *total curvature* of Γ is given by

$$C(\Gamma) = \sum_{i=1}^{n-1} \int_{\Gamma_i} \kappa(s)\,ds + \sum_{i=1}^{n-1} \theta_i. \tag{4.25}$$

We have included θ_1 in the sum in case the curve is closed; of course, θ_1 is 0 otherwise, for example it is 0 in the curve of Figure 4.8.

The sign convention means that if we traverse a simple closed curve in a counterclockwise manner, the total curvature is 2π. (A curve is simple if it does not cross itself.) This is sometimes called the "turning tangents" theorem.

Let S be a compact subset of $\mathbb{R}^2$, with a boundary ∂S made up entirely of curves such as Γ described above. If S is connected but contains m voids then the total curvature of ∂S equals $2\pi - 2\pi m$, since integration around the boundary of each void requires a circuit in the clockwise direction (see Figure 4.9). If S is made up of m_1 disjoint components and contains m_2 voids then, adding over all components, we see that the total curvature of S equals $2\pi m_1 - 2\pi m_2 = 2\pi x(S)$, where $x(S) \equiv m_1 - m_2$ is the *Euler characteristic* of S. This is one form of the Gauss-Bonnet theorem in differential geometry [see O'Neill (1966, p. 372ff)]. The Euler characteristic of the set in Figure 4.9 equals -1.

A region S such as that depicted in Figure 4.9 is called an *oriented polygonal region* (O'Neill 1966, p. 386). The boundary of such an S is assumed to be comprised entirely of a finite number of simple

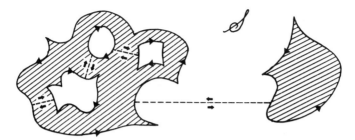

Figure 4.9. Integrating curvature around boundary ∂S of set S. Total curvature of ∂S equals $2 \times 2\pi - 3 \times 2\pi = 2\pi x(S)$, where $x(S) = -1$ is the Euler characteristic of S.

closed curves, which are such that the turning angles at junctions are clearly defined. The boundary is the formal sum of a finite number of smooth curves Γ_i. Curvature along these smooth curves is defined as the rate of change of the angle of the tangent, and the curvature at the junction of two smooth curves equals the angle between the two tangents.

Consider a Boolean model $\mathcal{C}$ in which shapes distributed as S are centered at points of a stationary Poisson process $\mathcal{P}$ having intensity λ. Assume that all realizations of S are oriented polygonal regions, and that $\alpha \equiv E(\|S\|_2)$, $\beta \equiv E(\|\partial S\|_1)$ and $E\{|x(S)|\}$ are all finite. Let $\chi \equiv E\{x(S)\}$ be the mean Euler characteristic of S. We shall compute the expected curvature per unit area resulting from this model.

Let $\xi + S$ be an arbitrary set from $\mathcal{C}$. (We could take $\xi = 0$ if we wished. See Section 4.1 for a definition of "arbitrary.") Denote by $I(s)$ the indicator of the event that the point $P(s)$ on the boundary $\partial(\xi + S)$ of $\xi + S$ is not covered by other sets from the Boolean model. Assume that $\partial(\xi + S)$ is made up of smooth curves Γ_i with junctions at points $P(s_i)$, and that the turning angle at $P(s_i)$ equals θ_i. The total curvature of $\xi + S$ computed from parts of the boundary that are visible (i.e., not covered by other sets) is expressible as an analogue of formula (4.25):

$$\sum_i \int_{\Gamma_i} \kappa(s) I(s)\, ds + \sum_i \theta_i I(s_i).$$

We call this the *visible curvature* of $\xi + S$. Taking expectations conditional on $\xi + S$, using properties of the Palm (or conditional) distribution of a Boolean model described in Section 4.1, and recalling that $E\{I(s)\} = e^{-\alpha\lambda}$ [formula (4.7)], we see that the expected visible curvature of $\xi + S$, conditional on $\xi + S$, equals

$$2\pi x(\xi + S)e^{-\alpha\lambda} = 2\pi x(S)e^{-\alpha\lambda},$$

where $x(S)$ denotes the Euler characteristic of the set S. Now take expectations with respect to S, obtaining $2\pi E\{x(S)e^{-\alpha\lambda}\} = 2\pi\chi e^{-\alpha\lambda}$ for the expected visible curvature of an arbitrary random set. The intensity of sets per unit area equals λ, and so the expected visible curvature per unit area is

$$2\pi\chi\lambda e^{-\alpha\lambda}. \tag{4.26}$$

That is, the expected visible curvature of random sets centered within any Borel subset $\mathcal{R}$ of $\mathbb{R}^2$ equals

$$2\pi\chi\lambda e^{-\alpha\lambda}\|\mathcal{R}\|.$$

We should stress that formula (4.26) does *not* require the assumption of isotropy.

The last step in the derivation of formula (4.26) uses the notion of marked point processes and the Campbell theorem. In more detail, the argument runs as follows. Associate with each point ξ_i of the driving Poisson process a mark M_i equal to the total visible curvature of $\xi_i + S_i$. Let $\mathcal{R}$ be a subset of $\mathbb{R}^2$, and put $f(\mathbf{x}, m) = m$ if $\mathbf{x} \in \mathcal{R}$, $f(\mathbf{x}, m) = 0$ otherwise. Then

$$\sum_{i=1}^{\infty} f(\xi_i, M_i)$$

equals the total visible curvature of random sets centered within $\mathcal{R}$. Apply the form (4.4) of the Campbell theorem to deduce that the expected value of this series equals $\lambda\|\mathcal{R}\|E(M)$, where M has the "mark distribution." Then use formula (4.8) to identify the distribution of M, obtaining $E(M) = 2\pi\chi e^{-\alpha\lambda}$.

The value of expected curvature displayed in formula (4.26) ignores "uncovered crossings," where the boundary of a second set covers the boundary of $\xi + S$ and the point of intersection is not covered by a third set. The angles between intersecting boundaries make significant contributions to total curvature, as we shall show in the next paragraph. Therefore $2\pi\chi\lambda e^{-\alpha\lambda}$ is not the expected total curvature per unit area in the Boolean model. However, expected visible curvature can occasionally be measured directly. If we know that shapes have smooth boundaries (e.g., random radius disks) then any discontinuity in the boundary of the Boolean model pattern can in theory be identified as an uncovered crossing, and deliberately omitted from a calculation of total curvature. If $K_v(\mathcal{R})$ denotes visible curvature within a large region $\mathcal{R}$, then the result in the previous paragraph establishes that

$$E\{K_v(\mathcal{R})\} = 2\pi\chi\lambda\|\mathcal{R}\|e^{-\alpha\lambda}.$$

The ergodic argument used to prove Theorem 3.6 in Chapter 3 gives in the present case,

$$K_v(\mathcal{R})/\|\mathcal{R}\| \to 2\pi\chi\lambda e^{-\alpha\lambda}$$

almost surely as $\mathcal{R}$ increases. We know that the vacancy $V = V(\mathcal{R})$ within $\mathcal{R}$ satisfies $E\{V(\mathcal{R})\} = \|\mathcal{R}\|e^{-\alpha\lambda}$ [formula (3.5)], and

$$V(\mathcal{R})/\|\mathcal{R}\| \to e^{-\alpha\lambda}$$

almost surely as $\mathcal{R}$ increases (Theorem 3.6). Therefore

$$\hat{\lambda}_3 \equiv K_v(\mathcal{R})/\{2\pi\chi V(\mathcal{R})\}$$

is a consistent estimator of λ. In most cases $\chi = 1$ (e.g., when S consists of single a connected component having no voids). However, the estimator $\hat{\lambda}_3$ is rarely useful in practice, owing to difficulties in computing $K_v(\mathcal{R})$.

Next we calculate the contribution to curvature from uncovered crossings. Assume shapes in the Boolean model are isotropic, and as before let $\xi + S$ be an arbitrary random set in the Boolean model. By Theorem 4.3, the expected total number of intersections of an infinitesimal interval of length ds of $\partial(\xi + S)$ with parts of boundaries of other sets in the Boolean model equals $2\pi^{-1}\lambda ds\beta$. The probability that any given crossing is uncovered equals $e^{-\alpha\lambda}$, and the expected angle between the boundaries of two intersecting sets must be $-\pi/2$, by symmetry. (The minus sign comes from our convention that the curvature of a closed curve is $+2\pi$ if the curve is traced in an *anticlockwise* direction. The turning angle of each uncovered crossing lies between 0 and $-\pi$; see Figure 4.10.) Therefore the expected contribution to total curvature from uncovered crossings that occur within the interval of length ds equals

$$2\pi^{-1}\lambda ds\beta e^{-\alpha\lambda}(-\pi/2) = -\beta\lambda e^{-\alpha\lambda}ds.$$

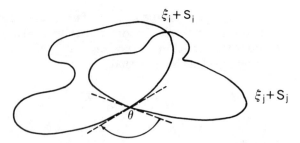

Figure 4.10. Curvature at uncovered crossing. Angle θ is negative with mean $-\pi/2$.

Integrating around the boundary of $\xi + S$ we see that, conditional on $\xi + S$, the expected contribution to total curvature from uncovered crossings of $\partial(\xi + S)$ is

$$-\beta\lambda e^{-\alpha\lambda}\|\partial(\xi + S)\|_1 = -\beta\lambda e^{-\alpha\lambda}\|\partial S\|_1.$$

Adding this contribution over all sets in the Boolean model, and noting that each contribution will be counted twice (once for each set intersecting at an uncovered crossing), we deduce that the expected total curvature from uncovered crossings per unit area equals

$$\tfrac{1}{2}\lambda E\left\{-\beta\lambda e^{-\alpha\lambda}\|\partial S\|_1\right\} = -\tfrac{1}{2}(\beta\lambda)^2 e^{-\alpha\lambda}.$$

[We have again used the Campbell theorem in the form (4.4), employing (4.8) to identify the "mark distribution."] More explicitly, the expected total curvature from uncovered crossings of random sets centered within the Borel set $\mathcal{R}$ equals

$$-\tfrac{1}{2}(\beta\lambda)^2 e^{-\alpha\lambda}\|\mathcal{R}\|.$$

In this formulation, the full turning angle is included for any intersection between two sets both centered within $\mathcal{R}$; only half the turning angle is included if one of the sets is centered within $\mathcal{R}$ and the other outside $\mathcal{R}$.

This curvature should be added to the expected visible curvature per unit area calculated at (4.26). Therefore the expected total curvature per unit area in the Boolean model is

$$2\pi\chi\lambda e^{-\alpha\lambda} - \tfrac{1}{2}(\beta\lambda)^2 e^{-\alpha\lambda}. \tag{4.27}$$

More precisely, the expected total curvature of random sets centered within $\mathcal{R}$ equals

$$\left\{2\pi\chi\lambda e^{-\alpha\lambda} - \tfrac{1}{2}(\beta\lambda)^2 e^{-\alpha\lambda}\right\}\|\mathcal{R}\|,$$

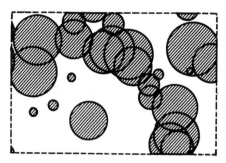

Figure 4.11. Pattern of clumps and voids in a Boolean model of random-radius disks. [From Kellerer (1983).]

using the rule in the previous paragraph to decide on turning angles. We pause to combine these results into a theorem.

Theorem 4.5. *Take $k = 2$. Consider the Boolean model in which each shape is distributed as S and the driving Poisson process has intensity λ. Assume that realizations of S are isotropic orientable polygonal regions and that $\alpha \equiv E(\|S\|_2)$, $\beta \equiv E(\|\partial S\|_1)$, and $\chi \equiv E\{x(S)\}$ are all well-defined and finite. Let $\mathcal{R}$ denote a Borel subset of $\mathbb{R}^2$. Then the expected visible curvature of those sets centered within $\mathcal{R}$ equals*

$$2\pi\chi\lambda\|\mathcal{R}\|e^{-\alpha\lambda}.$$

The expected total curvature of the pattern within $\mathcal{R}$ equals

$$2\pi\left\{\chi\lambda - (4\pi)^{-1}(\beta\lambda)^2\right\}\|\mathcal{R}\|e^{-\alpha\lambda}.$$

A realization of a Boolean model is a pattern of clumps and voids, formed from the overlapping random sets (see Figure 4.11). From the second part of Theorem 4.5 we see that the mean Euler characteristic per unit area of this pattern is

$$\left\{\chi\lambda - (4\pi)^{-1}(\beta\lambda)^2\right\}e^{-\alpha\lambda}. \tag{4.28}$$

Remembering that the Euler characteristic of a figure equals the number of disjoint components minus the number of voids, we conclude that the expected number of clumps minus voids per unit area in the Boolean model pattern is given by (4.28). Note that this formula requires the assumption that shapes are isotropic.

Sometimes it is of interest to compute the expected number of clumps minus voids in the set $\mathcal{R}(\mathcal{C})$ formed by intersection of sets from the Boolean model $\mathcal{C} = \{\xi_i + S_i, i \geq 1\}$ with a given region $\mathcal{R}$:

$$\mathcal{R}(\mathcal{C}) \equiv \mathcal{R} \cap \left\{ \bigcup_{i=1}^{\infty} (\xi_i + S_i) \right\}. \qquad (4.29)$$

In the case of the pattern depicted in Figure 4.11, this corresponds to $(2\pi)^{-1}$ times the expected curvature of the shaded set. It is different from the quantity $\{\chi\lambda - (4\pi)^{-1}(\beta\lambda)^2\}\|\mathcal{R}\|e^{-\alpha\lambda}$, obtainable from (4.28), in that it includes contributions to curvature arising from the boundary of $\mathcal{R}$.

Assume that shapes generating the Boolean model are isotropic and that $\mathcal{R}$ is a connected oriented polygonal region without voids (e.g., a convex set). We first compute the expected contribution to curvature of $\mathcal{R}(\mathcal{C})$ from uncovered crossings of the boundary of $\mathcal{R}$ by boundaries of random sets. Using Theorem 4.3 we see that the expected number of uncovered crossings of an infinitesimal segment of length dt of the boundary of $\mathcal{R}$ equals $2\pi^{-1}\lambda\,dt\beta e^{-\alpha\lambda}$. The mean turning angle at any such crossing is $\pi/2$. (All angles lie between 0 and π, and the mean is $\pi/2$ by symmetry.) Therefore the expected contribution to total curvature arising from uncovered crossings of the boundary of $\mathcal{R}$ by boundaries of random sets is

$$\int_{\partial\mathcal{R}} 2\pi^{-1}\lambda\,dt\,\beta e^{-\alpha\lambda}\pi/2 = \beta\lambda\|\partial\mathcal{R}\|_1 e^{-\alpha\lambda}.$$

Next we compute the expected contribution to curvature of $\mathcal{R}(\mathcal{C})$ from curvature of the boundary of $\mathcal{R}$. The curvature at the position of the infinitesimal segment of length dt on $\partial\mathcal{R}$ makes a contribution if and only if it is *covered*. (In Figure 4.11, only the bottom left-hand corner makes a nonzero contribution.) The probability that the infinitesimal segment is covered equals $(1 - e^{-\alpha\lambda})$. Integrating over all such segments, we see that the total contribution to expected curvature from this source is $(1 - e^{-\alpha\lambda})$ times the total curvature of $\mathcal{R}$; the latter equals 2π. The total contribution to curvature from the boundary of $\mathcal{R}$ comes from adding this term to the one derived in the previous paragraph:

$$\beta\lambda\|\partial\mathcal{R}\|_1 e^{-\alpha\lambda} + (1 - e^{-\alpha\lambda})2\pi.$$

Finally we add this quantity to the expected total curvature *inside* $\mathcal{R}$, available from Theorem 4.5, getting the mean total curvature of the set $\mathcal{R}(\mathcal{C})$:

$$\beta\lambda\|\partial\mathcal{R}\|_1 e^{-\alpha\lambda} + (1 - e^{-\alpha\lambda})2\pi + 2\pi\left\{\chi\lambda - (4\pi)^{-1}(\beta\lambda)^2\right\}\|\mathcal{R}\|_2 e^{-\alpha\lambda}.$$

(We are guilty of a sleight of hand here, in that we have equated expected curvature occurring inside $\mathcal{R}$ with expected curvature of sets centered inside $\mathcal{R}$. The reader may check that these two means are identical.) The mean number of clumps minus voids in $\mathcal{R}(\mathcal{C})$ is obtainable by dividing through by 2π. We have proved the following theorem.

Theorem 4.6. *Take $k = 2$. Consider the Boolean model $\mathcal{C} = \{\xi_i + S_i, i \geq 1\}$ in which each shape S_i is distributed as the isotropic random shape S and the driving Poisson process $\{\xi_i, i \geq 1\}$ has intensity λ. Assume that realizations of S are oriented polygonal regions, and that $\alpha \equiv E(\|S\|_2)$, $\beta \equiv E(\|\partial S\|_1)$, and $\chi \equiv E\{x(S)\}$ are all well-defined and finite. Let $\mathcal{R}$ denote a connected oriented polygonal region without voids. Then the expected number of clumps minus voids in the set $\mathcal{R}(\mathcal{C})$ defined at (4.9) equals*

$$\left\{\chi\lambda - (4\pi)^{-1}(\beta\lambda)^2\right\}\|\mathcal{R}\|_2 e^{-\alpha\lambda} + (2\pi)^{-1}\beta\lambda\|\partial\mathcal{R}\|_1 e^{-\alpha\lambda} + 1 - e^{-\alpha\lambda}.$$

By way of comparison, the expected number of clumps minus voids per unit area in the Boolean model equals

$$\left\{\chi\lambda - (4\pi)^{-1}(\beta\lambda)^2\right\}e^{-\alpha\lambda}. \tag{4.30}$$

We conclude this discussion by considering four examples. The first two illustrate particular cases of formula (4.30) for mean number of clumps minus voids per unit area. Examples (iii) and (iv) indicate how that formula should be modified if shapes are not isotropic.

(i) *Fixed radius disks.* If shapes generating the Boolean model are fixed radius disks with radius r then $\chi = 1$, $\alpha = \pi r^2$, $\beta = 2\pi r$, and

$$(4\pi)^{-1}(\beta\lambda)^2 = (4\pi)^{-1}(2\pi r\lambda)^2 = \pi r^2\lambda^2 = \alpha\lambda^2.$$

Therefore by (4.30), the mean number of clumps minus voids per unit area equals

$$\lambda(1 - \alpha\lambda)e^{-\alpha\lambda}.$$

(ii) *Random length, isotropic line segments.* If shapes are isotropic line segments with mean length l then $\chi = 1$, $\alpha = 0$, and $\beta = 2l$. Therefore by (4.30), the mean number of clumps minus voids per unit area equals

$$\lambda(1 - \pi^{-1}l^2\lambda).$$

(iii) Parallelograms with fixed orientation. Assume shapes generating the Boolean model are all identical to a fixed parallelogram S with area α. If $\xi + S$ is an arbitrary set from the Boolean model, the probability that n crossings of a given side σ of $\xi + S$ occur equals the chance that n points from the driving Poisson process fall within the union of two particular disjoint parallelograms congruent to S. Therefore the expected number of crossings of σ equals $2\alpha\lambda$. The probability that a given crossing is uncovered is $e^{-\alpha\lambda}$. The turning angle at each crossing of σ depends only on choice of σ. If the angle is $-\theta$, then the turning angle for crossings of the side opposite to σ equals $-(\pi - \theta)$. Therefore the mean turning angle for crossings of opposite sides is

$$\tfrac{1}{2}\{-\theta - (\pi - \theta)\} = -\tfrac{1}{2}\pi,$$

and the expected value of the total of all turning angles of uncovered crossings of the arbitrary set $\xi + S$ is

$$4.2\alpha\lambda e^{-\alpha\lambda}(-\pi/2) = -4\pi\alpha\lambda e^{-\alpha\lambda}.$$

Thus, the contribution of uncovered crossings to expected total curvature per unit area in the Boolean model, equals

$$\tfrac{1}{2}\lambda(-4\pi\alpha\lambda e^{-\alpha\lambda}) = -2\pi\alpha\lambda^2 e^{-\alpha\lambda}.$$

(The factor $\tfrac{1}{2}$ is included because we have counted each crossing twice, once for each set intersecting.) Adding this to the expected visible curvature per unit area derived at (4.26), and noting that $\chi = 1$, we see that expected total curvature per unit area of the union of all sets in the Boolean model equals

$$2\pi\lambda(1 - \alpha\lambda)e^{-\alpha\lambda}.$$

Therefore the mean number of clumps minus voids per unit area equals

$$\lambda(1 - \alpha\lambda)e^{-\alpha\lambda},$$

the same as in Example (i).

(iv) Triangles with fixed orientations. Assume shapes are identical to a fixed triangle S with area α. Arguing as in Example (iii) we obtain the following results: the expected number of crossings of any given side equals $4\alpha\lambda$; the probability that a given crossing is uncovered equals $e^{-\alpha\lambda}$; all turning angles have the form $-(\pi - \theta)$, where θ is one of the angles of the triangle; the average over all turning angles is $-2\pi/3$; the expected value of the total of all

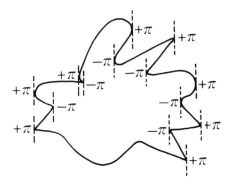

Figure 4.12. Tangent count of an irregular blob. Total count equals $8\pi - 6\pi = 2\pi$.

turning angles of uncovered crossings of all sides of a given set $\xi + S$ is

$$3 \cdot 4\alpha\lambda \cdot e^{-\alpha\lambda} \cdot (-2\pi/3) = -8\pi\alpha\lambda e^{-\alpha\lambda};$$

the contribution of uncovered crossings to expected total curvature per unit area equals

$$\tfrac{1}{2}\lambda(-8\pi\alpha\lambda e^{-\alpha\lambda}) = -4\pi\alpha\lambda^2 e^{-\alpha\lambda};$$

the expected total curvature per unit area equals

$$2\pi\lambda e^{-\alpha\lambda} - 4\pi\alpha\lambda^2 e^{-\alpha\lambda} = 2\pi\lambda(1 - 2\alpha\lambda)e^{-\alpha\lambda};$$

and the mean number of clumps minus voids per unit area equals

$$\lambda(1 - 2\alpha\lambda)e^{-\alpha\lambda}.$$

If the boundary of each random set in a Boolean model can be identified, and if random sets are connected and without voids, then an estimate of λ may be based on the fact that each set has curvature equal to 2π. Draw all those tangents to a set that are inclined at a given angle, count $+\pi$ if in the neighborhood of the point where the tangent touches the set the tangent lies outside the set, and count $-\pi$ if it lies inside (see Figure 4.12). Then the total count equals 2π for each set. The value of λ may be estimated from observation of the Boolean model within a given region $\mathcal{R}$ by computing the total tangent count within $\mathcal{R}$ and dividing by 2π (see Figure 4.13). The result is called the *tangent count estimator*,

$$\hat{\lambda}_4 \equiv (\text{total tangent count within } \mathcal{R}$$
$$\text{for tangents in a given direction})/(2\pi\|\mathcal{R}\|),$$

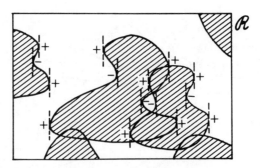

Figure 4.13. Tangent count within region $\mathcal{R}$. Total tangent count equals $10\pi - 5\pi = 5\pi$; $\hat{\lambda}_4 = 5\pi/(2\pi\|\mathcal{R}\|) = 5/(2\|\mathcal{R}\|)$.

and is unbiased for λ. It is consistent for λ as $\mathcal{R}$ increases. Unbiasedness does not require the point process driving the coverage process to be Poisson; it need only be stationary with intensity λ. Neither is isotropy needed. Consistency holds if the point process is stationary and ergodic, and may be proved by modifying the proof of Theorem 3.6 in Chapter 3.

Accuracy of the tangent count estimator is likely to be increased if $\hat{\lambda}_4$ is averaged over as many tangent inclinations as possible. If we average (i.e., integrate) over all possible inclinations then the resulting estimator is based on total curvature of all sets and parts of sets lying within $\mathcal{R}$. Notice that total curvature of sets in the coverage process is different from total curvature of their union; the latter quantity was examined in Theorems 4.5 and 4.6.

4.4. ISOLATED SETS

Let $\mathcal{C} = \{\xi_i + S_i, i \geq 1\}$ be a Boolean model in $\mathbb{R}^k$, where the driving Poisson process $\mathcal{P} = \{\xi_i, i \geq 1\}$ is stationary with intensity λ and independent of the independent and identically distributed shapes S, S_1, S_2, ... We say that a set $\xi_i + S_i$ is *isolated*, or part of a clump of order 1, if it intersects no other sets in the Boolean model, that is, if $(\xi_i + S_i) \cap (\xi_j + S_j) = \phi$ for all $j \neq i$. The chance that an arbitrary set is isolated equals the probability that the random set S, independent of $\mathcal{C}$, intersects no sets from $\mathcal{C}$. We describe the latter event by saying that "S is isolated." (Recall the discussion of Palm distributions in Section 4.1. In particular, remember that the distribution of a stationary Poisson process $\mathcal{P}$ conditional on a point being sited at $\mathbf{x}$ is just that of $\mathcal{P} \cup \{\mathbf{x}\}$. Without loss of generality, the arbitrary set is centered at $\mathbf{x} = 0$, in which case it is distributed as S.)

We need the notion of two "arbitrary sets" in the Boolean model $\mathcal{C}$. Of course it makes no sense to talk about two arbitrary sets from $\mathcal{C}$ within $\mathbb{R}^k$, just as it is nonsensical to discuss a uniform distribution on $\mathbb{R}^k$. But we may properly define two arbitrary sets centered within a bounded region $\mathcal{R}$. Let M equal the number of points of $\mathcal{P}$ that lie within $\mathcal{R}$, and conditional on M let the associated random sets in $\mathcal{C}$ be $\xi^{(1)} + S^{(1)}, \ldots, \xi^{(M)} + S^{(M)}$, numbered in completely random order. Define

$$p_1 \equiv P(\xi^{(1)} + S^{(1)} \text{ isolated} \,|\, M \geq 1) = P(S \text{ isolated}) \qquad (4.31)$$

and

$$p_2 = p_2(\mathcal{R}) \equiv P(\xi^{(1)} + S^{(1)}, \; \xi^{(2)} + S^{(2)} \text{ both isolated} \,|\, M \geq 2). \quad (4.32)$$

Of course, the event that $\xi^{(1)} + S^{(1)}$ and $\xi^{(2)} + S^{(2)}$ are both isolated means that these two sets do not intersect one another or any other sets in $\mathcal{C}$.

Let $N = N(\mathcal{R})$ equal the total number of isolated sets $\xi_i + S_i$ whose centers ξ_i lie with the region $\mathcal{R}$, a Borel subset of $\mathbb{R}^k$. Then

$$E(N) = \lambda \|\mathcal{R}\| P(S \text{ isolated}) = \lambda \|\mathcal{R}\| p_1. \qquad (4.33)$$

This intuitively obvious formula may be proved very quickly using the marked point process argument, as follows. Take the mark M_i at ξ_i to equal 1 if $\xi_i + S_i$ is isolated, 0 otherwise. Put $f(\mathbf{x}, m) = m$ if $\mathbf{x} \in \mathcal{R}$, $f(\mathbf{x}, m) = 0$ otherwise. Then

$$N = \sum_{i=1}^{\infty} f(\xi_i, M_i).$$

The mean of N may be found by applying formula (4.4), resulting in $E(N) = \lambda \|\mathcal{R}\| E(M)$ where by (4.8), $M \equiv I(S \text{ isolated})$. Formula (4.33) is immediate. This argument does not require the driving point process to be Poisson, although computation of the value of p_1 in (4.33) does need that type of information.

A longer but in some ways more direct approach to (4.33) could go as follows. Assume that the shape S is essentially bounded, with $P(\text{rad} S < r) = 1$ for some $r > 0$. Suppose the observation region $\mathcal{R}$ is bounded, and let $\mathcal{A}$ be a set sufficiently large to contain

$$\mathcal{R}^{2r} \equiv \left\{ \mathbf{y} \in \mathbb{R}^k : |\mathbf{x} - \mathbf{y}| < 2r \text{ for some } \mathbf{x} \in \mathcal{R} \right\}.$$

Any random set in the Boolean model that intersects a random set centered within $\mathcal{R}$ must be centered within $\mathcal{A}$. Let M' be a Poisson random variable with mean $\lambda\|\mathcal{A}\|$. Conditional on M', let $\xi_{(1)}, \ldots, \xi_{(M')}$ be independent and uniformly distributed over $\mathcal{A}$, and let $S_{(1)}, \ldots, S_{(M')}$ be independent copies of S, independent also of the $\xi_{(i)}$'s. Then N has the same distribution as

$$N' \equiv \sum_{i=1}^{M'} I(\xi_{(i)} + S_{(i)} \text{ isolated}, \xi_{(i)} \in \mathcal{R}),$$

where the series is interpreted as 0 if $M' = 0$. Since $E(M') = \lambda\|\mathcal{A}\|$ and $P(\xi_{(i)} \in \mathcal{R}) = \|\mathcal{R}\|/\|\mathcal{A}\|$ then

$$E(N) = E(N') = E\left\{ \sum_{i=1}^{M'} P(\xi_{(i)} + S_{(i)} \text{ isolated} \mid \xi_{(i)} \in \mathcal{R}) P(\xi_{(i)} \in \mathcal{R}) \right\}$$

$$= E(M' p_1 \|\mathcal{R}\|/\|\mathcal{A}\|)$$

$$= \lambda p_1 \|\mathcal{R}\|.$$

The argument above quickly gives a formula for the variance of N. For if $a \equiv \|\mathcal{R}\|/\|\mathcal{A}\|$ then, since $\text{var}(M') = E(M') = \lambda\|\mathcal{A}\|$,

$$\text{var}(N) = \text{var}(N')$$

$$= E\{\text{var}(N'|M')\} + \text{var}\{E(N'|M')\}$$

$$= E\left\{ M'(ap_1 - a^2 p_1^2) + M'(M'-1)(a^2 p_2 - a^2 p_1^2) \right\} + \text{var}(M' a p_1)$$

$$= E(M')(ap_1 - a^2 p_1^2) + (EM')^2 (a^2 p_2 - a^2 p_1^2) + E(M') a^2 p_1^2$$

$$= \lambda\|\mathcal{R}\|p_1 + (\lambda\|\mathcal{R}\|)^2 (p_2 - p_1^2).$$

That is,

$$\text{var}(N) = \lambda\|\mathcal{R}\|p_1 + (\lambda\|\mathcal{R}\|)^2 (p_2 - p_1^2). \tag{4.34}$$

This result may be written down directly from formulas for second moment measures of marked point processes, but since those results would be of only limited use to us in the future we do not develop them here. Of course, a truncation argument shows that formula (4.34) applies generally, not just in the case where $\text{rad}(S)$ is bounded and $\mathcal{R}$ is bounded.

Similar formulas may be developed for the mean and variance of the number of clumps of any fixed order centered within a region $\mathcal{R}$. (In Section 4.5 we shall treat the problem of defining the center of a higher-order clump.) Under appropriate regularity conditions on shape distribution, both the mean and variance of these clump counts grow like $\|\mathcal{R}\|$ as $\mathcal{R}$ increases. In fact, the clump counts satisfy a central limit theorem with asymptotic variance equal to const.$\|\mathcal{R}\|$. We confine ourselves here to proving an approximate formula for the large-region variance of the number of first-order clumps in a Boolean model. The variances of numbers of higher-order clump counts are exceedingly complicated.

To formalize the argument we reintroduce the equivalent scale-changed model described in Chapter 3, Section 3.4. We work in k-dimensional space. Assume that the Poisson process driving the Boolean model has intensity λ, and that each shape is distributed as

$$\delta S = \{\delta \mathbf{x} : \mathbf{x} \in S\},$$

where S is a given random shape and $\delta = \delta(\lambda) \to 0$ as $\lambda \to \infty$. The quantities δ and λ are connected by the formula $\delta^k \lambda \to \rho$ as $\lambda \to \infty$, where $0 \leq \rho < \infty$. Let $\mathcal{C}(\delta, \lambda)$ denote this model, and let P be the probability measure associated with $\mathcal{C}(\delta, \lambda)$. Write $N = N(\mathcal{R})$ for the number of isolated sets from $\mathcal{C}(\delta, \lambda)$ whose centers lie in $\mathcal{R}$. Formulas (4.33) and (4.34) hold in the present case, with p_1 and p_2 given by

$$p_1(\delta, \lambda) = P(\delta S \text{ is isolated})$$

and

$$p_2(\delta, \lambda) = p_2(\delta, \lambda; \mathcal{R})$$
$$= P(\xi^{(1)} + \delta S^{(1)}, \, \xi^{(2)} + \delta S^{(2)} \text{ both isolated} | M \geq 2).$$

[Compare these expressions with (4.31) and (4.32).] We keep $\mathcal{R}$ fixed as λ increases and δ decreases.

If we scale up the Boolean model $\mathcal{C}(\delta, \lambda)$ by multiplying each dimension by the factor δ^{-1}, we obtain a new model in which each shape is distributed as S and the driving Poisson process has intensity $\delta^k \lambda$, which converges to ρ as $\lambda \to \infty$. Let $\mathcal{C}_0$ be a Boolean model in which shapes are distributed as S and the driving Poisson process has intensity ρ, and let P_0 be the probability measure associated with $\mathcal{C}_0$. Let $S_{(1)}$ and $S_{(2)}$ be independent copies of S, independent of $\mathcal{C}_0$. We say that $\mathbf{x} + S_{(1)}$ is isolated if it intersects none of the sets in $\mathcal{C}_0$, and that $\mathbf{x}_{(1)} + S_{(1)}$ and $\mathbf{x}_{(2)} + S_{(2)}$ are both isolated if these sets do not intersect one another or any of the sets in $\mathcal{C}_0$. It is easy to see that as $\lambda \to \infty$ and $\delta \to 0$ in such a manner that $\delta^k \lambda \to \rho$,

$$p_1(\delta, \lambda) \to P_0(S_{(1)} \text{ isolated}).$$

It is a little more difficult to show that

$$\lambda \|\mathcal{R}\| \left\{ p_2(\delta, \lambda) - p_1^2(\delta, \lambda) \right\}$$

$$\to \rho \left[\int_{\mathbb{R}^k} \left\{ P_0(\mathbf{x} + S_{(1)}, S_{(2)} \text{ both isolated}) \right. \right.$$

$$\left. - P_0(S_{(1)} \text{ isolated})^2 \right\} d\mathbf{x}$$

$$+ 4 P_0(S_{(1)} \text{ isolated}) \int_{\mathbb{R}^k} P_0 \left\{ (\mathbf{x} + S_{(1)}) \cap S_{(2)} \neq \phi; \right.$$

$$\left. \left. S_{(2)} \text{ isolated} \right\} d\mathbf{x} \right]. \tag{4.35}$$

At the end of this section we shall prove (4.35) in the case where $\text{rad}(S)$ is essentially bounded. A derivation in the case $E\{\bar{v}(S)^2\} < \infty$, where

$$\bar{v}(S) \equiv \inf \{\|T\| : S \subseteq T, \text{ and } T \text{ is a sphere}\},$$

may be found in Hall (1986).

Assuming (4.35) for the present, we may deduce the following theorem from (4.33) and (4.34). Let $\mathcal{R}$ be a bounded Borel set with $\|\mathcal{R}\| > 0$.

Theorem 4.7. *Asssume $E\{\bar{v}(S)^2\} < \infty$, and let N denote the number of isolated sets in the Boolean model $\mathcal{C}(\delta, \lambda)$ defined just above whose centers lie within $\mathcal{R}$. If $\delta \to 0$ as $\lambda \to \infty$ in such a manner that $\delta^k \lambda \to \rho$ where $0 \le \rho < \infty$, then*

$$(\lambda \|\mathcal{R}\|)^{-1} E(N) \to P_0(S_{(1)} \text{ isolated})$$

and

$$(\lambda \|\mathcal{R}\|)^{-1} \text{var}(N) \to \kappa^2(S) \equiv P_0(S_{(1)} \text{ isolated})$$

$$+ \rho \int_{\mathbb{R}^k} \left\{ P_0(\mathbf{x} + S_{(1)}, S_{(2)} \text{ both isolated}) \right.$$

$$\left. - P_0(S_{(1)} \text{ isolated})^2 \right\} d\mathbf{x} + 4\rho P_0(S_{(1)} \text{ isolated})$$

$$\times \int_{\mathbb{R}^k} P_0 \left\{ (\mathbf{x} + S_{(1)}) \cap S_{(2)} \neq \phi; S_{(2)} \text{ isolated} \right\} d\mathbf{x}$$

as $\lambda \to \infty$. Both integrals here are absolutely convergent.

The scale-changed Boolean model $C(\delta, \lambda)$ is a flexible and powerful tool for developing approximations in a variety of circumstances, from high-intensity models (Chapter 3, Sections 3.5 and 3.6) through moderate-intensity models (the work above and Chapter 3, Section 3.4) to low-intensity models (end of Section 3.6 and next section). By adjusting the values of δ and λ we may study many different situations. For example, consider a Boolean model in which shapes distributed as S are centered at points of a stationary Poisson process with intensity ρ (fixed). Suppose we observe the Boolean model pattern within the large region

$$l\mathcal{R} \equiv \{l\mathbf{x} : \mathbf{x} \in \mathcal{R}\},$$

which is simply $\mathcal{R}$ expanded by the factor l along each dimension. Let N_l denote the number of isolated sets centered within $l\mathcal{R}$. By (4.33),

$$E(N_l) = \lambda \|l\mathcal{R}\| P_0(S_{(1)} \text{ isolated}),$$

and by Theorem 4.7,

$$\mathrm{var}(N_l) \sim \lambda \|l\mathcal{R}\| \kappa^2(S)$$

as l increases.

Before passing to a proof of (4.35) we provide a little information about calculation of the quantities above. In the case $k = 2$, if S is convex and isotropic with mean area $\alpha \equiv E(\|S\|_2)$ and mean perimeter $\beta \equiv E(\|\partial S\|_1)$, then

$$P_0(S_{(1)} \text{ isolated}) = e^{-\alpha\rho} E\left[\exp\left\{-\rho(\|S\|_2 + (2\pi)^{-1}\beta\|\partial S\|_1)\right\}\right],$$

and the quantity p_1 defined at (4.31) is given by

$$p_1 \equiv e^{-\alpha\lambda} E\left[\exp\left\{-\lambda(\|S\|_2 + (2\pi)^{-1}\beta\|\partial S\|_1)\right\}\right]$$

[see (4.16) in Section 4.2]. Calculation of $\kappa^2(S)$ is more difficult. We treat only the case where S is a fixed sphere of radius r. Let v_k and s_k denote, respectively, the k-dimensional content and the $(k-1)$-dimensional surface content of a unit k-dimensional sphere.

Then

$$P_0(S_{(1)} \text{ isolated}) = \exp\left\{-\rho(2r)^k v_k\right\},$$

$$P_0\left\{(\mathbf{x}+S_{(1)})\cap S_{(2)} \neq \phi;\ S_{(2)} \text{ isolated}\right\}$$
$$= \begin{cases} P_0(S_{(1)} \text{ isolated}) & \text{if } |\mathbf{x}| \leq 2r \\ 0 & \text{if } |\mathbf{x}| > 2r, \end{cases}$$

and

$$P_0(\mathbf{x}+S_{(1)},\ S_{(2)} \text{ both isolated})$$
$$= \begin{cases} 0 & \text{if } |\mathbf{x}| \leq 2r \\ P_0(S_{(1)} \text{ isolated})^2 \exp\{\rho(2r)^k B(|\mathbf{x}|/4r)\} & \text{if } 2r < |\mathbf{x}| \leq 4r \\ P_0(S_{(1)} \text{ isolated})^2 & \text{if } |\mathbf{x}| > 4r, \end{cases}$$

where

$$B(x) = \begin{cases} 2\pi^{(k-1)/2}\{\Gamma(\tfrac{1}{2}+\tfrac{1}{2}k)\}^{-1}\int_x^1 (1-y^2)^{(k-1)/2}\,dy & \text{if } 0 \leq x \leq 1 \\ 0 & \text{if } x > 1 \end{cases}$$

$$(4.36)$$

denotes the content of the lens of intersection of two unit k-dimensional spheres whose centers are distant $2x$ apart. Therefore

$$\int_{\mathbb{R}^k} \left\{ P_0(\mathbf{x}+S_{(1)}, S_{(2)} \text{ both isolated}) - P_0(S_{(1)} \text{ isolated})^2 \right\} d\mathbf{x}$$

$$= -P_0(S_{(1)} \text{ isolated})^2 \int_{|\mathbf{x}| \leq 2r} d\mathbf{x}$$

$$+ P_0(S_{(1)} \text{ isolated})^2 \int_{2r < |\mathbf{x}| \leq 4r} \left[\exp\left\{\rho(2r)^k B(|\mathbf{x}|/4r)\right\} - 1\right] d\mathbf{x}$$

$$= -P_0(S_{(1)} \text{ isolated})^2 (2r)^k v_k$$

$$+ P_0(S_{(1)} \text{ isolated})^2 (4r)^k s_k \int_{\frac{1}{2}}^1 x^{k-1} \left[\exp\left\{\rho(2r)^k B(x)\right\} - 1\right] dx$$

and

$$\int_{\mathbb{R}^k} P_0\left\{(\mathbf{x}+S_{(1)})\cap S_{(2)} \neq \phi;\ S_{(2)} \text{ isolated}\right\} d\mathbf{x}$$

$$= P_0(S_{(1)} \text{ isolated}) \int_{|\mathbf{x}| \leq 2r} d\mathbf{x}$$

$$= \exp\left\{-\rho(2r)^k v_k\right\}(2r)^k v_k.$$

Combining these results we see that if S is a fixed sphere of radius r,

$$\kappa^2(S) = \exp\left\{-\rho(2r)^k v_k\right\} + \rho(4r)^k s_k \exp\left\{-2\rho(2r)^k v_k\right\}$$
$$\times \int_{\frac{1}{2}}^{1} x^{k-1}\left[\exp\left\{\rho(2r)^k B(x)\right\} - 1\right] dx$$
$$+ 3\rho(2r)^k v_k \exp\left\{-2\rho(2r)^k v_k\right\}.$$

Proof of (4.35) in the case where rad(S) *is essentially bounded.*
First note that if rad$(S) < r$ with probability 1 then the integrands of both integrals on the right-hand side of (4.35) are 0 if $|\mathbf{x}| > 4r$. Therefore both integrals converge.

Returning to the definitions of p_1 and p_2 in formulas (4.31) and (4.32), let M equal the number of points from the driving Poisson process that lie within $\mathcal{R}$, and conditional on M let the associated random sets in $\mathcal{C}(\delta, \lambda)$ be $\xi^{(1)} + \delta S^{(1)}, \dots, \xi^{(M)} + \delta S^{(M)}$, numbered in completely random order. Then as $\lambda \to \infty$,

$$P(M \geq 2) = 1 - (1 + \lambda\|\mathcal{R}\|)\exp(-\lambda\|\mathcal{R}\|) = 1 + o(\lambda^{-1})$$

[see formula (4.6), Section 4.1],

$$p_2(\delta, \lambda)$$
$$= P(\xi^{(1)} + \delta S^{(1)}, \xi^{(2)} + \delta S^{(2)} \text{ both isolated}; M \geq 2)/P(M \geq 2)$$
$$= P(\xi^{(1)} + \delta S^{(1)}, \xi^{(2)} + \delta S^{(2)} \text{ both isolated}; M \geq 2) + o(\lambda^{-1}),$$

and

$$P(\xi^{(1)} + \delta S^{(1)}, \xi^{(2)} + \delta S^{(2)} \text{ both isolated}; M \geq 2)$$
$$= \sum_{m=2}^{\infty} P(\xi^{(1)} + \delta S^{(1)}, \xi^{(2)} + \delta S^{(2)}$$
$$\text{both isolated}|M = m)P(M = m)$$
$$= \sum_{m=2}^{\infty} P(\xi_{(1)} + \delta S_{(1)}, \xi_{(2)} + \delta S_{(2)}$$
$$\text{both isolated}|M = m - 2)P(M = m),$$

where $\xi_{(1)}$ and $\xi_{(2)}$ denote independent random variables uniformly distributed on $\mathcal{R}$, and $S_{(1)}$ and $S_{(2)}$ are independent copies of S, all being independent of one another and of $\mathcal{C}(\delta, \lambda)$. Combining these results we conclude that

$$p_2(\delta,\lambda)+o(\lambda^{-1})=\sum_{m=2}^{\infty}P(\xi_{(1)}+\delta S_{(1)},\xi_{(2)}+\delta S_{(2)}$$

$$\text{both isolated}|M=m-2)P(M=m)$$

$$=\sum_{m=2}^{\infty}P(\xi_{(1)}+\delta S_{(1)},\xi_{(2)}+\delta S_{(2)}$$

$$\text{both isolated}|M=m-2)P(M=m-2)$$

$$+\sum_{m=2}^{\infty}P(\xi_{(1)}+\delta S_{(1)},\xi_{(2)}+\delta S_{(2)}\text{ both isolated}|M=m-2)$$

$$\times\{P(M=m)-P(M=m-2)\}$$

$$=P(\xi_{(1)}+\delta S_{(1)},\xi_{(2)}+\delta S_{(2)}\text{ both isolated})+A,$$

where

$$A=A(\delta,\lambda)$$

$$\equiv\sum_{m=0}^{\infty}P(\xi_{(1)}+\delta S_{(1)},\xi_{(2)}+\delta S_{(2)}\text{ both isolated}|M=m)P(M=m)$$

$$\times\left\{\frac{\mu^2}{(m+1)(m+2)}-1\right\}$$

and $\mu\equiv E(M)=\lambda\|\mathcal{R}\|$. Therefore (4.35) follows from the relations

$$\lambda\|\mathcal{R}\|\left\{P(\xi_{(1)}+\delta S_{(1)},\xi_{(2)}+\delta S_{(2)}\text{ both isolated})\right.$$

$$\left.-\prod_{i=1}^{2}P(\xi_{(i)}+\delta S_{(i)}\text{ isolated})\right\}$$

$$\to\rho\int_{\mathbb{R}^k}\left\{P_0(\mathbf{x}+S_{(1)},S_{(2)}\text{ both isolated})\right.$$

$$\left.-P_0(S_{(1)}\text{ isolated})^2\right\}d\mathbf{x}\qquad(4.37)$$

and

$$\lambda\|\mathcal{R}\|A(\delta,\lambda)\to4\rho P_0(S_{(1)}\text{ isolated})\int_{\mathbb{R}^k}P_0\{(\mathbf{x}+S_{(1)})\cap S_{(2)}\neq\phi;$$

$$S_{(2)}\text{ isolated}\}d\mathbf{x}.\qquad(4.38)$$

We prove these in Steps (i) and (ii) below.

Step (i). Proof of (4.37). Rescale the Boolean model by the factor δ^{-1} along each dimension, resulting in a new model in which the driving Poisson process has intensity $\delta^k \lambda$ and each shape is distributed as S. Write P_1 for the probability measure associated with this new coverage process. The left-hand side of (4.37) equals

$$\lambda \|\mathcal{R}\| \|\delta^{-1}\mathcal{R}\|^{-2} \iint_{(\delta^{-1}\mathcal{R})^2} \Big\{ P_1\big(\mathbf{x}_{(1)} + S_{(1)}, \mathbf{x}_{(2)} + S_{(2)} \text{ both isolated}\big)$$

$$- \prod_{i=1}^{2} P_1\big(\mathbf{x}_{(i)} + S_{(i)} \text{ isolated}\big) \Big\} d\mathbf{x}_{(1)} \, d\mathbf{x}_{(2)}$$

$$= \delta^k \lambda \int_{\mathcal{R}_\delta} \delta^k \|\mathcal{R}\|^{-1} \|\delta^{-1}\mathcal{R} \cap (\delta^{-1}\mathcal{R} - \mathbf{x})\|$$

$$\times \Big\{ P_1\big(\mathbf{x} + S_{(1)}, S_{(2)} \text{ both isolated}\big) - P_1\big(S_{(1)} \text{ isolated}\big)^2 \Big\} d\mathbf{x},$$

where $\mathcal{R}_\delta \equiv \{\mathbf{x}_1 - \mathbf{x}_2 : \mathbf{x}_1, \mathbf{x}_2 \in \delta^{-1}\mathcal{R}\}$. The integrand of the last-written integral equals 0 if $|\mathbf{x}| > 4r$. For fixed $\mathbf{x}$, $\|\delta^{-1}\mathcal{R} \cap (\delta^{-1}\mathcal{R} - \mathbf{x})\| \sim \delta^{-k}\|\mathcal{R}\|$ as $\delta \to 0$, and so for fixed $\mathbf{x}$ the integrand converges to

$$P_0\big(\mathbf{x} + S_{(1)}, S_{(2)} \text{ both isolated}\big) - P_0\big(S_{(1)} \text{ isolated}\big)^2.$$

The set $\mathcal{R}_\delta$ increases to $\mathbb{R}^k$ as $\delta \to 0$. [See Halmos (1950, Lemma B, page 68) for the case $k = 1$.] Therefore (4.37) follows by the dominated convergence theorem.

Step (ii). Proof of (4.38). Using Markov's inequality we obtain for any $\epsilon, l > 0$:

$$P\big(|M - \mu| > \mu^{(1/2)+\epsilon}\big) \leq E\big(|M - \mu|/\mu^{(1/2)+\epsilon}\big)^l$$
$$= O\big(\mu^{(1/2)l}/\mu^{((1/2)+\epsilon)l}\big) = O\big(\mu^{-\epsilon l}\big).$$

Therefore if $\nu \equiv \mu^{(1/2)+\epsilon}$ for some positive ϵ,

$$A(\delta, \lambda) = \sum_{m:|m-\mu|\leq\nu} P\big(\boldsymbol{\xi}_{(1)} + \delta S_{(1)}, \boldsymbol{\xi}_{(2)} + \delta S_{(2)} \text{ both isolated}|M = m\big)$$

$$\times P(M = m)\Big\{ \frac{\mu^2}{(m+1)(m+2)} - 1 \Big\} + o(\mu^{-1}). \qquad (4.39)$$

Let $n = m - \mu$. Then

$$\mu^2/(m+1)(m+2) = \{1 + (n+1)/\mu\}^{-1}\{1 + (n+2)/\mu\}^{-1}$$
$$= 1 - (2n+3)/\mu + 3(n/\mu)^2$$
$$+ O\big\{|n|^3/\mu^3 + (|n|+1)/\mu^2\big\}.$$

Substituting into (4.39) and choosing $0 < \epsilon < \frac{1}{6}$, we see that

$$
\begin{aligned}
A(\delta,\lambda) &= \sum_{m:\,|m-\mu|\le\nu} P(\xi_{(1)} + \delta S_{(1)}, \xi_{(2)} + \delta S_{(2)} \\
&\qquad \text{both isolated;}\ M = m) \\
&\quad \times \left\{ -(2n+3)/\mu + 3(n/\mu)^2 \right\} + o(\mu^{-1}) \\
&= \sum_{m=0}^{\infty} P(\xi_{(1)} + \delta S_{(1)}, \xi_{(2)} + \delta S_{(2)}\ \text{both isolated;}\ M = m) \\
&\quad \times \left\{ -(2n+3)/\mu + 3(n/\mu)^2 \right\} + o(\mu^{-1}) \\
&= -3\mu^{-1} P(\xi_{(1)} + \delta S_{(1)}, \xi_{(2)} + \delta S_{(2)}\ \text{both isolated}) \\
&\quad - 2\mu^{-1} \sum_{m=0}^{\infty} P(\xi_{(1)} + \delta S_{(1)}, \xi_{(2)} + \delta S_{(2)} \\
&\qquad \text{both isolated;}\ M = m)(m - \mu) \\
&\quad + 3\mu^{-2} \sum_{m=0}^{\infty} P(\xi_{(1)} + \delta S_{(1)}, \xi_{(2)} + \delta S_{(2)} \\
&\qquad \text{both isolated;}\ M = m)(m - \mu)^2 \\
&\quad + o(\mu^{-1}).
\end{aligned}
$$

By (4.37),

$$
\begin{aligned}
P(\xi_{(1)} + \delta S_{(1)}, \xi_{(2)} + \delta S_{(2)}\ \text{both isolated}) \\
= P(\delta S_{(1)}\ \text{isolated})^2 + O(\lambda^{-1}) \\
= P_0(S_{(1)}\ \text{isolated})^2 + o(1),
\end{aligned}
$$

and similarly it may be shown that

$$
\begin{aligned}
\sum_{m=0}^{\infty} P(\xi_{(1)} + \delta S_{(1)}, \xi_{(2)} + \delta S_{(2)} \\
\text{both isolated}|M = m)P(M = m)(m - \mu)^2 \\
\sim \sum_{m=0}^{\infty} P_0(S_{(1)}\ \text{isolated})^2 P(M = m)(m - \mu)^2 \\
= P_0(S_{(1)}\ \text{isolated})^2 E(M - \mu)^2 \\
= \mu P_0(S_{(1)}\ \text{isolated})^2.
\end{aligned}
$$

Therefore

$$A(\delta, \lambda) \equiv 2A_1(\delta, \lambda) + o(\mu^{-1}), \tag{4.40}$$

where

$$A_1(\delta, \lambda) = -\mu^{-1} \sum_{m=0}^{\infty} P(\boldsymbol{\xi}_{(1)} + \delta S_{(1)}, \boldsymbol{\xi}_{(2)} + \delta S_{(2)}$$

$$\text{both isolated}; \ M = m)(m - \mu).$$

An application of Abel's method of summation (Whittaker and Watson 1927, p. 17) gives

$$A_1(\delta, \lambda) = \sum_{m=0}^{\infty} P(\boldsymbol{\xi}_{(1)} + \delta S_{(1)}, \boldsymbol{\xi}_{(2)} + \delta S_{(2)} \text{ both isolated} | M = m)$$

$$\times \{P(M = m) - P(M = m - 1)\}$$

$$= \sum_{m=0}^{\infty} P(M = m) \Big\{ P(\boldsymbol{\xi}_{(1)} + \delta S_{(1)}, \boldsymbol{\xi}_{(2)} + \delta S_{(2)}$$

$$\text{both isolated} | M = m)$$

$$- P(\boldsymbol{\xi}_{(1)} + \delta S_{(1)}, \boldsymbol{\xi}_{(2)} + \delta S_{(2)}$$

$$\text{both isolated} | M = m + 1) \Big\}.$$

Let $\boldsymbol{\xi}_{(3)} + \delta S_{(3)}$ denote an independent copy of $\boldsymbol{\xi}_{(1)} + \delta S_{(1)}$, independent of everything defined so far. Then

$$P(\boldsymbol{\xi}_{(1)} + \delta S_{(1)}, \boldsymbol{\xi}_{(2)} + \delta S_{(2)} \text{ both isolated} | M = m + 1)$$

$$= P(\boldsymbol{\xi}_{(1)} + \delta S_{(1)}, \boldsymbol{\xi}_{(2)} + \delta S_{(2)}$$

$$\text{both isolated}; \ \boldsymbol{\xi}_{(3)} + \delta S_{(3)} \text{ intersects}$$

$$\text{neither } \boldsymbol{\xi}_{(1)} + \delta S_{(1)} \text{ nor } \boldsymbol{\xi}_{(2)} + \delta S_{(2)} | M = m),$$

and so

$$A_1(\delta,\lambda) = \sum_{m=0}^{\infty} P(M=m)P(\boldsymbol{\xi}_{(1)}+\delta S_{(1)}, \boldsymbol{\xi}_{(2)}+\delta S_{(2)} \text{ both isolated};$$

$$\boldsymbol{\xi}_{(3)}+\delta S_{(3)} \text{ intersects at least one of}$$

$$\boldsymbol{\xi}_{(1)}+\delta S_{(1)}, \boldsymbol{\xi}_{(2)}+\delta S_{(2)}|M=m)$$

$$= P(\boldsymbol{\xi}_{(1)}+\delta S_{(1)}, \boldsymbol{\xi}_{(2)}+\delta S_{(2)} \text{ both isolated}; \boldsymbol{\xi}_{(3)}+\delta S_{(3)}$$

$$\text{intersects at least one of } \boldsymbol{\xi}_{(1)}+\delta S_{(1)}, \boldsymbol{\xi}_{(2)}+\delta S_{(2)})$$

$$\sim 2P\left\{\boldsymbol{\xi}_{(1)}+\delta S_{(1)}, \boldsymbol{\xi}_{(2)}+\delta S_{(2)} \text{ both isolated};\right.$$

$$\left.(\boldsymbol{\xi}_{(3)}+\delta S_{(3)})\cap(\boldsymbol{\xi}_{(2)}+\delta S_{(2)}) \neq \phi\right\}$$

$$\sim 2P(\boldsymbol{\xi}_{(1)}+\delta S_{(1)} \text{ isolated})$$

$$\times P\left\{(\boldsymbol{\xi}_{(3)}+\delta S_{(3)})\cap(\boldsymbol{\xi}_{(2)}+\delta S_{(2)}) \neq \phi; \boldsymbol{\xi}_{(2)}+\delta S_{(2)} \text{ isolated}\right\}$$

$$= 2P(\delta S_{(1)} \text{ isolated})\|\mathcal{R}\|^{-2}$$

$$\times \iint_{\mathcal{R}^2} P\{(\mathbf{x}_{(1)}+\delta S_{(1)})\cap(\mathbf{x}_{(2)}+\delta S_{(2)}) \neq \phi;$$

$$\mathbf{x}_{(2)}+\delta S_{(2)} \text{ isolated}\}\, d\mathbf{x}_{(1)}\, d\mathbf{x}_{(2)}$$

$$\sim 2P_0(S_{(1)} \text{ isolated})\delta^k\|\mathcal{R}\|^{-1}$$

$$\times \int_{\mathbb{R}^k} P_0\{(\mathbf{x}+S_{(1)})\cap S_{(2)} \neq \phi; S_{(2)} \text{ isolated}\}\, d\mathbf{x},$$

the last line following as in Step (i) above. The desired result (4.38) follows on combining this estimate with (4.40). $\square$

4.5. CLUMPS OF VARIOUS ORDERS IN SPARSE BOOLEAN MODELS

The previous section described properties of "isolated sets," or clumps of order one. Now we examine clumps of arbitrary finite order. To ensure reasonable mathematical tractability we concentrate on the case of low-intensity, or sparse Boolean models, in which the covered portion of a region $\mathcal{R}$, expressed as a proportion of the total content of the region, is small. We may describe sparse models using the scale change device of the previous section, in which the model $\mathcal{C}(\delta,\lambda) \equiv \{\boldsymbol{\xi}_i+\delta S_i, \ i \geq 1\}$ is driven by a Poisson process $\mathcal{P} = \{\boldsymbol{\xi}_i, i \geq 1\}$ with intensity λ and each shape δS_i has the distribution of

δS. In this formulation S is a given random set, and $\delta = \delta(\lambda)$ and λ vary in such a manner that

$$\eta \equiv \delta^k \lambda \to 0.$$

Our initial results depend on δ and λ only through η, and so we impose conditions on δ and λ only through η.

The hallmark of sparse Boolean models is that most of the random sets are isolated, with smaller numbers of sets occurring in clumps of orders 2, 3, and so on. Sparse Boolean models can arise in dust-particle counting experiments, for example, as mathematical models behind statistical tests for compliance with hygiene standards. In these circumstances it may be difficult to distinguish between isolated and nonisolated sets, particularly if the sizes and configurations of shapes are irregular. On the other hand, it is quite easy to count the total number of clumps or parts of clumps occurring within a given region. We may wish to compare clump count (known) with particle count (unknown), and correct for any systematic error.

Throughout this section we assume that the coverage process is a Boolean model and that the random shape S is connected with probability 1. A random set $\xi_i + \delta S_i$ is said to be a member of a clump of order n if there exist $n - 1$ different sets $\xi_{j_l} + \delta S_{j_l}$ $(1 \le l \le n - 1)$, with indices all different from one another and from i, such that the union

$$(\xi_i + \delta S_i) \cup \left\{ \bigcup_{l=1}^{n-1} (\xi_{j_l} + \delta S_{j_l}) \right\}$$

is a connected set and is disjoint from each set $\xi_m + \delta S_m$ whose index is not included in the set $\{i, j_1, \ldots, j_{n-1}\}$ (see Figure 4.14). Thus, a clump of order n cannot be part of a clump of order $n + 1$, in our notation. Some authors use a different definition of clump order.

The notion of an "arbitrary" random set was defined in Section 4.1. Let $p(n)$ denote the probability that an arbitrary random set is part of a clump of order n. We shall develop approximations to $p(n)$ in the case of sparse Boolean models. The theory is much simpler if S is a sphere in $\mathbb{R}^k$ of random radius R, centered at the origin, and so we begin there. The three numbered points below describe only that case.

Consider the Boolean model $\mathcal{C}(\delta, \lambda) \equiv \{\xi_i + \delta S_i, \, i \ge 1\}$ in $\mathbb{R}^k$, in which the driving Poisson process has intensity λ and shapes are

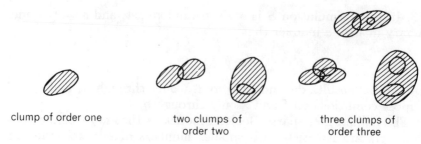

clump of order one   two clumps of   three clumps of
                                    order two   order three

Figure 4.14. Clumps of various orders.

distributed as δS, where S is the sphere

$$S \equiv \left\{ \mathbf{x} \in \mathbb{R}^k : |\mathbf{x}| \leq R \right\}$$

of random radius R. For ease of notation, take S to be independent of $\mathcal{C}(\delta, \lambda)$. We say that δS is isolated if it intersects no sets $\boldsymbol{\xi}_i + \delta S_i$ from $\mathcal{C}(\delta, \lambda)$.

(i) Clumps of Order One. The probability that an arbitrary set in the Boolean model is isolated equals the chance that δS is isolated (see Section 4.1). Given that $\mathrm{rad}\,(S) = r$, the chance that δS is isolated equals

$$\exp\left\{ -\lambda v_k E(\delta R + \delta r)^k \right\} = \exp\left\{ -\eta v_k E(R + r)^k \right\},$$

where $\eta = \delta^k \lambda$. [To prove this result in detail, use a minor modification of the marked point process argument leading to (4.7).] Therefore the probability that an arbitrary set in $\mathcal{C}(\delta, \lambda)$ is (part of) a clump of order one equals

$$p(1) = P(S \text{ isolated}) = E\left\{ P(S \text{ isolated} \mid \mathrm{rad}\,S = R) \right\}$$
$$= \int_0^\infty \exp\left\{ -\eta v_k E(R + r)^k \right\} dP(R \leq r). \qquad (4.41)$$

Assume that $E(\|S\|^3) < \infty$, or equivalently, that $E(R^{3k}) < \infty$. Under the assumption that $\eta \to 0$ as δ and λ vary, we may expand $p(1)$

as a power series in η at least as far as terms of order η^3:

$$
\begin{aligned}
p(1) &= \int_0^\infty \left[1 - \eta v_k E(R+r)^k \right. \\
&\qquad \left. + \tfrac{1}{2} \left\{ \eta v_k E(R+r)^k \right\}^2 \right] dP(R \leq r) + O(\eta^3) \\
&= 1 - \eta v_k E(R_1 + R_2)^k \\
&\qquad + \tfrac{1}{2} \eta^2 v_k^2 E \left[E \left\{ (R_1 + R_2)^k \mid R_2 \right\} \right]^2 + O(\eta^3), \qquad (4.42)
\end{aligned}
$$

where R_1 and R_2 are independent copies of R. In the special case of $k = 2$ dimensions we have

$$
\begin{aligned}
p(1) &= 1 - 2\pi(\mu_1^2 + \mu_2)\eta \\
&\qquad + \tfrac{1}{2}\pi^2 (8\mu_1^2 \mu_2 + 4\mu_1 \mu_3 + 3\mu_2^2 + \mu_4)\eta^2 + O(\eta^3),
\end{aligned}
$$

where $\mu_i \equiv E(R^i)$. If sphere radius is fixed and equal to r then for general k,

$$
p(1) = 1 - v_k(2r)^k \eta + \tfrac{1}{2} v_k^2 (2r)^{2k} \eta^2 + O(\eta^3). \qquad (4.43)
$$

(ii) Clumps of Order Two. Let $T(r)$ denote the closed sphere of radius r centered at the origin, and let $\mathbf{x}$ be any k-vector of length $x > 0$. Define

$$
f(r_1, r_2; x) \equiv E\left[\| T(r_1 + R) \cup \{\mathbf{x} + T(r_2 + R)\} \| \right]
$$

to be the mean content of the union of two spheres whose centers are distant x apart and whose respective radii are $r_1 + R$ and $r_2 + R$. The probability that no random spheres from the Boolean model intersect the set $T(\delta r_1) \cup \{\delta \mathbf{x} + T(\delta r_2)\}$ equals

$$
\exp\left\{ -\delta^k \lambda f(r_1, r_2; x) \right\}.
$$

The chance that a given fixed sphere T_1 of radius δr_1 intersects a sphere in the Boolean model whose radius lies between δr_2 and $\delta(r_2 + dr_2)$, and whose center is distant between δx and $\delta(x + dx)$ from the center of T_1, equals

$$
P(r_2 < R \leq r_2 + dr_2) \lambda s_k (\delta x)^{k-1} \delta \, dx
$$

if $x \leq r_1 + r_2$; 0 otherwise. Therefore the probability that an arbitrary random set is part of a clump of order two equals

$$
p(2) \equiv \int_0^\infty dP(R_1 \leq r_1) \int_0^\infty dP(R_2 \leq r_2)
$$

$$
\times \int_0^{r_1+r_2} \lambda s_k \delta^k x^{k-1} \exp\left\{-\delta^k \lambda f(r_1, r_2; x)\right\} dx
$$

$$
= \eta s_k E\left[\int_0^{R_1+R_2} x^{k-1} \exp\left\{-\eta f(R_1, R_2; x)\right\} dx\right].
$$

If $E(\|S\|^3) < \infty$ then, under our assumption that $\eta \equiv \delta^k \lambda \to 0$ as δ and λ vary, we may expand $p(2)$ at least as far as terms of order η^3:

$$
p(2) = \eta s_k E\left[\int_0^{R_1+R_2} x^{k-1}\left\{1 - \eta f(R_1, R_2; x)\right\} dx\right] + O(\eta^3)
$$

$$
= \eta v_k E(R_1 + R_2)^k
$$

$$
- \eta^2 s_k E\left\{\int_0^{R_1+R_2} x^{k-1} f(R_1, R_2; x)\, dx\right\} + O(\eta^3). \quad (4.44)
$$

(Note that $k^{-1} s_k = v_k$.)

The integral appearing in the term of order η^2 in (4.44) is most easily calculated in the case where sphere radius is fixed and equal to r, say. Then

$$
E\left\{\int_0^{R_1+R_2} x^{k-1} f(R_1, R_2; x)\, dx\right\}
$$

$$
= \int_0^{2r} x^{k-1} (2r)^k \left\{2v_k - B(x/4r)\right\} dx
$$

$$
= (2r)^{2k}\left\{2k^{-1} v_k - \int_0^1 x^{k-1} B(x/2)\, dx\right\},
$$

where $B(x)$ denotes the content of the lens of intersection of two unit k-dimensional spheres whose centers are distant $2x$ apart [see (4.36)]. Substituting into (4.44) and noting that $k^{-1} s_k = v_k$, we see that in the case where the spheres are of fixed radius r,

$$
p(2) = v_k (2r)^k \eta - v_k^2 (2r)^{2k}\left\{2 - k v_k^{-1} \int_0^1 x^{k-1} B(x/2)\, dx\right\} \eta^2 + O(\eta^3).
$$

$$
(4.45)
$$

(iii) Clumps of Order Three. We claim that if $E(\|S\|^3) < \infty$ then the probability that an arbitrary random sphere T_1 is part of a clump of order four or more equals $O(\eta^3)$ as $\eta \to 0$. To prove this, observe that the chance that the radius of T_1 is in the range $((n_1 - 1)\delta, n_1\delta]$ equals $q_1(n_1) \equiv P(n_1 - 1 < R \le n_1)$. Given that T_1 has radius in this range, the chance that at least one sphere T_2 with radius in the range $((n_2 - 1)\delta, n_2\delta]$ intersects T_2 is no greater than

$$1 - \exp\left[-\lambda P(n_2 - 1 \le R \le n_2)\{(n_1 + n_2)\delta\}^k v_k\right]$$
$$\le \eta v_k(n_1 + n_2)^k P(n_2 - 1 < R \le n_2) \equiv q_2(n_1, n_2),$$

say. (Use the marked point process and "Poisson-ness" arguments.) Given that spheres T_1 and T_2 have radii in these ranges and intersect, the chance that some sphere T_3 with radius in the range $((n_3 - 1)\delta, n_3\delta]$ intersects $T_1 \cup T_2$, is not greater than

$$\eta v_k(n_1 + 2n_2 + n_3)^k P(n_3 - 1 < R \le n_3) \equiv q_3(n_1, n_2, n_3).$$

Given that spheres T_1, T_2, and T_3 with radii in these ranges intersect, the chance that some sphere T_4 with radius in the range $((n_4 - 1)\delta, n_4\delta]$ intersects $T_1 \cup T_2 \cup T_3$ is not greater than

$$\eta v_k(n_1 + 2n_2 + 2n_3 + n_4)^k P(n_4 - 1 < R \le n_4) \equiv q_4(n_1, n_2, n_3, n_4).$$

The chance that an arbitrary sphere T_1 is part of a clump of order four or more is not greater than

$$\sum_{n_1=1}^{\infty} \sum_{n_2=1}^{\infty} \sum_{n_3=1}^{\infty} \sum_{n_4=1}^{\infty} q_1(n_1)q_2(n_1,n_2)q_3(n_1,n_2,n_3)q_4(n_1,n_2,n_3,n_4)$$
$$\le C_1\eta^3 \sum_{n_1=1}^{\infty} \sum_{n_2=1}^{\infty} \sum_{n_3=1}^{\infty} \sum_{n_4=1}^{\infty} (n_1^k + n_2^k)$$
$$\times (n_1^k + n_2^k + n_3^k)(n_1^k + n_2^k + n_3^k + n_4^k)$$
$$\times \prod_{i=1}^{4} P(n_i - 1 < R \le n_i)$$
$$\le C_2\eta^3$$

if $E(\|S\|^3) = v_k^3 E(R^{3k}) < \infty$, where C_1 and C_2 are positive constants not depending on η. This establishes the claim made in the first sentence of this paragraph.

In consequence, if $E(\|S\|^3) < \infty$ then

$$1 - p(1) - p(2) - p(3) = O(\eta^3). \qquad (4.46)$$

We may now deduce from (4.42) and (4.44) that

$$p(3) = \left[s_k E\left\{ \int_0^{R_1+R_2} x^{k-1} f(R_1, R_2; x)\, dx \right\} \right.$$
$$\left. - \tfrac{1}{2} v_k^2 E\left\{ E((R_1 + R_2)^k \mid R_2) \right\}^2 \right] \eta^2 + O(\eta^3).$$

If the spheres in the Boolean model have fixed radius r then this formula reduces to

$$p(3) = v_k^2 (2r)^{2k} \left\{ \tfrac{3}{2} - k v_k^{-1} \int_0^1 x^{k-1} B(x/2)\, dx \right\} \eta^2 + O(\eta^3). \qquad (4.47)$$

In the special case $k = 2$,

$$\int_0^1 x^{k-1} B(x/2)\, dx = 4 \int_0^1 x\, dx \int_{x/2}^1 (1 - y^2)^{1/2}\, dy$$
$$= \tfrac{1}{8}(4\pi - 3\sqrt{3}).$$

Substituting this result into (4.43), (4.45), (4.46), and (4.47) we see that when $k = 2$ and the "spheres" are disks of fixed radius r,

$$p(1) = 1 - 4\pi r^2 \eta + 8\pi^2 r^4 \eta^2 + O(\eta^3), \qquad (4.48)$$
$$p(2) = 4\pi r^2 \eta - 4\pi(4\pi + 3\sqrt{3}) r^4 \eta^2 + O(\eta^3), \qquad (4.49)$$
$$p(3) = 4\pi(2\pi + 3\sqrt{3}) r^4 \eta^2 + O(\eta^3), \qquad (4.50)$$

and

$$\sum_{n=4}^{\infty} p(n) = O(\eta^3) \qquad (4.51)$$

as $\eta \to 0$. We may deduce from work on infinite clumping in Section 4.7 that the probability that an arbitrary sphere is part of an infinite clump equals 0 for all sufficiently small η. For such values of η, $\sum_{n \geq 4} p(n)$ equals the probability that the clump containing an arbitrary sphere is of order four or more.

The expected number of spheres per unit content that are part of a clump of order n equals $\lambda p(n)$. If we count clumps once for each sphere they contain, then a clump of order n is counted exactly n times. Therefore the expected number of clumps of order n per unit content of $\mathbb{R}^k$ equals $\lambda n^{-1} p(n)$. Assuming that infinite clumps have

0 probability, the expected total number of clumps per unit content equals

$$\nu \equiv \lambda \sum_{n=1}^{\infty} n^{-1} p(n). \qquad (4.52)$$

More explicitly, suppose we have a reasonable way of defining the "center" of a clump. (The "right-hand center" definition introduced following Theorem 4.8 below is just one of many possibilities.) Then the expected number of clumps centered within any Borel-measurable set $\mathcal{R} \subseteq \mathbb{R}^k$ equals $\nu \|\mathcal{R}\|$. If the spheres are disks of fixed radius r in $k = 2$ dimensions, then it follows from (4.48)–(4.51) that

$$\nu = \lambda \left\{ 1 - 2\pi r^2 \eta + \tfrac{2}{3}\pi (4\pi - 3\sqrt{3}) r^4 \eta^2 + O(\eta^3) \right\}. \qquad (4.53)$$

If the spheres are of random radius distributed as R in $k \geq 1$ dimensions, then by (4.42) and (4.44),

$$\nu = \lambda \left\{ 1 - \tfrac{1}{2} v_k E(R_1 + R_2)^k \eta + O(\eta^2) \right\}, \qquad (4.54)$$

where R_1 and R_2 are independent copies of R. Formulas for ν will be discussed further in Section 4.6.

Now we drop our assumption that shapes are random radius spheres, and permit them to have more general configurations. Shapes generating the Boolean model $\mathcal{C}$ are distributed as δS, and the driving Poisson process has intensity λ. In this circumstance it can be quite tedious to derive power series expansions for clumping probabilities, like those in (4.42) and (4.44). The assumption that S be convex hardly simplifies matters, since a clump of order two or more is usually not convex.

One fundamental principle that emerges from our study of the special case where shapes are spheres is that the probability of a shape being part of a clump of order n appears to be $O(\eta^{n-1})$ as $\eta \to 0$. This property does in fact hold for general shape configurations. We begin by considering the case $n = 1$, which is by far the easiest.

Let $\mathcal{C}^\dagger$ denote the Boolean model in which the driving Poisson process has intensity λ and shapes are k-dimensional spheres centered at the origin with radii distributed as $\delta\mathrm{rad}(S)$. (Shapes generating $\mathcal{C}$ are distributed as δS.) Write $p(1)$ or $p^\dagger(1)$, respectively, for the probability that an arbitrary set in $\mathcal{C}$ or $\mathcal{C}^\dagger$ is isolated. Then

$$p^\dagger(1) \leq p(1) \leq 1. \qquad (4.55)$$

From formula (4.41) we see that

$$p^\dagger(1) = \int_0^\infty \exp\left\{-\eta v_k E(R+r)^k\right\} dP(R \le r),$$

where $R = \mathrm{rad}(S)$. Therefore if $E\{\bar{v}(S)\} < \infty$ then

$$p^\dagger(1) = 1 + O(\eta),$$

and so by (4.55),

$$p(1) = 1 + O(\eta)$$

as $\eta \to 0$.

We turn next to the case $n \ge 2$. The theorem below declares that if $E\{\bar{v}(S)^n\} < \infty$ then $p(n)$ is asymptotic to a constant multiple of η^{n-1} as $\eta \to 0$. Let $S_0, S_1, S_2, \ldots$ be independent and identically distributed copies of S, assume S is connected with probability 1, and define

$$f(\mathbf{x}_1, \ldots, \mathbf{x}_{n-1}) \equiv P\left\{\text{the set } S_0 \cup \bigcup_{j=1}^{n-1}(\mathbf{x}_j + S_j) \text{ is connected}\right\}.$$

$$(4.56)$$

Theorem 4.8. *Assume $E\{\bar{v}(S)^n\} < \infty$, and let $p(n)$ denote the probability that an arbitrary set in the Boolean model $\mathcal{C}(\delta, \lambda)$ is part of a clump of order n, where $n \ge 2$. If δ and λ vary in such a manner that $\eta \equiv \delta^k \lambda \to 0$, then*

$$p(n) = \eta^{n-1} \frac{1}{(n-1)!}$$
$$\times \int_{\ldots} \int_{(\mathbb{R}^k)^{n-1}} f(\mathbf{x}_1, \ldots, \mathbf{x}_{n-1}) \, d\mathbf{x}_1 \ldots d\mathbf{x}_{n-1} + o(\eta^{n-1}) \qquad (4.57)$$

and

$$\sum_{i=n+1}^\infty p(i) = o(\eta^{n-1}). \qquad (4.58)$$

The integral on the right-hand side of (4.57) is finite.

At the end of this section we prove Theorem 4.8 in the case where $\mathrm{rad}(S)$ is essentially bounded.

If we are to count the number of clumps centered within a given region, we must specify what we mean by the "center" of a clump.

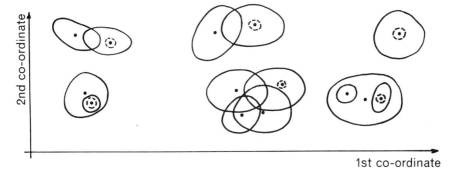

Figure 4.15. Right-hand centers when $k = 2$. Centers of sets are marked by dots, and centers (i.e., right-hand centers) of clumps are ringed.

Each clump of order n may be expressed as a connected union,

$$C \equiv \bigcup_{j=1}^{n} (\xi_{i_j} + \delta S_{i_j})$$

say. In a sense it has n centers, $\xi_{i_1}, \ldots, \xi_{i_n}$. The *right-hand center* of the clump is defined to be that vector ξ_{i_j} whose first coordinate is greatest. The "center" of any given clump will be taken to be its right-hand center. Figure 4.15 illustrates the case $k = 2$. This definition of clump center is only appropriate to sparse Boolean models, where infinite clumps cannot occur.

Let $\mathcal{R}$ be a Borel subset of $\mathbb{R}^k$, and let $N(n) = N(n; \mathcal{R})$ denote the number of clumps of order n centered within $\mathcal{R}$. We have already seen that the expected number of clumps of order n per unit content of $\mathbb{R}^k$ equals $\lambda n^{-1} p(n)$. Therefore

$$E\{N(n)\} = \lambda \|\mathcal{R}\| n^{-1} p(n).$$

(A detailed proof uses the marked point process argument.) Theorem 4.8 may be interpreted as providing information about properties of mean number of nth order clumps per unit area in a sparse Boolean model. For example, it implies that as $\eta \to 0$,

$$(\lambda \eta^{n-1})^{-1} E\{N(n)\} \to \mu(n)$$
$$\equiv \frac{\|\mathcal{R}\|}{n!} \int \ldots \int_{(\mathbb{R}^k)^{n-1}} f(\mathbf{x}_1, \ldots, \mathbf{x}_{n-1}) \, d\mathbf{x}_1 \ldots d\mathbf{x}_{n-1},$$

provided $E\{\bar{v}(S)^n\} < \infty$. A similar but longer argument may be used to show that the variance of $N(n)$ obeys a similar formula:

$$(\lambda \eta^{n-1})^{-1} \text{var}\{N(n)\} \to \mu(n).$$

These two properties suggest that in a sparse Boolean model, $N(n)$ is approximately Poisson-distributed. (Note that a Poisson random variable has its mean equal to its variance.)

Our next theorem confirms that $N(n)$ is asymptotically Poisson-distributed as $\eta \to 0$. We assume (for the first time in this section) that $\delta = \delta(\lambda) \to 0$ and $\lambda \to \infty$, as well as that $\eta \equiv \delta^k \lambda \to 0$. This is equivalent to studying sparse Boolean models within increasingly large regions. The problem breaks naturally into two parts: case (i), in which $\lambda \eta^{n-1} \to a$ (say) and $a < \infty$; and case (ii), where $\lambda \eta^{n-1} \to \infty$. In the latter case the number of clumps of order n centered within $\mathcal{R}$ diverges to $+\infty$, and so the Poisson approximation is really a normal approximation. [If X is Poisson with mean γ then $(X - \gamma)/\gamma^{1/2} \to N(0,1)$ in distribution as $\gamma \to \infty$.]

We continue to assume that S is connected with probability 1, and define f as in (4.56). The region $\mathcal{R}$ is taken to be Riemann measurable with $\|\mathcal{R}\| > 0$.

Theorem 4.9. *Assume that* rad(S) *is essentially bounded, and let $N(n)$ denote the number of clumps of order n from the Boolean model $C(\delta, \lambda)$ centered within $\mathcal{R}$, where $n \geq 1$. Put $\eta \equiv \delta^k \lambda$. If $\delta = \delta(\lambda) \to 0$ and $\lambda \to \infty$ such that $\lambda \eta^{n-1} \to a \ (< \infty)$ then $N(n)$ is asymptotically Poisson-distributed with mean $a\mu(n)$, where*

$$\mu(n) \equiv \frac{\|\mathcal{R}\|}{n!} \int \cdots \int_{(\mathbb{R}^k)^{n-1}} f(\mathbf{x}_1, \ldots, \mathbf{x}_{n-1}) d\mathbf{x}_1 \ldots d\mathbf{x}_{n-1}.$$

If $\lambda \eta^{n-1} \to \infty$ then $\{N(n) - EN(n)\}/\{\text{var} N(n)\}^{1/2}$ is asymptotically normal $N(0,1)$, and both $E\{N(n)\}$ and $\text{var}\{N(n)\}$ are asymptotic to $\lambda \eta^{n-1} \mu(n)$.

A proof is given at the end of this section.

The total number of clumps of all orders centered within $\mathcal{R}$ is given by

$$N_{tot} = N_{tot}(\mathcal{R}) \equiv \sum_{n=1}^{\infty} N(n),$$

and has mean

$$E(N_{tot}) = \lambda \|\mathcal{R}\| \sum_{n=1}^{\infty} n^{-1} p(n).$$

(Again we assume that η is so small that the probability that an infinite clump occurs equals 0; see Section 4.7.) The results stated so far are sufficient to give us a central limit theorem for N_{tot}. Assume that for some sufficiently large $m \geq 1$, $\lambda\eta^{2m}$ is bounded. Observe that

$$N_{tot} - E(N_{tot})$$
$$= \sum_{n=1}^{m} \{N(n) - EN(n)\} + \left\{ \sum_{n=m+1}^{\infty} N(n) - \sum_{n=m+1}^{\infty} EN(m) \right\}.$$
$$(4.59)$$

By (4.58),

$$0 \leq E\left\{ \sum_{n=m+1}^{\infty} N(n) \right\} = \lambda\|\mathcal{R}\| \sum_{n=m+1}^{\infty} n^{-1}p(n)$$
$$\leq \lambda\|\mathcal{R}\| \sum_{n=m+1}^{\infty} p(n)$$
$$= o(\lambda\eta^{m}) = o(\lambda^{1/2}),$$

and by Theorem 4.9,

$$N(n) - E\{N(n)\} = O_p\left\{ (\lambda\eta^{n-1})^{1/2} \right\} = o_p(\lambda^{1/2})$$

for $2 \leq n \leq m$. Substituting into (4.59) we conclude that

$$N_{tot} - E(N_{tot}) = N(1) - E\{N(1)\} + o_p(\lambda^{1/2}).$$

Theorem 4.9 implies that $N(1) - E\{N(1)\}$ is asymptotically normal $N(0, \lambda\|\mathcal{R}\|)$, and so this asymptotic distribution must be shared by $N_{tot} - E(N_{tot})$.

Of course, the total number M of random sets centered within $\mathcal{R}$ is Poisson-distributed with mean $\lambda\|\mathcal{R}\|$. Therefore $M - E(M)$ is also asymptotically normal $N(0, \lambda\|\mathcal{R}\|)$. The fact that the limit behaviors of $N(1) - EN(1)$, $N_{tot} - E(N_{tot})$, and $M - E(M)$ are identical reflects the essential character of sparse Boolean models: few sets overlap, and so most sets are present as clumps of order one.

Let $\mathcal{R}$ be a Riemann-measurable set with $\|\mathcal{R}\| > 0$, and put $l\mathcal{R} \equiv \{l\mathbf{x} : \mathbf{x} \in \mathcal{R}\}$. Consider the Boolean model in $\mathbb{R}^k$ where sets are distributed as S (not δS) and the driving Poisson process has intensity λ. The expected order of an arbitrary clump may be defined by an ergodic argument as the almost sure limit

$$\nu^* \equiv \lim_{l \to \infty} \frac{\text{number of sets centered within } l\mathcal{R}}{\text{number of clumps centered within } l\mathcal{R}}.$$

Trivially,

$$\|l\mathcal{R}\|^{-1} \text{ (number of sets centered within } l\mathcal{R}) \to \lambda$$

almost surely as $l \to \infty$. The ergodic argument used to prove Theorem 3.6 in Chapter 3 readily gives

$$\|l\mathcal{R}\|^{-1} \text{ (number of clumps centered within } l\mathcal{R}) \to \nu$$

$$\equiv \lambda \sum_{n=1}^{\infty} n^{-1} p(n)$$

almost surely, provided $E\{\bar{v}(S)\} < \infty$ and infinite clumps have zero probability of forming [see (4.52)]. Therefore

$$\nu^* = \lambda/\nu = \left\{ \sum_{n=1}^{\infty} n^{-1} p(n) \right\}^{-1},$$

which of course is independent of our choice of $\mathcal{R}$. Formulas for ν in the case where S is a sphere are given by (4.53) and (4.54). In particular, if $k = 2$ and S is a disk of fixed radius r then by (4.53),

$$\nu^* = 1 + 2\pi r^2 \eta + \tfrac{2}{3}\pi(2\pi + 3\sqrt{3})r^4 \eta^2 + O(\eta^3)$$

as $\eta \to 0$.

Armitage (1949) reports the results of an empirical comparison of the approximation

$$\tilde{\nu}^* \equiv 1 + 2\pi r^2 \eta + \tfrac{2}{3}\pi(2\pi + 3\sqrt{3})r^4 \eta^2 \qquad (4.60)$$

with estimates of ν^* obtained by simulation. Table 4.1 summarizes some of his results.

Proof of Theorem 4.8 in the case* rad(S) *essentially bounded.
[See Hall (1986) for the general case, $E\{\bar{v}(S)^n\} < \infty$.]

Let $R \equiv \text{rad}(S)$, and assume $R \leq c$ with probability 1, where c is a fixed positive constant. Then for some random k-vector $\mathbf{X}$, $S \subseteq \mathbf{X} + \mathcal{T}(R)$ with probability 1, where $\mathcal{T}(R)$ is the closed sphere of radius R centered at the origin. Without loss of generality, $\mathbf{X} = \mathbf{0}$ (see Section 4.1). In our usual notation, let the Boolean model be driven by the Poisson process $\mathcal{P} \equiv \{\xi_i, i \geq 1\}$ and generated by shapes $\{\delta S_i, i \geq 1\}$.

TABLE 4.1. Comparison of Approximation, $\tilde{\nu}^*$, to Expected Order, ν^*, of an Arbitrary Clump

$a\lambda$	$\tilde{\nu}^*$	Estimated ν^* $\pm$ Standard Error
0.016	1.032	1.026±0.015
0.024	1.05	1.05 ±0.02
0.035	1.07	1.08 ±0.04
0.063	1.14	1.15 ±0.03
0.098	1.22	1.28 ±0.05
0.141	1.34	1.38 ±0.07
0.192	1.49	1.47 ±0.08
0.251	1.68	1.71 ±0.12

Approximation $\tilde{\nu}^*$ is defined by (4.60). Shapes are disks of fixed area $a = \pi r^2$, and Poisson intensity of disk centers equals λ. [From Armitage (1949).]

The chance that an arbitrary set in the Boolean model is part of a clump of order $n+1$ or more is bounded by the probability that n or more points of $\mathcal{P}$ lie within a given sphere of radius $2nc\delta$. The latter probability equals

$$\sum_{m=n}^{\infty} \{\lambda\|\mathcal{T}(2nc\delta)\|\}^m \exp\{-\lambda\|\mathcal{T}(2nc\delta)\|\}/m!$$

$$\leq \{\lambda\|\mathcal{T}(2nc\delta)\|\}^n$$

$$= \eta^n \left\{(2nc)^k v_k\right\}^n.$$

Therefore

$$\sum_{m=n+1}^{\infty} p(m) \leq \eta^n \left\{(2nc)^k v_k\right\}^n,$$

from which follows (4.58) under the assumption $\mathrm{rad}\,(S) \leq c$.

Define $t \equiv 2(n-1)c$. If the arbitrary set $\xi_i + \delta S_i$ is part of a clump

$$(\xi_i + \delta S_i) \cup \left\{\bigcup_{l=1}^{n-1}(\xi_{j_l} + S_{j_l})\right\}$$

of order n, then all the points ξ_{j_l} must lie within the sphere $\xi_i + \mathcal{T}(t\delta)$. Let the random variable M have the Poisson distribution with mean $\mu \equiv \lambda\|\mathcal{T}(t\delta)\| = \eta\|\mathcal{T}(t)\|$, and conditional on M distribute M points $\xi_{(1)}, \ldots, \xi_{(M)}$ independently and uniformly within $\mathcal{T}(t\delta)$. Let $S_{(0)}, S_{(1)}, \ldots$ be independent copies of S, independent also of M and of $\xi_{(1)}, \ldots, \xi_{(M)}$. In view of the remark two sentences above,

$$p(n) = P\left(\begin{array}{l}\text{the clump containing } \delta S_{(0)} \\ \text{and formed only from the random} \\ \text{sets } \delta S_{(0)}, \xi_{(1)} + \delta S_{(1)}, \ldots, \xi_{(M)} + \delta S_{(M)} \text{ is of size } n\end{array}\right)$$

$$= \sum_{m=n-1}^{\infty} P(M=m) \|T(t)\|^{-m}$$

$$\times \int \ldots \int_{\{T(t)\}^m} g_n(\mathbf{x}_1, \ldots, \mathbf{x}_m) \, d\mathbf{x}_1 \ldots d\mathbf{x}_m, \qquad (4.61)$$

where

$$g_n(\mathbf{x}_1, \ldots, \mathbf{x}_m) \equiv P\left(\begin{array}{l}\text{the clump containing } S_{(0)} \text{ and} \\ \text{formed only from the random sets} \\ S_{(0)}, \mathbf{x}_1 + S_{(1)}, \ldots, \mathbf{x}_m + S_{(m)} \text{ is of size } n.\end{array}\right).$$

Since $g_n(\mathbf{x}_1, \ldots, \mathbf{x}_{n-1}) = f(\mathbf{x}_1, \ldots, \mathbf{x}_{n-1})$ then the first term in the series on the right-hand side of (4.61) equals

$$\frac{\{\eta \|T(t)\|\}^{n-1}}{(n-1)!} \exp\{-\eta \|T(t)\|\} \|T(t)\|^{-(n-1)}$$

$$\times \int \ldots \int_{(\mathbb{R}^k)^{n-1}} f(\mathbf{x}_1, \ldots, \mathbf{x}_{n-1}) \, d\mathbf{x}_1 \ldots d\mathbf{x}_{n-1},$$

which equals the right-hand side of (4.57) plus terms of order η^n. The sum of all but the first term in the series on the right-hand side of (4.61) is dominated by

$$\sum_{m=n}^{\infty} P(M=m) = \sum_{m=n}^{\infty} \{\eta \|T(t)\|\}^m \exp\{-\eta \|T(t)\|\}/m!$$

$$= O(\eta^n).$$

Result (4.57) is immediate. $\square$

Proof of Theorem 4.9.   Our proof is similar to that of Theorem 3.5 in Section 3.4. Assume that shapes generating the Boolean model are distributed as δS, which is entirely contained within a sphere of radius $c\delta$ centered at the origin. Let d be a large positive number.

Divide $\mathbb{R}^k$ into a lattice of k-dimensional cubes of side length $cd\delta$ and with nearest faces of adjacent cubes separated by spacings of width $4nc\delta$. Let $\mathcal{D}_1,\ldots,\mathcal{D}_\nu$ denote those cubes that are contained entirely within $\mathcal{R}$. (During the proof of Theorem 3.5 we used the symbol n instead of ν, and required the spacings between adjacent cubes to be $2c\delta$ instead of $4nc\delta$.) Let $\mathcal{A}_1 \equiv \bigcup_{1\le i\le\nu}\mathcal{D}_i$, let $\mathcal{A}_2$ denote the union of all those rectangular boxes forming the spacings between cubes and that are contained wholly within $\mathcal{R}$, and let $\mathcal{A}_3$ be the region formed from the intersection of $\mathcal{R}$ with all those cubes $\mathcal{D}_i$ or rectangular spacings that are not completely within or without $\mathcal{R}$. Then $\mathcal{A}_1,\mathcal{A}_2$, and $\mathcal{A}_3$ are as defined in the proof of Theorem 3.5, and are illustrated by Figure 3.4 in the case $k=2$. Write N_i for the number of clumps of order n whose right-hand centers lie within $\mathcal{A}_i$. Then

$$N(n) = N_1 + N_2 + N_3.$$

In step (i) of the proof we establish a Poisson limit (if $\lambda\eta^{n-1} \to a$) or a normal limit (if $\lambda\eta^{n-1} \to \infty$) for N_1. Then in step (ii) we prove that for large values of d, N_2 and N_3 are negligible in comparison with N_1. The desired limit theorems for $N(n)$ follow on combining these results.

Step (i). Let $\mathcal{D}_i'$ be the k-dimensional cube of side length $(d+2n)c\delta$, concentric to $\mathcal{D}_i$. Any clump of order n whose right-hand center lies within $\mathcal{D}_i$, is composed entirely of sets centered within $\mathcal{D}_i'$. Let M_i equal the number of clumps of order n centered within $\mathcal{D}_i$, and let K_i equal the total number of points of the driving Poisson process within $\mathcal{D}_i'$. We begin by listing properties of the variables M_i.

Since nth order clumps are of radius at most $nc\delta$, and since all cubes are distant at least $4nc\delta$ apart, then the variables $M_1,\ldots,M_\nu$ are mutually independent. Furthermore,

$$E\left\{M_i^2 I(M_i \ge 2)\right\} \le E\left\{K_i^2 I(K_i \ge 2n)\right\}$$

$$= \sum_{m=2n}^{\infty} m^2 \frac{(\lambda\|\mathcal{D}_1'\|)^m}{m!} \exp(-\lambda\|\mathcal{D}_1'\|)$$

$$\le (\lambda\|\mathcal{D}_1'\|)^{2n} \sum_{m=2n}^{\infty} \frac{m^2}{m!},$$

provided η is so small that $\lambda\|\mathcal{D}_1'\| = \eta\{(d+2n)c\}^k \leq 1$. Consequently,

$$E\left\{M_i^2 I(M_i \geq 2)\right\} = O(\eta^{2n}) \tag{4.62}$$

as $\eta \to 0$. A similar argument shows that

$$0 \leq P(M_i = 1) - P(M_i = 1, K_i = n) \leq P(K_i \geq n+1) = O(\eta^{n+1}). \tag{4.63}$$

Moreover,

$$
\begin{aligned}
P(M_i &= 1, K_i = n) \\
&= P(K_i = n) P(M_i = 1 \mid K_i = n) \\
&= \frac{1}{n!}(\lambda\|\mathcal{D}_1'\|)^n \exp(-\lambda\|\mathcal{D}_1'\|) \\
&\quad \times \|\mathcal{D}_1'\|^{-n} \int \ldots \int_{(\mathcal{D}_1')^n} P\left\{\bigcup_{i=1}^{n}(\mathbf{x}_i + \delta S_i) \text{ connected}\right\} d\mathbf{x}_1 \ldots d\mathbf{x}_n \\
&= \eta^n b + O(\eta^{n+1}),
\end{aligned}
\tag{4.64}
$$

where

$$b \equiv \frac{1}{n!} \int \ldots \int_{\mathcal{E}_1^n} P\left\{\bigcup_{i=1}^{n}(\mathbf{x}_i + S_i) \text{ connected}\right\} d\mathbf{x}_1 \ldots d\mathbf{x}_n$$

and $\mathcal{E}_1$ is any k-dimensional cube of side length $(d+2n)c$ with the same orientation as $\mathcal{D}_1$.

Combining (4.63) and (4.64) we conclude that

$$P(M_i = 1) = \eta^n b + O(\eta^{n+1}).$$

This estimate and (4.62) yield

$$E(M_i) = \eta^n b + O(\eta^{n+1}) \tag{4.65}$$

and $E(M_i^2) = \eta^n b + O(\eta^{n+1})$, so that

$$\text{var}(M_i) = \eta^n b + O(\eta^{n+1}). \tag{4.66}$$

Furthermore, if $x = x(\eta)$ is any sequence diverging to $+\infty$ as $\eta \to 0$, then for sufficiently small η,

$$E\left\{(M_i - EM_i)^2 I(|M_i - EM_i| > x)\right\}$$
$$\leq 2E\left\{M_i^2 I(M_i \geq 2)\right\} + 2\{E(M_i)\}^2 = O(\eta^{2n}). \quad (4.67)$$

We also need a little asymptotic theory for the quantity b. As $d \to \infty$ (meaning that the side length of the cube $\mathcal{E}_1$ diverges),

$$n! \, b \sim \|\mathcal{E}_1\| \int \cdots \int_{(\mathbb{R}^k)^{n-1}} P\left\{S_1 \cup \bigcup_{i=2}^{n} (\mathbf{x}_i + S_i) \text{ connected}\right\} d\mathbf{x}_2 \ldots d\mathbf{x}_n$$

$$= \|\mathcal{E}_1\| \int \cdots \int_{(\mathbb{R}^k)^{n-1}} f(\mathbf{x}_1, \ldots, \mathbf{x}_{n-1}) \, d\mathbf{x}_1 \ldots d\mathbf{x}_{n-1}.$$

But $\|\mathcal{E}_1\| \sim (cd)^k$ as $d \to \infty$, and so

$$b \sim \left\{(cd)^k / \|\mathcal{R}\|\right\} \mu(n) \quad (4.68)$$

as $d \to \infty$.

We are now in a position to prove a weak limit theorem for

$$N_1 = \sum_{i=1}^{\nu} M_i.$$

Note that the distribution of M_i does not depend on i, and that the value of ν is asymptotic to

$$\|\mathcal{R}\| / \{(d + 2n)c\delta\}^k = \lambda \eta^{-1} \|\mathcal{R}\| / \{(d + 2n)c\}^k$$

as $\eta \to 0$.

Case (a). $\lambda \eta^{n-1} \to a < \infty$. In this situation, as $\eta \to 0$,

$$\sum_{i=1}^{\nu} E\{M_i I(M_i \geq 2)\} = O(\lambda \eta^{-1} \eta^{2n}) \to 0,$$

by (4.62), and

$$\sum_{i=1}^{\nu} E(M_i) = \nu \eta^n b + O(\nu \eta^{n+1})$$

$$\to a \mu_1,$$

by (4.65), where

$$\mu_1 \equiv \|\mathcal{R}\| b / \{(d + 2n)c\}^k. \quad (4.69)$$

These two properties imply Poisson convergence for a sum of independent, identically distributed, nonnegative, integer-valued random variables M_i (see Appendix II). Therefore

$$N_1 \equiv \sum_{i=1}^{\nu} M_i \text{ is asymptotically Poisson-distributed}$$

$$\text{as } \eta \to 0, \text{ with mean } a\mu_1. \qquad (4.70)$$

Note that by (4.68) and (4.69),

$$\mu_1 \to \mu(n) \qquad (4.71)$$

as $d \to \infty$.

Case (b). $\lambda\eta^{n-1} \to \infty$. Here,

$$\text{var}(N_1) = \sum_{i=1}^{\nu} \text{var}(M_i) = \nu\eta^n b + O(\nu\eta^{n+1})$$

$$= \lambda\eta^{n-1}\mu_1 + o(\lambda\eta^{n-1}), \qquad (4.72)$$

using (4.66). If we take $\epsilon > 0$ and $x = \epsilon\{\text{var}(N_1)\}^{1/2}$ in (4.67), we may deduce that Lindeberg's condition holds for the series $\sum_{i=1}^{\nu} (M_i - EM_i)$:

$$\{\text{var}(N_1)\}^{-1} \sum_{i=1}^{\nu} E\left[(M_i - EM_i)^2 I\left\{|M_i - EM_i| > \epsilon(\text{var}N_1)^{1/2}\right\}\right]$$

$$= O\left\{(\nu\eta^n)^{-1}(\nu\eta^{2n})\right\}$$

$$= O(\eta^n) \to 0$$

as $\eta \to 0$. Lindeberg's central limit theorem (Appendix II) now ensures that

$$\{N_1 - E(N_1)\} / \{\text{var}(N_1)\}^{1/2} \to N(0,1) \qquad (4.73)$$

in distribution.

Step (ii). Here we show that for large d, the variables N_2 and N_3 appearing in the formula $N(n) = N_1 + N_2 + N_3$ are negligible in comparison with N_1.

The expected number of clumps of order n, per unit content of $\mathbb{R}^k$, equals $\lambda n^{-1} p(n)$. Therefore the expected number of nth order clumps centered within $\mathcal{A}_i$ equals

$$E(N_i) = \lambda n^{-1} p(n) \|\mathcal{A}_i\|.$$

Theorem 4.8 asserts that $p(n) = C_1\eta^{n-1} + o(\eta^{n-1})$, where the constant C_1 depends only upon n and the distribution of S. Hence

$$E(N_i) = \lambda\eta^{n-1} n^{-1} C_1 \|\mathcal{A}_i\| + o(\lambda\eta^{n-1}) \qquad (4.74)$$

as $\eta \to 0$.

Case (a). $\lambda\eta^{n-1} \to a < \infty$. When $i = 2$, $\|\mathcal{A}_i\| \leq C_2/d$ for a constant C_2 not depending on d [see (3.37)], and when $i = 3$, $\|\mathcal{A}_i\| \to 0$ as $\eta \to 0$ [see (3.36)]. Therefore by (4.74),

$$\lim_{d\to\infty} \limsup_{\eta\to 0} E(N_i) = 0$$

for $i = 2, 3$. This result, together with (4.70) and (4.71), implies that $N(n) \equiv N_1 + N_2 + N_3$ is asymptotically Poisson-distributed with mean $a\mu(n)$, as had to be proved.

Case (b). $\lambda\eta^{n-1} \to 0$. Both N_2 and N_3 may be written in the form

$$N_i = \sum_{j=1}^{m_i} \sum_{l=1}^{n_{ij}} M_{ijl}, \qquad (4.75)$$

where $m_i \leq 2^k$, the M_{ijl}'s are the numbers of nth order clumps centered within respective disjoint regions $\mathcal{A}_{ijl}$ of dimensions no more than $\{(d+2n)c\delta\} \times \cdots \times \{(d+2n)c\delta\}$, and for fixed i and j, the variables $M_{ij1}, \ldots, M_{ijn_{ij}}$ are mutually independent. The total number of terms in the double series (4.75) is $O(\delta^{-k})$ as $\delta \to 0$.

Independence implies that

$$\text{var}(N_i) \leq C_3 \sum_{j=1}^{m_i} \sum_{l=1}^{n_{ij}} \text{var}(M_{ijl}) \leq C_3 \sum_{j=1}^{m_i} \sum_{l=1}^{n_{ij}} E(M_{ijl}^2),$$

where C_3 depends only on k. Now,

$$E(M_{ijl}^2) \leq E(M_{ijl}) + E\left\{M_{ijl}^2 I(M_{ijl} \geq 2)\right\}.$$

An argument like that leading to (4.62) shows that

$$\sup_{i,j,l} E\left\{M_{ijl}^2 I(M_{ijl} \geq 2)\right\} = O(\eta^{2n})$$

as $\eta \to 0$. Combining these estimates we conclude that for a constant $C_4 > 0$,

$$\text{var}(N_i) \leq C_3 \sum_{j=1}^{m_i} \sum_{l=1}^{n_{ij}} E(M_{ijl}) + C_4 \sum_{j=1}^{m_i} \sum_{l=1}^{n_{ij}} \eta^{2n}$$

$$= C_3 E(N_i) + O(\delta^{-k}\eta^{2n}).$$

Referring to formula (4.72) for $\mathrm{var}(N_1)$, and to formula (4.74) for $E(N_i)$, we conclude that

$$\mathrm{var}(N_i)/\mathrm{var}(N_1) \leq C_1 C_3 \|\mathcal{A}_i\|/(n\mu_1) + o(1)$$

for $i = 2$ and 3.

In the case $i = 2$ we have $\|\mathcal{A}_i\| \leq C_2/d$, and when $i = 3$, $\|\mathcal{A}_i\| \to 0$ as $\eta \to 0$ [see (3.36) and (3.37)]. Therefore

$$\lim_{d \to \infty} \limsup_{\eta \to 0} \mathrm{var}(N_i)/\mathrm{var}(N_1) = 0$$

for $i = 2$ and 3. This result, together with (4.71)–(4.73), implies that $N(n) - E\{N(n)\}$ is asymptotically normally distributed with 0 mean and asymptotic variance $\lambda \eta^{n-1} \mu(n)$. $\square$

4.6. EXPECTED TOTAL NUMBER OF CLUMPS

One of the problems in determining the expected total number of clumps within a given region $\mathcal{R}$ is defining just what it is that we mean by this statement. We shall show in the next section that under certain quite reasonable conditions, clumps of infinite order can form. How are infinite clumps to contribute to the total number of clumps in $\mathcal{R}$? The "right-hand center" (see Section 4.5) of an infinite clump is not necessarily well-defined. One concise way of posing the problem is to argue as follows. Given a coverage process in $\mathbb{R}^k$, let $\mathcal{A}(\mathbf{x})$ (a subset of $\mathbb{R}^k$) denote the clump containing $\mathbf{x} \in \mathbb{R}^k$, assuming $\mathbf{x}$ is covered. Define

$$m(\mathbf{x}) \equiv \begin{cases} \|\mathcal{A}(\mathbf{x})\|^{-1} & \text{if } \mathbf{x} \text{ is covered} \\ 0 & \text{otherwise,} \end{cases}$$

for $\mathbf{x} \in \mathcal{R}$. We may take the "weighted average,"

$$M(\mathcal{R}) \equiv \int_{\mathcal{R}} m(\mathbf{x})\, d\mathbf{x},$$

to equal the total number of clumps "within" $\mathcal{R}$.

For each $\mathbf{x}$,

$$E\{m(\mathbf{x})\} = E\left\{ C_0^{-1} I(C_0 > 0) \right\},$$

where C_0 denotes the content of the clump containing the origin (and equals 0 if the origin is not covered). Therefore we might define

$$\nu^\dagger \equiv E\left\{C_0^{-1}I(C_0 > 0)\right\} \qquad (4.76)$$

to equal the expected number of clumps per unit content in the coverage process. In this case, $E\{M(\mathcal{R})\} = \nu^\dagger\|\mathcal{R}\|$ would be the expected number of clumps within $\mathcal{R}$. Even in the absence of infinite clumps this definition can be unsatisfactory. Consider for example the case where all shapes are line segments or "sticks," so that $C_0 \equiv 0$ and $\nu^\dagger = 0$. Nevertheless, in many cases of practical interest (e.g., for random spheres, polygons, etc.) we have $\nu = \nu^\dagger$. The definition of $\nu^\dagger$ and the right-hand center definition of ν both ignore infinite clumps.

Formula (4.76) may be employed to estimate ν from simulations, but it is not very useful from a theoretical viewpoint. Apparently it is only in the case of $k = 1$ dimension that ν admits a simple formula. There, if shapes are line segments with mean length α and if the driving point process is Poisson with intensity λ, then $\nu = \lambda e^{-\alpha\lambda}$ (see Exercise 4.8). In other situations the best available results are approximations, such as those developed in Section 4.5, or bounds. In this section we derive an upper bound ν_1 to ν in the case of convex isotropic sets distributed in a $k = 2$ dimensional Boolean model $\mathcal{C}$. Recall that Theorem 4.6 [formula (4.30)] gives the expected number of clumps minus voids per unit area.

Suppose shapes generating the Boolean model $\mathcal{C}$ are distributed as the isotropic random convex set S with mean area $\alpha \equiv E(\|S\|_2)$ and mean perimeter $\beta \equiv E(\|\partial S\|_1)$. Let λ denote the intensity of the driving Poisson process. The expected number of sets in $\mathcal{C}$ that intersect a fixed convex set S in the plane is given in Theorem 4.1; it is

$$\nu(S) \equiv \lambda\left\{\|S\|_2 + \alpha + (2\pi)^{-1}\|\partial S\|_1\beta\right\}.$$

Let S^* denote the isotropic image of S after a random rotation, independent of $\mathcal{C}$. Since $\nu(S)$ depends on S only through its area and perimeter, then the expected number of random sets from $\mathcal{C}$ that intersect S^* equals $\nu(S)$. By symmetry, the expected number of sets that intersect S^* to the left of its center must equal the expected number of sets intersecting to the right of center [see Chapter 3, Section 3.1 for a definition of "center"]. Therefore the expected number of sets intersecting S^* to the right of center must equal $\frac{1}{2}\nu(S)$. (The expectation here is taken over the random rotation, as well as over the number of intersecting sets.) The "Poisson-ness argument" (see Section 4.1) implies that this number is Poisson-distributed. Without loss of generality, the center of S^* is the origin $\mathbf{0}$. Therefore the chance that no sets from $\mathcal{C}$ intersect S^* and have their centers to the right of $\mathbf{0}$ is

$$\exp\left\{-\tfrac{1}{2}\nu(\mathcal{S})\right\}.$$

Take S (having the distribution of an arbitrary shape) to be independent of $\mathcal{C}$. Then S is the isotropic image of some set S' after a random rotation. The probability that an arbitrary set $\xi_i + S_i$ from $\mathcal{C}$ has no intersections with sets from $\mathcal{C}$ whose centers lie to the right of ξ_i equals the chance that no sets from $\mathcal{C}$ that intersect S have their centers to the right of the origin. By first conditioning on S' and then taking expectations with respect to S', we see that this probability is

$$E\left[\exp\left\{-\tfrac{1}{2}\nu(S')\right\}\right] = E\left[\exp\left\{-\tfrac{1}{2}\nu(S)\right\}\right]$$
$$= e^{-\frac{1}{2}\alpha\lambda}E\left[\exp\left\{-\tfrac{1}{2}\lambda(\|S\|_2 + (2\pi)^{-1}\beta\|\partial S\|_1)\right\}\right].$$

Therefore the mean number of random sets $\xi_i + S_i$ per unit area with the property that no sets intersecting them have centers to the right of ξ_i, equals

$$\nu_1 \equiv \lambda e^{-(1/2)\alpha\lambda}E\left[\exp\left\{-\tfrac{1}{2}\lambda(\|S\|_2 + (2\pi)^{-1}\beta\|\partial S\|_1)\right\}\right].$$

(Use the marked point process argument.) Call this property Π.

A clump that consists of just one or two random sets is counted precisely once in the argument above, since it contains exactly one set with property Π. Clumps containing three or more sets will be counted at least once, and may be counted two or more times, depending on their configurations (see Figure 4.16).

Therefore the quantity ν_1 forms an upper bound to the expected number ν of clumps per unit area. (The "right-hand center" of a two-dimensional set was defined in Section 4.5, following Theorem 4.8. An argument based on right-hand centers runs like this: Any clump whose right-hand center lies within a region $\mathcal{R}$ has that center satisfying property Π. Therefore the expected number of right-hand centers of clumps falling within $\mathcal{R}$ is dominated by the expected number of sets whose centers fall within $\mathcal{R}$ and that satisfy property Π: $\nu\|\mathcal{R}\| \leq \nu_1\|\mathcal{R}\|$.) Consequently,

$$\nu \leq \nu_1. \tag{4.77}$$

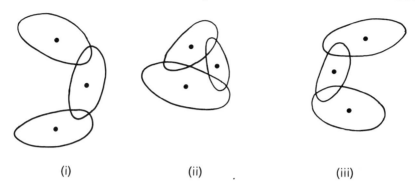

(i)   (ii)   (iii)

Figure 4.16. Counting clumps of order three in the formula for ν_1. Clumps of types (i) and (ii) are counted once only, since each contains only one shape that intersects no shape whose center lies to the right. However, clumps of type (iii) are counted twice, since they contain two shapes that intersect no shapes with centers to the right. (Shape centers are denoted by dots.)

It is instructive to interpret this formula in the case of sparse Boolean models studied in Section 4.5. A two-dimensional sparse Boolean model $\mathcal{C}(\delta, \lambda)$ is one where shapes are distributed as δS and the driving Poisson process has intensity λ, and where δ and λ vary in such a manner that $\eta \equiv \delta^2 \lambda \to 0$. In this context formula (4.77) reduces to

$$\nu \leq \nu_1 \equiv \lambda e^{-(1/2)\alpha\eta} E\left[\exp\left\{-(1/2)\eta(\|S\|_2 + (2\pi)^{-1}\beta\|\partial S\|_1)\right\}\right].$$
$$(4.78)$$

The technique used to derive this inequality counts clumps of orders one and two correctly, and counts each clump of order n no more than $n-1$ times if $n \geq 3$. Therefore the amount by which ν_1 overestimates ν is less than

$$\sum_{n=3}^{\infty} (n-2)\,(\text{expected number of } n\text{th order clumps per unit area})$$

$$= \lambda \sum_{n=3}^{\infty} (n-2)n^{-1}p(n)$$

$$< \lambda \sum_{n=3}^{\infty} p(n),$$

provided infinite clumps have 0 probability of occurring. [We define $p(n)$ to be the probability that an arbitrary random set in the Boolean model $\mathcal{C}(\delta, \lambda)$ is part of a clump of order n; see Section 4.5.]

In view of Theorem 4.8,

$$\sum_{n=3}^{\infty} p(n) = O(\eta^2)$$

under the assumption that $E\{\bar{v}(S)^3\} < \infty$. Consequently,

$$\nu_1 + O(\lambda \eta^2) \le \nu \le \nu_1. \tag{4.79}$$

We may expand the expression (4.78) for ν_1 to obtain the first few terms in a power series in η:

$$
\begin{aligned}
\nu_1 &= \lambda E \left[1 - \tfrac{1}{2}\eta \left\{ \alpha + \|S\|_2 + (2\pi)^{-1}\beta \|\partial S\|_1 \right\} \right. \\
&\quad \left. + \tfrac{1}{2}(\tfrac{1}{2}\eta)^2 \left\{ \alpha + \|S\|_2 + (2\pi)^{-1}\beta \|\partial S\|_1 \right\}^2 + O(\eta^3) \right] \\
&= \lambda \left[1 - \eta \left\{ \alpha + (4\pi)^{-1}\beta^2 \right\} \right. \\
&\quad + \tfrac{1}{8}\eta^2 \left\{ 4\alpha^2 + \mathrm{var}(\|S\|_2) + (2\pi)^{-2}\beta^4 + (2\pi)^{-2}\beta^2 \mathrm{var}(\|\partial S\|_1) \right. \\
&\quad \left. \left. + 2\pi^{-1}\alpha\beta^2 + \pi^{-1}\beta \mathrm{cov}(\|S\|_2, \|\partial S\|_1) \right\} + O(\eta^3) \right],
\end{aligned}
$$

provided $E\{\bar{v}(S)^3\} < \infty$. [To check that this regularity condition is sufficient, notice that since S is convex then $\|S\|_2 \le \bar{v}(S)$ and $\|\partial S\|_1^2 \le 4\pi\bar{v}(S)$.] Combining this estimate with (4.79) we conclude that

$$\nu = \lambda \left[1 - \eta \left\{ \alpha + (4\pi)^{-1}\beta^2 \right\} + O(\eta^2) \right].$$

In the case where S is a disk of fixed radius r, the expansion for ν_1 above and inequality (4.77) give

$$\nu \le \lambda \left\{ 1 - 2\pi r^2 \eta + 2\pi^2 r^4 \eta^2 + O(\eta^3) \right\}.$$

Formula (4.53) gives ν accurately up to terms of order η^3:

$$\nu = \lambda \left\{ 1 - 2\pi r^2 \eta + (2/3)\pi(4\pi - 3\sqrt{3})r^4 \eta^2 + O(\eta^3) \right\}.$$

Since

$$\left\{ \tfrac{2}{3}\pi(4\pi - 3\sqrt{3}) \right\} / 2\pi^2 \doteq 0.78,$$

then the terms of order η^2 in these expansions are in reasonable agreement.

We conclude by summarizing our main results.

Theorem 4.10. *Let* $\mathcal{C}(\delta, \lambda)$ *denote a Boolean model in* $\mathbb{R}^2$ *where the driving Poisson process has intensity* λ *and shapes are distributed as* δS. *Assume* S *is convex and isotropic, and put* $\alpha \equiv E(\|S\|_2)$, $\beta \equiv E(\|\partial S\|_1)$, *and* $\eta \equiv \delta^2 \lambda$. *Then the expected number of clumps per unit area,* ν, *is dominated by*

$$\nu_1 \equiv \lambda e^{-(1/2)\alpha \eta} E \left[\exp \left\{ -\tfrac{1}{2}\eta (\|S\|_2 + (2\pi)^{-1}\beta\|\partial S\|_1) \right\} \right].$$

If $E\{\bar{v}(S)^3\} < \infty$ *then*

$$
\begin{aligned}
\nu_1 = \lambda \Big[1 &- \eta \left\{ \alpha + (4\pi)^{-1}\beta^2 \right\} \\
&+ \tfrac{1}{8}\eta^2 \left\{ 4\alpha^2 + \mathrm{var}(\|S\|_2) + (2\pi)^{-2}\beta^4 + (2\pi)^{-2}\mathrm{var}(\|\partial S\|_1) \right. \\
&\left. + 2\pi^{-1}\alpha\beta^2 + \pi^{-1}\beta\,\mathrm{cov}(\|S\|_2, \|\partial S\|_1) \right\} + O(\eta^3) \Big]
\end{aligned}
$$

as $\eta \to 0$. *This expansion gives an accurate description of* ν *up to terms of order* $\lambda\eta$, *in the sense that*

$$\nu = \lambda \left[1 - \eta \left\{ \alpha + (4\pi)^{-1}\beta^2 \right\} + O(\eta^2) \right]$$

as $\eta \to 0$.

We showed in Section 4.5 that the expected order of an arbitrary clump in a Boolean model equals

$$\nu^* \equiv \lambda / \nu.$$

If $\nu = \nu^\dagger$, the latter defined at (4.76), then formula (4.76) gives us an expression that can be used to estimate ν^* from simulations:

$$\nu^* \equiv \lambda \left[E \left\{ C_0^{-1} I(C_0 > 0) \right\} \right]^{-1}, \qquad (4.80)$$

where C_0 is the content of the clump containing the origin. Theorem 4.10 gives us a lower bound in the case of a two-dimensional Boolean model generated by isotropic convex shapes:

$$\nu^* \geq \lambda / \nu_1.$$

Formula (4.80) is available even when clump order can be infinite with positive probability. If the Boolean model is generated by shapes with the distribution of S, and if $E\{\bar{v}(S)\} < \infty$, then the denominator in (4.80) is nonzero for all finite, positive values of Poisson intensity λ. In this case ν^* never takes the value infinity. Therefore the expected order of an arbitrary clump is finite, even when clumps can have infinite order with positive probability! This paradox highlights the difficulty of counting the number of clumps within a given region. The paradox may be resolved by noting that the "weighted average" definition (4.76) of the expected number of clumps per unit area gives infinite clumps zero weight.

4.7. CLUMPS OF INFINITE ORDER

Consider the k-dimensional Boolean model $\mathcal{C} \equiv \{\xi_i + S_i, i \geq 1\}$ in which the driving Poisson process $\mathcal{P} \equiv \{\xi_i, i \geq 1\}$ has intensity λ and shapes S_i are distributed as S. Assume throughout this section that S is connected with probability 1. The order of the clump containing an arbitrary set $\xi_i + S_i$ is the supremum of all values of n for which there exist distinct indices $j_l, l \geq 1$, such that the set

$$(\xi_i + S_i) \cup \left\{ \bigcup_{l=1}^{n-1} (\xi_{j_l} + S_{j_l}) \right\}$$

is connected. Let this supremum be M, so that M has the distribution of the order of the clump containing an arbitrary set. (See Section 4.1 for a definition of "arbitrary.") In the present section we investigate circumstances under which $P(M = \infty)$ is strictly positive, or where $E(M) = \infty$.

The phenomenon of "infinite clumping" is known as *percolation*. In the present context we often speak of *continuum* percolation, as distinct from lattice percolation in which the components of clumps are sited on some sort of lattice (e.g., vertices or edges of a combinatorial graph). Appendix VI takes a brief look at lattice percolation.

Infinite clumps of connected sets cannot form in one dimension unless the whole line is to be covered. To see this, notice that the only connected sets in one dimension are line segments. Let λ denote the intensity of the Poisson process driving a linear Boolean model, and suppose shapes generating the model are random-length intervals with mean length α. If $\alpha = \infty$ then with probability 1 the entire real line is covered (Exercise 2.6). If $\alpha < \infty$ then expected clump length equals $\lambda^{-1}(e^{\alpha\lambda} - 1)$ (Theorem 2.2) and the expected number of segments in an arbitrary clump is $e^{\alpha\lambda}$ (Exercise 4.8; see Chapter 2, Section 2.1 for a definition of "order of an arbitrary clump"). Therefore if $\alpha < \infty$ then with probability 1 no infinite clumps occur, and *in a certain sense* expected clump order is finite. Nevertheless, the expected order of the clump containing an arbitrary segment can be infinite, even when $\alpha < \infty$. Indeed, this expected order is finite if and only if segment length has finite *variance* (see Theorem 4.11(ii) below).

If we drop the assumption that sets are connected then it is possible for infinite-order clumps to form in a linear Boolean model without the entire real line being covered (Exercise 4.11). However for our purposes we may regard such curiosities as pathological.

The propensity of shapes to form clumps has little to do with shape content. Random isotropic line segments in $k = 2$ dimensions have zero two-dimensional content, but can form infinite clumps if the driving Poisson process has sufficiently high intensity. It is possible to construct a random sphere-like shape with finite mean k-dimensional content and such that each set in any Boolean model generated by such shapes is part of an infinite clump, no matter what the value of Poisson intensity (see Exercise 4.12). The contents $\overline{v}(S)$ of the "smallest sphere containing S," and $\underline{v}(S)$ of "the largest sphere contained inside S," are often of more importance than the content of S itself.

We begin by considering the simplest case—a k-dimensional Boolean model of spheres. Suppose shapes generating the model are distributed as S, and the driving Poisson process has intensity λ.

Theorem 4.11.   *Take $k \geq 1$. Assume S is a k-dimensional sphere with $E(\|S\|) < \infty$.*

(i) The condition $E(\|S\|^{2-(1/k)}) < \infty$ is sufficient for the number of spheres in each clump to be finite with probability 1 for all sufficiently small λ.

(ii) The condition $E(\|S\|^2) < \infty$ is necessary and sufficient for the expected number of spheres in the clump containing an arbitrary sphere to be finite for all sufficiently small λ.

(iii) If $E(\|S\|^2) = \infty$ then the expected number of spheres in the same clump as a given sphere and distant no more than one sphere away from that sphere is infinite for all positive values of λ.

In parts (i) and (ii) of Theorem 4.11, the assumption that λ be "sufficiently small" is essential; the stated results do not hold for large λ. Indeed, there is a critical value of λ beyond which infinite clumping occurs with positive probability. See Corollaries 4.14 and 4.15, and the discussion following those results.

The proof of Theorem 4.11 is deferred until the end of this section. As an immediate corollary, we obtain sufficient conditions for finiteness of clump order when shapes have a general distribution.

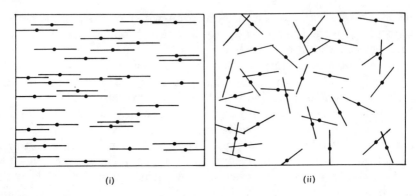

Figure 4.17. Random sticks with (i) fixed orientation, (ii) orientation uniformly distributed on $[0,\pi)$.

Corollary 4.12.   *Take $k \geq 1$.*
   (i) If $E\{\overline{v}(S)^{2-(1/k)}\} < \infty$ then the number of sets in each clump is finite with probability 1 for all sufficiently small λ.
   (ii) If $E\{\overline{v}(S)^2\} < \infty$ then the expected number of sets in the clump containing an arbitrary set is finite for all sufficiently small λ.

These two results take care of the case of low intensity. Our next task is to determine whether infinite clumps can form under conditions of high intensity. This requires constraints quite different from those imposed above.

To illustrate the contingencies that can arise, let us consider the case $k = 2$ and examine the distribution of "sticks" (line segments) in a plane. Suppose each stick is of unit length, and the ith stick is centered at the ith point $\boldsymbol{\xi}_i$ of a Poisson process. If all the sticks have the same orientation then no intersections ever occur, and so there is no possibility of even finite clumping. However, if each stick has an independent orientation uniformly distributed on $[0,\pi)$, then infinite clumps of overlapping sticks can form under conditions of high intensity. Indeed, the condition of isotropy is stronger than necessary. The only requirement for clumping is that the distribution of orientation be nondegenerate, or equivalently, that it have at least two distinct points of support. See Figure 4.17.

A k-dimensional analogue of the stick example may be obtained by replacing the sticks by small sections of hyperplanes, centered at points of a Poisson process and oriented independently of one another. Place into $\mathbb{R}^k$ a Cartesian coordinate system with components $(x_1,\ldots,x_k)$, and let $S^{(0)}$ denote the closed unit $(k-1)$-

dimensional sphere lying in the hyperplane perpendicular to the x_k-axis and centered at the origin. Given a unit vector $\boldsymbol{\theta}$ from the space Ω of all such vectors, let $S_{\boldsymbol{\theta}}^{(0)}$ denote the image of $S^{(0)}$ after rotation to a hyperplane with normal vector $\boldsymbol{\theta}$, still centered at the origin. Let Θ be a random vector distributed on Ω (not necessarily uniformly distributed), and define the random shape $S^{(0)}$ to be $S_{\Theta}^{(0)}$. Consider the Boolean model $C^{(0)}$ in which each shape is distributed as $S^{(0)}$ and the driving Poisson process has intensity λ.

Theorem 4.13.   *Take $k \geq 2$. Assume that the distribution of Θ has at least two points of support, $\boldsymbol{\theta}_1$ and $\boldsymbol{\theta}_2$, with $\boldsymbol{\theta}_1 \neq \boldsymbol{\theta}_2$ and $\boldsymbol{\theta}_1 \neq -\boldsymbol{\theta}_2$. Then there exists a constant λ_0, depending on the distribution of Θ, such that the probability that an arbitrary random set from $C^{(0)}$ is part of an infinite clump is strictly positive for all $\lambda > \lambda_0$.*

Theorem 4.13 admits many generalizations. For example, the set $S^{(0)}$ does not have to be hyperplanar—it could be a portion of the surface of a k-dimensional sphere of positive, even random radius. Such general results may be used to obtain sufficient conditions for high-intensity percolation in a wide variety of Boolean models. Roughly speaking, those conditions require that with positive probability S contains a random set with properties like the shape $S^{(0)}$ above. The corollary below is a convenient but relatively weak result of this type. Its proof is given at the very end of the section.

Consider the Boolean model in which shapes are distributed as S and the driving Poisson process has intensity λ. Recall that $\underline{v}(S)$ denotes the content of the largest sphere contained within S.

Corollary 4.14.   *Take $k \geq 2$. If $E\{\underline{v}(S)\} > 0$ then the probability that an arbitrary random set in the Boolean model is part of a clump of infinite order is positive for all sufficiently large λ.*

The probability $p(\lambda)$ that an arbitrary set is part of an infinite clump is a nondecreasing function of λ. (By an infinite clump we mean of course a clump composed of an infinite number of sets, not simply a clump for which the expected number of sets is infinite.) Combining Corollaries 4.12 and 4.14 we see that for many distributions of random shapes, there exists a critical intensity $\lambda_c > 0$ with the property that $p(\lambda) = 0$ if $\lambda < \lambda_c$ and $p(\lambda) > 0$ if $\lambda > \lambda_c$. Likewise there exists an intensity λ_c' beyond which the expected order of the clump containing an arbitrary set is infinite. The following corollary makes this explicit.

Corollary 4.15. *Take $k \geq 2$. If*

$$0 < E\{\underline{v}(S)\} \quad \text{and} \quad E\left\{\overline{v}(S)^{2-(1/k)}\right\} < \infty$$

then there exists $\lambda_c > 0$, depending on the distribution of S, such that the probability that an arbitrary random set in the Boolean model is part of a clump of infinite order is 0 for $\lambda < \lambda_c$ and positive for $\lambda > \lambda_c$. If

$$0 < E\{\underline{v}(S)\} \quad \text{and} \quad E\left\{\overline{v}(S)^2\right\} < \infty$$

then there exists $\lambda_c' > 0$, depending on the distribution of S, such that the expected order of the clump containing an arbitrary random set is finite for $\lambda < \lambda_c'$ and infinite for $\lambda > \lambda_c'$.

Clearly $0 < \lambda_c' \leq \lambda_c < \infty$, provided $0 < E\{\overline{v}(S)\}$ and $E\{\overline{v}(S)^2\} < \infty$.

We should stress that saying that the expected order of the clump containing an arbitrary set is infinite is quite different from saying that the expected order of an arbitrary clump is infinite. Indeed, the latter is *never* infinite if $E\{\overline{v}(S)\} < \infty$, even when clump order can be infinite with positive probability (see the closing paragraph of Section 4.6). This paradox arises partly because of the difficulty of defining an "arbitrary clump" when clump order can be infinite.

The notion of "criticality" described in Corollary 4.15 is familiar in several areas of probability theory. For example, in the case of a simple Galton-Watson branching process where offspring distribution has mean μ, the population will become extinct with probability 1 if $\mu \leq 1$ and explode with positive probability if $\mu > 1$ (Karlin and Taylor 1975, p. 396ff). Percolative processes on regular lattices are governed by critical probabilities, above which infinite clusters of bonds or sites occur with positive probability (Kesten 1982, Chapter 2). But unlike these areas, the continuum percolation model lacks much by way of orderly mathematical structure. Methods based on enumeration, which prove so useful in determining critical probabilities for lattice percolation, lose much of their power in the continuum case. Bounds for λ_c and λ_c' obtainable by theoretical arguments are quite poor, and one has to turn to Monte Carlo simulations or methods involving extrapolation from the lattice case for more explicit results.

It is clear that in some pathological cases we may have $0 = \lambda_c' < \lambda_c < \infty$; for example, take S to be a k-dimensional sphere with $E(\|S\|^{2-(1/k)}) < \infty$ and $E(\|S\|^2) = \infty$. No rigorous information is

TABLE 4.2. Estimated Critical Intensities for Boolean
Models Generated by Fixed-Radius Disks and Spheres in $k = 2$
and $k = 3$ Dimensions

Reference	Disks ($k = 2$)	Spheres ($k = 3$)
Gawlinski and Stanley (1981)	1.13	
Vicsek and Kertész (1981)	1.16	
Haan and Zwanzig (1977)	1.15	0.35
Fremlin (1976)	1.10	0.34
Gayda and Ottavi (1974)		0.33
Kurkijärvi (1974)		0.35
Ottavi and Gayda (1974)	1.02	
Pike and Seager (1974)	1.12	0.37
Ottavi and Gayda (1973)		0.38
Skal and Shkolovskii (1973)		0.38
Holcomb, Iwasawa, and Roberts (1972)		0.29
Domb (1972), based on Dalton, Domb, and Sykes (1964)	1.13 1.10	0.34
Holcomb and Rehr (1969)		0.29
Roberts and Storey (1968)		0.37
Roberts (1967)	0.96	
Gilbert (1961)	0.80	

In each case the results are standardized for shapes with unit content. Original results were standardized in a variety of ways, and some rounding errors may have been incurred by transformation. We list the estimated values in inverse chronological order, from the most recent (and presumably most accurate) to the oldest.

available in other situations. Some simulation studies (e.g., Roberts 1967, p. 625) make the explicit assumption that $\lambda_c = \lambda'_c$, while others make it implicitly. It is usually difficult to estimate λ_c from Monte Carlo experiments, but relatively easy to estimate λ'_c. A common approach to the latter problem is to estimate the asymptote of the graph of mean clump size against Poisson intensity.

Perhaps the most convincing evidence that λ_c and λ'_c are close or identical in many cases comes from approximation by the lattice case (e.g., Dalton, Domb, and Sykes 1964; Shante and Kirkpatrick 1971, Section 7). Table 4.2 summarizes various estimates of critical intensity (λ_c and/or λ'_c) in the case of Boolean models generated by fixed-radius spheres in $k = 2$ or 3 dimensions.

Continuum percolation can be used as a model for electrical conductivity in doped crystalline semiconductors, for polymer condensation in chemical solutions, for communication in random networks, for spread of contagious disease, etc. In the first of these examples, sets in the Boolean model (usually disks or spheres) represent electron-bearing impurities. As voltage across the crystal increases, so does the proportion of impurities available for conduction. Conduction occurs at a threshold voltage that depends on physical properties of the crystal and the impurities, and on the density of impurities.

We close this section with proofs of the main results stated above.

Proof of Theorem 4.11.   We may assume without loss of generality that the sphere S is centered at the origin. Let $R \equiv \operatorname{rad}(S)$.

Part (i). Let $R' - 1$ equal the integer part of R. The conditions $E(\|S\|^{2-(1/k)}) < \infty$, $E(R^{2k-1}) < \infty$, and $E\{(R')^{2k-1}\} < \infty$ are equivalent. If sphere radius is given the distribution of R' rather than R then the probability of an infinite clump occurring will not decrease. Therefore it suffices to establish part (i) of the theorem in the special case where R takes only positive integer values. We shall prove that if $E(R^{2k-1}) < \infty$ then the expected number of spheres comprising the perimeter of the clump containing an arbitrary set is finite, provided λ is sufficiently small. From this it follows that the order of the clump is almost surely finite, and (since there is only a countable number of clumps) that with probability 1 all clumps are of finite order.

We shall construct a multitype branching process (see Appendix V) to bound the number of spheres in the clump perimeter. There will be a countable infinity of types, indexed by positive integers corresponding to sphere radii. Our first step is to determine the distribution of types.

Suppose an initial sphere of radius i is centered at a point $\mathbf{z}$. The number of spheres of radius j in our coverage process that intersect the initial sphere, and that protrude at least partially beyond that sphere, has a Poisson distribution with parameter

$$\mu_{ij} \equiv \lambda v_k \left[(i+j)^k - \{\max(0, i-j)\}^k \right] p_j, \qquad (4.81)$$

where $p_j \equiv P(R = j)$. Let this number be N_j. The variables N_j, $j \geq 1$, are stochastically independent. We ignore spheres that are wholly contained within the initial sphere, since they cannot contribute to the perimeter of the clump. If an individual of type i is present in the nth generation, then the vector of numbers of types of his progeny in the $(n+1)$th generation will be given the same distribution as $(N_1, N_2, \ldots)$. Here N_j represents the number of children of type j.

Using this distribution of types, we construct the branching process as below. The individuals in the process are points in $\mathbb{R}^k$. The individual in the 0th generation is the center of the "arbitrary" sphere, which we take without loss of generality to be the origin. Given individuals $\varsigma_{n1}, \ldots, \varsigma_{nN_n}$ in the nth generation, we define the $(n+1)$th generation as follows. Suppose ς_{nl} is of type i. Let $\mathcal{P}_{nl}$ be a Poisson process in $\mathbb{R}^k$ of intensity λ, independent of the previous history of the process and also of $\mathcal{P}_{nl'}$ for $l' \neq l$. Center spheres at the points of $\mathcal{P}_{nl}$, the radii being independent and distributed as R. The progeny of ς_{nl} of type j in the $(n+1)$th generation are those points of $\mathcal{P}_{nl}$ whose associated spheres are of radius j and that intersect the sphere of radius i centered at Z_{nl}, but are not wholly contained within that sphere. The expected number of individuals in all generations of this multitype process is greater than or equal to the expected number of spheres that protrude from the clump containing the initial sphere in the coverage process. [This observation uses the "lack of memory" property of a Poisson process. Specifically, if $\mathcal{P}$ is homogeneous Poisson and if $x_1, \ldots, x_m$ are arbitrary fixed points, then the conditional, or Palm, distribution of $\mathcal{P} \setminus \{x_1, \ldots, x_m\}$, given that points of $\mathcal{P}$ occur at $x_1, \ldots, x_m$, is the same as the unconditional distribution of $\mathcal{P}$. See Chapter 1, Section 1.7. This property, and the fact that our construction adds extra points in certain cases, ensures that the conditional distribution of the $(k+1)$st generation of the branching process dominates that of the coverage process.]

We prove next that the expected total population size is finite for all sufficiently small λ. The expected number of immediate type j progeny born to a type i individual equals μ_{ij}. Define $v_i^{(n)}$ to be the expected number of type i individuals in the nth generation, and let $\mathbf{v}^{(n)}$ be the infinite row vector whose ith element is $v_i^{(n)}$. If the initial individual was of type i, then

$$\mathbf{v}^{(n)} = \mathbf{i}\mathbf{M}^n,$$

where $\mathbf{i}$ is the row vector whose ith element is 1 and all other elements are 0, and where $\mathbf{M} = (\mu_{ij})$ is a matrix with an infinite number of rows and columns (see Appendix V). Therefore the expected total number of individuals in the nth generation, given that the initial individual was of type i, is

$$\sum_{j=1}^{\infty} v_j^{(n)} = \sum_{j=1}^{\infty} \mu_{ij}^{(n)},$$

where $\mu_{ij}^{(n)}$ is the (i,j)th element of $\mathbf{M}^n$. The expected total number of individuals in all generations, given that the initial individual was of type i, is

$$\mu_i \equiv \sum_{n=1}^{\infty} \sum_{j=1}^{\infty} \mu_{ij}^{(n)}. \tag{4.82}$$

In view of formula (4.81), for $i \le j$ we have

$$\mu_{ij} = \lambda v_k (i+j)^k p_j \le 2^k \lambda v_k j^k p_j,$$

while for $i > j$,

$$\mu_{ij} = \lambda v_k \left\{ (i+j)^k - (i-j)^k \right\} p_j \le \text{const.} \, \lambda i^{k-1} j p_j,$$

where the constant depends on none of i, j, or λ. Therefore in general,

$$\mu_{ij} \le c \lambda j \{ \max(i,j) \}^{k-1} p_j \le c \lambda i^{k-1} j^k p_j$$

for all i, j, and λ, where c is chosen $\ge 2^k v_k$ and depends on none of i, j, and λ. Consequently,

$$\mu_{ij}^{(2)} = \sum_{l=1}^{\infty} \mu_{il} \mu_{lj}$$

$$\le (c\lambda)^2 i^{k-1} j^k p_j \sum_{l=1}^{\infty} l^{2k-1} p_l$$

$$= (c\lambda)^2 \mu i^{k-1} j^k p_j$$

$$\le (c\lambda \mu)^2 i^{k-1} j^k p_j,$$

where $\mu \equiv \sum_{l \ge 1} l^{2k-1} p_l < \infty$, the last inequality following from the fact that $E(R^{2k-1}) < \infty$. If $\mu_{ij}^{(n-1)} \le (c\lambda\mu)^{n-1} i^{k-1} j^k p_j$ for all i, j, and λ, then it follows easily that $\mu_{ij}^{(n)} \le (c\lambda\mu)^n i^{k-1} j^k p_j$, and so the latter formula must be true for all i, j, n, and λ, using mathematical induction. Substituting this estimate into (4.82) we see that

$$\mu_i \le i^{k-1} \sum_{n=1}^{\infty} (c\lambda\mu)^n \sum_{j=1}^{\infty} j^k p_j < \infty,$$

provided only that λ is chosen so small that $c\lambda\mu < 1$. In this case, the expected number of spheres that form the perimeter of the clump containing an arbitrary sphere of radius i is finite. Since each of these spheres has finite radius then the dimensions of the clump are finite with probability 1. Therefore the total number of spheres making up the clump must be finite with probability 1.

Part (ii). First we prove that if $E(\|S\|^2) < \infty$, or equivalently, if the distribution of R satisfies $E(R^{2k}) < \infty$, then the expected number of spheres in an arbitrary clump is finite provided λ is chosen sufficiently small. This requires only a modification of the branching process argument given above. We assume as before that R takes only integer values. On the present occasion we must bound the *total number* of spheres in the clump, not just the number of spheres protruding from the clump.

Suppose a sphere of radius i is centered at a point $\mathbf{z}$. Instead of N_j we consider N'_j, equal to the number of spheres of radius j in our coverage process that intersect the initial sphere. The variables N'_j are stochastically independent, and N'_j is Poisson-distributed with mean

$$\mu'_{ij} \equiv \lambda v_k (i+j)^k p_j. \tag{4.83}$$

The branching process is defined as before, except that the type j progeny of the type i individual ς_{nl} in the $(n+1)$th generation are taken to be those points in $\mathcal{P}_{nl}$ whose associated spheres are of radius j and intersect the sphere of radius i centered at ς_{nl}. We may derive an analogue of formula (4.82), and so the proof of this part of the theorem will be complete if we show that

$$\sum_{n=1}^{\infty}\sum_{j=1}^{\infty} \mu'^{(n)}_{ij} < \infty, \tag{4.84}$$

where $\mu'^{(n)}_{ij}$ is the (i,j)th element of the nth power of the matrix $\mathbf{M}' \equiv (\mu'_{ij})$.

In view of (4.83) we have

$$\mu'_{ij} \leq c\lambda\{\max(i,j)\}^k p_j \leq c\lambda i^k j^k p_j$$

for all i, j, and λ, where $c = 2^k v_k$. It now follows as before that for $n \geq 1$,

$$\mu'^{(n)}_{ij} \leq (c\lambda\mu')^n i^k j^k p_j,$$

where $\mu' \equiv \sum_l l^{2k} p_l < \infty$. Therefore (4.84) will hold if λ is chosen so small that $c\lambda\mu' < \infty$.

Completion of Part (ii) and Part (iii). Here we show that if $E(\|S\|^2)$ $= \infty$ then the expected number of spheres that are in the same clump as a given sphere and are distant no more than one sphere away from that sphere is infinite for all values of λ. Let R have the distribution of the radius of S, define R'' to be the integer part of R, and note that $E(R'')^{2k} = \infty$. If sphere radius is given the distribution of R'' then the expected number of spheres distant one or more spheres away from a given sphere will not exceed the expected value in the case where radius has the distribution of R. Therefore we may assume without loss of generality that R takes only nonnegative integer values. (The sphere of radius zero centered at $\mathbf{z}$ is taken to be the singleton $\{\mathbf{z}\}$.)

Suppose a sphere $S^{(1)}$ of radius $i \geq 0$ is centered at a point $\mathbf{z}$. The number, N_j, of spheres of radius j that intersect $S^{(1)}$, is Poisson-distributed with parameter

$$\mu_j \equiv \lambda v_k (i+j)^k p_j,$$

where $p_j \equiv P(R = j)$. Let M denote the largest value of j for which $N_j > 0$, except that we define $M = -1$ if no spheres intersect $S^{(1)}$. The variables N_j are independent, and so

$$P(M = m) = P(N_m > 0) \prod_{j=m+1}^{\infty} P(N_j = 0)$$

$$= \{1 - \exp(-\mu_m)\} \exp\left(-\sum_{j=m+1}^{\infty} \mu_j\right), \qquad m \geq 0,$$

with

$$P(M = -1) = 1 - \sum_{m=0}^{\infty} P(M = m) = \exp\left(-\sum_{j=0}^{\infty} \mu_j\right).$$

Note that the entire Boolean model may be regarded as the superposition of *independent* Boolean models $C_0, C_1, \ldots,$ where C_j is generated by spheres of fixed radius j centered at points of a Poisson process of intensity λp_j. The event $\{M = m\}$ is the same as the event $\{N_m > 0; N_j = 0 \text{ for } j \geq m+1\}$, and so is measurable in the σ-field generated by the processes $C_m, C_{m+1}, \ldots$. Thus, for any $m > 0$ the events $\{M = m\}$ and $\{M \geq m\}$ are stochastically independent of any event in the σ-field generated by $C_0, C_1, \ldots, C_{m-1}$. Define

$$t \equiv \inf\{n \geq 1 : p_n > 0\}.$$

Conditional on $M \geq t+1$, let $S^{(2)}$ be any sphere of radius M intersecting $S^{(1)}$. In view of the preceding discussion, the following is true. Conditional on $M = m$, where $m \geq t+1$, the number of spheres of radius t that intersect $S^{(2)}$ is Poisson-distributed with mean

$$\nu_m \equiv \lambda v_k (m+t)^k p_t.$$

Therefore the expected number of spheres of radius t distant one sphere or less away from the initial sphere of radius i is not less than

$$\rho \equiv \sum_{m=t+1}^{\infty} P(M=m)\nu_m$$

$$\geq \lambda v_k p_t \exp\left(-\sum_{j=0}^{\infty} \mu_j\right) \sum_{m=t+1}^{\infty} \{1 - \exp(-\mu_m)\} m^k. \quad (4.85)$$

Since $E(\|S\|) < \infty$ then for fixed i,

$$\sum_{j=0}^{\infty} \mu_j \leq \text{const.} \sum_{j=1}^{\infty} j^k p_j < \infty,$$

and also,

$$1 - \exp(-\mu_m) \geq \mu_m \exp(-\mu_m)$$
$$\geq \lambda v_k m^k p_m \exp(-\mu_m)$$
$$\geq \text{const.} m^k p_m$$

for all $m \geq 1$, where "const." denotes a generic positive constant not depending on m. Substituting these estimates into (4.85), we see that

$$\rho \geq \text{const.} \sum_{m=t+1}^{\infty} m^{2k} p_m = \infty,$$

since $E(R^{2k}) = \infty$. Therefore $\rho = \infty$ for each value of i, which completes the proof of Theorem 4.11. $\square$

Proof of Corollary 4.12. The Corollary is trivial when $k = 1$. Therefore we may confine attention to the case $k \geq 2$. Let $\mathcal{C}$ be a Boolean model in which shapes are distributed as S. Given the random shape S, let T be the closed k-dimensional sphere centered at the origin and such that $\|T\| = \bar{v}(S)$, and let $\mathbf{Y}$ be a random vector in $\mathbb{R}^k$ such that $S \subseteq \mathbf{Y} + T$. Consider the Boolean model $\mathcal{C}'$

in which shapes have the distribution of $\mathbf{Y} + T$ and are centered at points of a Poisson process of the same intensity as that driving $\mathcal{C}$. The expected number of sets per clump, and the probability of an arbitrary set being part of an infinite clump, are not greater for C than for C'. The Corollary follows on applying Theorem 4.11 to the Boolean model C'. $\square$

Proof of Theorem 4.13.   We derive a lower bound to the probability that an arbitrary set is part of an infinite clump by comparing the Boolean model to a site percolation process on a rectangular lattice in $\mathbb{R}^k$.

Let $\epsilon > 0$ be so small that the sets

$$T_i(3\epsilon) \equiv \{\boldsymbol{\theta} \in \Omega : |\boldsymbol{\theta} - \boldsymbol{\theta}_i| \le 3\epsilon\}, \qquad \text{for} \quad i = 1 \text{ and } 2,$$

are disjoint, and the sets $T_1(3\epsilon)$ and $-T_2(3\epsilon)$ are disjoint. (The set Ω is the surface of the unit k-dimensional sphere centered at the origin.) Given $\mathbf{z} \in \mathbb{R}^k$ and $\boldsymbol{\theta} \in \Omega$, let $U(\mathbf{z}, \boldsymbol{\theta})$ denote the open $(k-1)$-dimensional sphere of unit radius centered at $\mathbf{z}$ and whose plane has its normal in the direction of $\boldsymbol{\theta}$. Let $\mathbb{Z}$ be the set of all integers. The sites in the percolation process will be the points of the lattice $(r_1 \mathbb{Z})^k$, for some $r_1 > 0$. Two sites will be said to be adjacent if they are distant exactly r_1 apart.

Choose r_1 so small that if $\mathbf{x}_1$ and $\mathbf{x}_2$ are any two adjacent sites of the lattice, the two $(k-1)$-dimensional spheres $\frac{1}{2}U(\mathbf{x}_1, \boldsymbol{\phi}_1)$ and $\frac{1}{2}U(\mathbf{x}_2, \boldsymbol{\phi}_2)$ have nonempty intersection whenever $\boldsymbol{\phi}_i \in T_i(2\epsilon)$ for $i = 1$ and 2. Next, choose $r_2 \in (0, \frac{1}{2}r_1)$ so small that the spheres $U(\mathbf{y}_1, \boldsymbol{\phi}_1)$ and $U(\mathbf{y}_2, \boldsymbol{\phi}_2)$ have nonempty intersection whenever $\mathbf{x}_1$ and $\mathbf{x}_2$ are adjacent sites, $|\mathbf{x}_i - \mathbf{y}_i| \le r_2$ for $i = 1$ and 2, and $\boldsymbol{\phi}_i \in T_i(\epsilon)$ for $i = 1$ and 2.

Let $T(\mathbf{x}, r_2)$ denote the closed k-dimensional sphere of radius r_2 centered at the point $\mathbf{x}$. Classify each site of the lattice $(r_1 \mathbb{Z})^k$ as either "type 1" or "type 2," in such a manner that no type i site is adjacent to another type i site, for $i = 1$ and 2. (Once any given site has been classified the types of all other sites are determined, so there are only two different possible classifications.) Distribute the points ξ_j of the Poisson process $\mathcal{P}$ throughout $\mathbb{R}^k$. Let $\mathbf{x} \in (r_1 \mathbb{Z})^k$ be a site of type i. We shall say that $\mathbf{x}$ is *occupied* if some point of $\mathcal{P}$ lies within the sphere $T(\mathbf{x}, r_2)$, and is such that the associated random shape $S_j^{(0)}$ [actually, a $(k-1)$-dimensional sphere $U(\mathbf{0}, \boldsymbol{\Theta}_j)$, for some random $\boldsymbol{\Theta}_j$] has orientation $\boldsymbol{\Theta}_j$ lying in the set $T_i(\epsilon)$. Write $\mathcal{P} = \{\xi_i, i \ge 1\}$ and $C^{(0)} = \{\xi_i + S_i^{(0)}, i \ge 1\}$. It

follows from our construction that the following properties hold. (i)
The probability p_i that a site of type i is occupied depends only
on i and not on other characteristics of the site. (ii) The value
of p_i increases to 1 as Poisson intensity, λ, increases to infinity.
(iii) The occupation of a given site $\mathbf{x}$ is stochastically independent
of the occupation pattern of any set A of sites for which $\mathbf{x} \notin A$.
(iv) If two adjacent sites $\mathbf{x}_1$ and $\mathbf{x}_2$ are occupied, then there exist
points $\xi_{j_l} \in \mathcal{T}(\mathbf{x}_l, r_2)$, for $l = 1$ and 2, such that the random sets
$\xi_{j_l} + S_{j_l}^{(0)}$, $l = 1$ and 2, have nonempty intersection. (v) Suppose we
delete from $\mathcal{C}^{(0)}$ all sets that are *not* of the form $\xi + U(\mathbf{0}, \Theta)$ for
some $\Theta \in \mathcal{T}_i(\epsilon)$ and $\xi \in \mathcal{T}(\mathbf{x}, r_2)$, where $\mathbf{x}$ is a type i site ($i = 1$ or
2). Let $\mathcal{C}^{(0)'}$ denote the resulting sequence of sets. If there exists
an infinite path comprised of bonds linking adjacent occupied sites,
then there occurs an infinite clump of sets from $\mathcal{C}^{(0)'}$.

In view of property (v), there will exist an infinite clump of
random sets $\xi_i + S_i^{(0)}$ if there exists an infinite path comprised of
bonds linking adjacent occupied sites. Consider the simpler site-
percolation process on the cubic lattice $\mathbb{Z}^k$, in which each site is
occupied with probability $p = \min(p_1, p_2)$ and vacant with probability $1 - p$, independently of all others. The probability that a given
site is part of an infinite path (or "cluster") for this process, is no
less than the probability that a site is part of an infinite path for
the former site-percolation process. It follows from the theory of
site percolation (Appendix VI) that if $p < 1$ is sufficiently large, the
probability of an infinite path occurring somewhere in $\mathbb{R}^k$ equals 1.
In view of property (ii) above, this means that the probability of an
infinite path occurring equals 1 for all sufficiently large λ. Therefore
if λ is sufficiently large then an infinite clump of sets from $\mathcal{C}^{(0)}$ occurs somewhere in $\mathbb{R}^k$. For such a value of λ, the probability that
an arbitrary set is part of an infinite clump must be positive. (If it
were 0, no infinite clump could occur anywhere in $\mathbb{R}^k$.) $\quad \square$

Proof of Corollary 4.14. Let T be an open sphere such that
$T \subseteq S$ and $\|T\| = \underline{v}(S)$. Without loss of generality, T is centered at
the origin. (This property may be achieved by giving S a random
translation, and such motions do not affect properties associated
with Boolean models; see Section 4.1.) Since $E\{\underline{v}(S)\} > 0$, we may
choose a fixed closed sphere $\mathcal{T}$ of radius $2r > 0$ and centered at the
origin, such that $P(\mathcal{T} \subseteq T) > 0$. Let $\mathcal{C} = \{\xi_i + S_i, i \geq 1\}$ denote the
Boolean model under study. Delete $\xi_i + S_i$ from $\mathcal{C}$ if $\mathcal{T} \not\subseteq S_i$, and
replace $\xi_i + S_i$ by $\xi_i + \mathcal{T}$ if $\mathcal{T} \subseteq S_i$. The result is a Boolean model $\mathcal{C}'$

in which shapes are all fixed and equal to $\mathcal{T}$ and the driving Poisson process has intensity $\lambda P(\mathcal{T} \subseteq S)$. If infinite clumps form in $\mathcal{C}'$ with positive probability, they also have positive probability of forming in $\mathcal{C}$. By rescaling $\mathcal{C}'$ in the obvious way we may assume that $\mathcal{T}$ has unit radius, and in that case $\mathcal{T}$ contains any number of $(k-1)$-dimensional unit spheres centered at the origin, of the type $\mathcal{S}^{(0)}$ considered in Theorem 4.13. It now follows from Theorem 4.13 that infinite clumping occurs in $\mathcal{C}'$ (and hence in $\mathcal{C}$) if λ is sufficiently large. $\square$

EXERCISES

4.1 Consider a coverage process in $\mathbb{R}^k$ where the driving point process is stationary with intensity λ, and shapes are distributed as the random set S, not necessarily bounded. Prove that the expected number of sets covering any given point $\mathbf{x} \in \mathbb{R}^k$ equals $\lambda E(\|S\|)$. [Note that it is possible to have $E(\|S\|) < \infty$, but to still have S unbounded with probability one.]

4.2 Consider a Boolean model $\mathcal{C}$ in the plane, in which the driving Poisson process $\mathcal{P}$ has intensity λ and each shape is distributed as the random isotropic convex set S. Suppose the pattern within a given convex region $\mathcal{R}$, a convex subset of $\mathbb{R}^2$, is recorded. Had we been able to observe the position of each point of $\mathcal{P}$ then a natural estimator of λ would have been

$$\hat{\lambda}^* \equiv (\text{number of points of } \mathcal{P} \text{ from within } \mathcal{R})/\|\mathcal{R}\|,$$

which has variance $\lambda\|\mathcal{R}\|^{-1}$. The estimator $\hat{\lambda}$ defined at (4.21), which does not require information about centers of random sets, actually does better than $\hat{\lambda}^*$ in terms of variance [see (4.23)]. Explain why, and discuss.

4.3 Let $\mathcal{C} \equiv \{\xi_i + S_i, i \geq 1\}$ be a Boolean model of random closed sets in the plane in which Poisson intensity equals λ and shapes are distributed as the random compact set S. The lowest right-hand endpoint of the set $\xi_i + S_i$ is said to lie within a region $\mathcal{R}$ if $\mathcal{R}$ contains a point $\mathbf{x} = (x_1, x_2)$ such that $\mathbf{x} \in \xi_i + S_i$ and

$$\sup\{y_1 : (y_1, y_2) \in \xi_i + S_i \text{ for some } y_2 \in \mathbb{R}\} = x_1,$$
$$\inf\{y_2 : (x_1, y_2) \in \xi_i + S_i\} = x_2.$$

Prove that the number of lowest right-hand endpoints within $\mathcal{R}$ is Poisson-distributed with mean $\lambda\|\mathcal{R}\|_2$.

4.4 Consider the Boolean model $\mathcal{C} \equiv \{\xi_i + S_i, i \geq 1\}$ of Exercise 4.3, and assume S is isotropic and an oriented polygonal region with probability 1. Let $\mathcal{R}$ be a connected orientable polygonal subset of $\mathbb{R}^2$, containing no voids. Prove that the expected Euler characteristic of the region

$$\mathcal{R} \cap \left\{ \bigcup_{i=1}^{\infty} (\xi_i + S_i) \right\}^{\sim}$$

equals

$$y \equiv \left\{ 1 + (4\pi)^{-1}(\beta\lambda)^2 \|\mathcal{R}\|_2 \right.$$
$$\left. + (2\pi)^{-1}\beta\lambda\|\partial\mathcal{R}\|_1 - \chi\lambda\|\mathcal{R}\|_2 \right\} e^{-\alpha\lambda},$$

where $\alpha \equiv E(\|S\|_2)$, $\beta \equiv E(\|\partial S\|_1)$ and χ equals the mean Euler characteristic of S. Give an intuitive argument suggesting that under conditions of high intensity, e^{-y} might be a good approximation to the probability that $\mathcal{R}$ is completely covered. Compare this with formula (3.68) derived in Chapter 3.

[Kellerer (1983)]

4.5 [Continuation] Suppose the set S is an isotropic random line segment. Assume that the distribution of segment length is bounded below by $2\,\mathrm{rad}\,(\mathcal{R})$, so that it is impossible for a segment to be contained entirely within the observation region $\mathcal{R}$. Use the result of Exercise 4.4 to deduce a formula for the expected number of cells into which $\mathcal{R}$ is divided by the random segments.

[Kellerer (1983)]

4.6 Show that under regularity conditions on shape distribution, the tangent count estimator $\hat{\lambda}_4$ is unbiased for λ if the driving point process is stationary. Prove strong consistency within an expanding region if the point process is Poisson.

4.7 Use marked point process and "Poisson-ness" arguments to give a detailed proof of formula (4.41) for the probability that an arbitrary sphere in a Boolean model of k-dimensional spheres is isolated.

4.8 Consider the one-dimensional Boolean model in which shapes are distributed as the line segment S with mean length $\alpha \equiv$

$E(\|S\|) < \infty$, and the driving Poisson process has intensity $\lambda > 0$. Show that the expected number of segments comprising an arbitrary clump equals $e^{\alpha\lambda}$. [Hint: There is a variety of proofs. One proceeds via the Laplace transform, as in Exercise 2.10. Alternatively, bypassing that result, use the ergodic theorem (Appendix II) and Theorem 2.2 to prove successively that the mean length of a clump/spacing pair is $\lambda^{-1}e^{\alpha\lambda}$, the mean number of clump/spacing pairs per unit length is $(\lambda^{-1}e^{\alpha\lambda})^{-1} = \lambda e^{-\alpha\lambda}$, and the mean number of segments per clump equals $\lambda/(\lambda e^{-\alpha\lambda}) = e^{\alpha\lambda}$.]

4.9 [Continuation] Suppose the line segments in Exercise 4.8 are of fixed length a. Let N_1 have the distribution of the number of segments in an arbitrary clump, and N_2 the distribution of the number of segments in the clump containing an arbitrary segment.

(i) Prove that $N_1 - 1$ has a geometric distribution. [If N has the geometric distribution with "probability" (i.e., parameter) p then $P(N = n) = p^n(1 - p)$, $n \geq 0$.]

(ii) Write down the exact distribution of N_2. [Hint: Consider that part of the clump lying to the right of the left-hand endpoint of the arbitrary segment, and then look in the opposite direction.]

4.10 Consider a one-dimensional Boolean model in which shapes are line segments of *random* length. As in Exercise 4.9, let N_1 and N_2 have the distribution of the number of segments in an arbitrary clump and the number of segments in the clump containing an arbitrary segment, respectively. Prove that

$$E(N_2) = E(N_1^2)/E(N_1).$$

The value of $E(N_1^2)$ may be found from Chapter 2, Exercise 2.10. In particular it follows that $E(N_2) < \infty$ if and only if segment length has finite variance.

4.11 Consider the one-dimensional Boolean model in which the driving Poisson process has intensity $\lambda > 0$ and shapes are distributed as the "discrete" random set S, where

$$P(S = \mathcal{S}_n) = cn^{-2}, \qquad n \geq 1,$$

$$\mathcal{S}_n \equiv \bigcup_{i=-n^2}^{n^2} (in^{-1} - (2n^2 + 1)^{-1}, in^{-1} + (2n^2 + 1)^{-1}), \quad n \geq 1,$$

and $c \equiv (\sum_{n \geq 1} n^{-2})^{-1}$. Show that $E(\|S\|) = 2$. Prove that each interval, no matter how small, is intersected with probability 1 by an infinite number of random sets from the Boolean model. This implies that with probability 1, each random set in the Boolean model has nonempty intersection with an infinite number of other random sets. [Hint: Let the random sets in the Boolean model be $\xi_i + S_i$, $i \geq 1$, where the S_i's are independent copies of S. Define N_i by $S_i = S_{N_i}$. Prove that the events

$$\mathcal{E}_j \equiv \{\text{for some } i, \ \xi_i \in (j-1, j) \text{ and } |\xi_i| \leq N_i\},$$
$$-\infty < j < \infty,$$

are independent and satisfy

$$\sum_{-\infty < j < \infty} P(\mathcal{E}_j) = \infty.]$$

4.12 Let R be a positive integer-valued random variable, and let S (a subset of $\mathbb{R}^k$) be the union of infinitesimally thin concentric shells of integer radius i, $1 \leq i \leq R$, centered at the origin:

$$S \equiv \left\{ \mathbf{x} \in \mathbb{R}^k : |\mathbf{x}| = i \text{ for some } 1 \leq i \leq R \right\}.$$

Prove that if $E(R^k) = \infty$ then in any k-dimensional Boolean model generated by shapes having this distribution, each set intersects an infinite number of other sets, with probability 1. Describe how to modify S so as to make it homeomorphic to a sphere, but still preserving the conclusion and having $E(\|S\|) < \infty$.

4.13 Consider the two-dimensional Boolean model in which random sets are isotropic line segments of mean length α, and the driving Poisson process has intensity λ. Prove that there exists $\lambda_0 > 0$ such that

$P(\text{an arbitrary set is part of a clump of infinite order})$

is positive for $\lambda > \lambda_0$, and 0 for $\lambda < \lambda_0$, if and only if $\alpha < \infty$.

4.14 Let $\mathcal{C} \equiv \{\xi_i + S_i, i \geq 1\}$ be a Boolean model in $\mathbb{R}^3$, where the driving Poisson process $\mathcal{P} \equiv \{\xi_i, i \geq 1\}$ has intensity λ and the random shapes S_i are distributed as the isotropic convex random set S. Let Π be a plane in $\mathbb{R}^3$. If $\xi_i + S_i$ intersects Π, let $\boldsymbol{\eta}_i$ denote the foot of the perpendicular from

ξ_i to Π. Prove that the set of such points η_i comprises a stationary Poisson process in Π with intensity $\delta\lambda$, where δ equals the mean distance between two parallel tangent planes to S. [Hint: Use the Poisson-ness and marked point process arguments. Note axioms (i) and (ii) for a stationary Poisson point process, given in Section 1.7.]

NOTES

Section 4.1

The notion of a marked point process goes back to Matthes (1963). See also Snyder (1975, Chapter 7), Matthes, Kerstan, and Mecke (1978), Franken, König, Arndt, and Schmidt (1981, Chapter 1), and Daley and Vere-Jones (1988, Chapter 12). Applications in stochastic geometry, for Boolean models and for more general coverage processes, appear in Stoyan (1979a,b), Hanisch and Stoyan (1979), Stoyan (1979c), Mecke and Stoyan (1980b), Mecke (1981), Stoyan and Mecke (1983), and Stoyan, Kendall, and Mecke (1987).

Some references for Campbell theorems were given in Section 4.1. See Moran (1968b, Chapter 9, Section 9.8) for a particularly lucid and informative account, with references to early work. The Campbell theorem has an associated spectral theory (Moran 1968b, p. 422), which could be of use in inference for coverage processes.

General and concise accounts of Palm distribution theory are given by Franken, König, Arndt, and Schmidt (1981, Section 1.2), Kallenberg (1983, Chapter 10), Karr (1986, Section 1.7), Stoyan, Kendall, and Mecke (1987, Section 4.4), and Daley and Vere-Jones (1988, Chapter 12).

Section 4.2

Theorems 4.1–4.3 in Section 4.2 are essentially special cases of general results in integral geometry, in particular Blaschke's "fundamental formula" (see Blaschke 1949, p. 49; Santaló 1953, p. 36; Miles 1974a; and Santaló 1976, p. 114, in the case $k = 2$). Our approach to Theorems 4.1 and 4.2 resembles that of Mack (1954, 1956). Theorem 4.3 may be proved from Cauchy's formula for the mean projection length of an isotropic line segment. See also Morton (1966). The theory of convex sets is discussed by Bonnesen and Fenchel (1948).

Discussion of ways of correcting counts for edge effects abounds in literature on microscopy and on industrial safety (e.g.,, Underwood 1970, Section 1.3.5; Gundersen 1977, 1978; Cooper, Feldman, and Chase 1978; Miles 1978; T. Schneider 1979). Miles (1974c) initiates a systematic approach to edge effect problems, introducing the associated point method. Related approaches include Gundersen's tiling method (Gundersen 1977) and tangent count techniques (e.g., Weibel 1973; DeHoff 1978; Ripley 1981, p. 202). Jensen and Sundberg (1986) have discussed the use of multiple associated points. Results in their Appendix A permit a very general account of the variance-reduction phenomenon discussed in our Exercise 4.2. Schwandtke, Ohser, and Stoyan (1987) introduce and discuss alternative low-variance estimators. See also Kellerer (1985). Laslett and Liow (1984) use transformations of the Boolean model pattern to estimate Poisson intensity. Their approach is also related to Miles' (1974c) associated point method. Parts of Theorem 4.4 are stated in Hall (1985f).

The fact that $\hat{\lambda}$ is unbiased for λ has long been known and used in various cases. In the case where random shapes are line segments (e.g., modeling asbestos fibers), edge-correction formulas may be derived via theory of homogeneous line segment processes; see e.g., Colman (1972), Parker and Cowan (1976), Fava and Santaló (1978), Cowan (1979), Mecke and Stoyan (1980a), Mecke (1981), and Stoyan and Stoyan (1986). Cox's appendix to Palmer (1948), and Cox (1969), are also relevant here. Lancaster (1950) discusses bias in the estimation of count variance.

Section 4.3
Theorems 4.5 and 4.6 in Section 4.3 are straightforward consequences of more general formulas of integral geometry. See for example Santaló (1976, p. 112ff). Our development is influenced by Kellerer (1983). See also Kellerer (1986). Properties of curvature play a major role in stereological problems (Underwood 1970; Miles 1976; Miles and Davy 1976; Davy and Miles 1977; Baddeley 1980b; Ohser 1980, last paragraph of Section 3). Applications of the tangent count estimator are surveyed by DeHoff (1978), Ripley (1981, p. 202) and Stoyan, Kendall, and Mecke (1987, pp. 201–203).

Advances in machine-based image analysis have inspired the development of integral geometric formulas that are suitable for "automatic" use. Some techniques based on tangent counts are in this category. More recent examples may be found in R. Schneider (1979a, 1979b, 1980), Weil (1983, 1984), Weil and Wieacker (1983), and Zähle (1984a, 1984b, 1986).

Sections 4.4–4.6
Mack (1954) derives the formula in Section 4.4 for the expected

number of isolated sets per unit content in the case of a two-dimensional Boolean model, and obtains the upper bound in Section 4.6 for the expected total number of clumps per unit area. See also Mack (1948, 1949) and Silberstein (1945). Mack (1956) elucidates some of the two-dimensional work in Mack (1954), and generalizes it to three dimensions. Hall (1986) gives a central limit theorem for counts of isolated sets. See also Lomnicki and Zaremba (1957) and Kester (1975). Armitage (1949) derives approximations to expected clump counts of various orders, for several types of shapes in a two- dimensional Boolean model. Section 4.5 further develops Armitage's work. See also Irwin, Armitage, and Davies (1949), Fullman (1953), Mack (1953), and Hasofer (1963).

Section 4.7
Physicists and mathematicians take distinctly different views of the percolation literature, in both the lattice and continuum cases. Recognition of the problem of continuum percolation dates from Gilbert (1961). Pike and Seager (1974) and Seager and Pike (1974) provide good introductions from a physicist's viewpoint. Calculations of rigorous bounds to critical percolative intensities are reported in Gilbert (1961), Pike and Seager (1974), Kirkwood and Wayne (1983), and Hall (1985g), but the range between upper and lower bounds is invariably considerable. In the case of fixed-radius, unit-area disks in $k = 2$ dimensions, the best currently available bounds appear to be $0.54 < \lambda_c < 2.78$ (Hall 1985g), compared with current estimates of $\lambda_c \doteq 1.1$ or 1.2 (Table 4.1).

Studies of continuum percolation with truly random shapes, as distinct from fixed shapes, include Kertész and Vicsek (1982), Gawlinski and Redner (1983), and Phani and Dhar (1984). Section 4.7 is based on Hall (1985g). Note that in Hall (1985g) an "arbitrary clump" is taken to be "a clump containing an arbitrary set." That definition is not really satisfactory—see Section 4.6—and in Section 4.7 we refer explicitly to "a clump containing an arbitrary set."

Work at the interface between applications and the mathematical theory of percolation includes Ambegaokar, Halperin, and Langer (1971), Kirkpatrick (1971), Coniglio, Stanley, and Klein (1979), and Kertész (1981), as well as several of the articles cited earlier.

CHAPTER FIVE
Elements of Inference

5.1. INTRODUCTION

A meaningful discussion of inference for coverage processes is hindered by the relatively restricted range of mathematically tractable stochastic models available for analysis. Processes more complex than even Boolean models of isotropic convex sets can be difficult to analyze without resorting to extensive simulation. This predicament should be recognized as one of the characteristics that distinguish inference for coverage processes from inference in classical areas of statistics.

Boolean models have several deficiencies when viewed as models for natural or physical phenomena: they permit no interactions between sets or set centers, and they admit no environmental heterogeneity. But in many cases they are a reasonable approximation, being "... a first step, where one admits only negligible interactions between the particles" (Serra 1982, p. 485). The planar random disk Boolean model epitomizes difficulties in modeling reality by coverage processes. In particular, situations where sets are very close to disks are rare, although they do exist (e.g., chemically etched fission particle tracks, lunar meteor craters). Nevertheless the random disk Boolean model has its place in applications, not least because its symmetry offers a plausible first-order approximation to phenomena as diverse as overlapping tree foliages in an aerial photograph of a forest, and interactive regions of electron-carrying impurities in a semiconductor.

Bearing all these caveats in mind, this chapter examines some of the theoretical aspects of inference in a Boolean model. *We treat only the case of a Boolean model in the plane.* It is imagined that the Boolean model divides $\mathbb{R}^2$ into just two regions, covered and uncovered. In practice, extra information about the process can sometimes be obtained by carefully scrutinizing covered regions, but information of that type is not used in any of the work in this chapter. Thus, covered areas are assumed to be completely black (opaque).

Recall that a set or point is said to be covered by the Boolean model $\mathcal{C} \equiv \{\xi_i + S_i, i \geq 1\}$ if it is completely contained in the union $\bigcup_{i \geq 1}(\xi_i + S_i)$. A set will be said to be *white* if it has no intersection with any set from the Boolean model. The converse to a white (completely uncovered) set is a nonwhite set. Only if the set is a singleton is this necessarily the same as a black (completely covered) set.

Some problems of inference were discussed in Chapter 4, particularly Sections 4.2 and 4.3, where they were connected directly to aspects of integral geometry developed there. Section 5.2 introduces porosity, point covariance, the variogram, and first contact distance, and discusses their uses in inference. Related techniques based on lattice counting and spatial correlation analysis are developed in Sections 5.3 and 5.4, respectively. Sections 5.5 and 5.6 discuss specialized techniques for inference in Boolean models where sets are "exactly" random radius disks. The present section briefly takes up the issues of image digitization, testing the Boolean model hypothesis, assessing isotropy, simulation-based methods, and multiphase models. Here and throughout this chapter, we use the term "isotropic set" for a set that is "isotropically randomly distributed." In two dimensions, that means a random set that has suffered a random, uniformly distributed rotation about some (any) center (see Chapter 3, Section 3.1 for details).

Coverage process data are often analyzed electronically, by digitizing the coverage pattern on a regular lattice or grid. Triangular, square, and hexagonal lattices are common [see Ahuja (1981a)]. Despite certain advantages of triangular and hexagonal lattices in depicting irregularly shaped patterns, square lattices are ubiquitous, due partly to their dominance of image processing technology. However, Serra's (1982) texture analyzer employs a triangular lattice.

Clearly the discretizing of a continuous coverage pattern on a lattice is likely to induce rounding errors. In some circumstances,

unbiased lattice-based estimates of various functions can be proposed, such as the counting estimates of covariance functions discussed in Section 5.2 below. But application of these quantities often requires interpolation, either explicit or implicit, to obtain in-
termediate values not available from the lattice. Thus, bias can be
a widespread (although not necessarily significant) problem. Some
theoretical analysis of bias and rounding error may proceed along
classical lines, using higher dimensional forms of results such as the
Euler-Maclaurin summation formula to obtain spatial analogues of
the error terms that Sheppard's corrections in classical statistical
theory (Kendall and Stuart 1977, p. 77) are designed to reduce.
Clearly some characteristics of coverage patterns are more suscepti-
ble to rounding errors and consequent bias than others; properties
related to connectivity, such as the mean Euler characteristic of a
coverage pattern, are a case in point. Serra (1982, p. 492ff) provides
an explicit asymptotic description of the sizes of errors in estimates
of mean Euler characteristic incurred in sampling on square and
hexagonal lattices.

In theory, although not always in practice, we may make some
progress toward testing the hypothesis that an observed coverage
process is a Boolean model with a certain shape distribution. Assume that the Poisson process driving the Boolean model has intensity λ, and that shapes generating the model are distributed as
the random convex isotropic set S. Let $\mathcal{S}_0$ be a given convex "test
set"; commonly $\mathcal{S}_0$ is a square if the digitizing lattice is square, and
a hexagon if the lattice is triangular or hexagonal. Theorem 4.1 in
Chapter 4 tells us that the probability that $\mathcal{S}_0$ is completely uncovered, or white, equals

$$q(\mathcal{S}_0) \equiv \exp\{-\nu(\mathcal{S}_0)\} = \exp\left[-\lambda\left\{\|\mathcal{S}_0\|_2 + \alpha + (2\pi)^{-1}\beta\|\partial\mathcal{S}_0\|_1\right\}\right],$$

where $\alpha \equiv E(\|S\|_2)$ and $\beta \equiv E(\|\partial S\|_1)$ are, respectively, the mean
area and mean perimeter of S. The quantity $1 - q(\mathcal{S}_0)$ is often called
the *capacity function*. It is easily estimated by counting. For example, if $\mathcal{S}_1, \ldots, \mathcal{S}_n$ are all congruent to $\mathcal{S}_0$ then

$$\hat{q}(\mathcal{S}_0) \equiv n^{-1}\sum_{i=1}^{n} I(\mathcal{S}_i \text{ white}) \qquad (5.1)$$

is an unbiased estimator of $q(\mathcal{S}_0)$, and $-\log\hat{q}(\mathcal{S}_0)$ is an estimator of
$\nu(\mathcal{S}_0)$. The estimators are consistent if as $n \to \infty$ the sets $\mathcal{S}_i$ range
over an indefinitely large subset of $\mathbb{R}^2$. If the shapes generating
the Boolean model are convex but not necessarily isotropic, then

$-\log\hat{q}(\mathcal{S}_0)$ remains an estimator of $\nu(\mathcal{S}_0)$ provided the test sets $\mathcal{S}_i$ are uniformly oriented images of $\mathcal{S}_0$. Now suppose we change $\mathcal{S}_0$ to $x\mathcal{S}_0$; that is, to $\mathcal{S}_0$ rescaled in each dimension by the positive number x. Then $\nu(x\mathcal{S}_0)$ is a quadratic in x:

$$\nu(x\mathcal{S}_0) = \lambda\left\{\alpha + \|\mathcal{S}_0\|x^2 + (2\pi)^{-1}\beta\|\partial\mathcal{S}_0\|x\right\}.$$

Therefore a plausible test of the Boolean model hypothesis is to compute $\log\hat{q}(x\mathcal{S}_0)$ for various values of x, and ascertain whether or not this function is approximately quadratic in x.

This approach is useful as a data-analytic tool for assessing the appropriateness of the Boolean model hypothesis. See for example Serra (1982, pp. 495–496 and 501). However, several difficulties impede a thorough application of the testing procedure. First of all, even the large-region (i.e., asymptotic) variance of $\log\hat{q}(x\mathcal{S}_0)$ is very complicated, let alone the variance of a statistic that measures the departure of $\log\hat{q}(x\mathcal{S}_0)$ from a quadratic. In most settings, values of $\hat{q}(x\mathcal{S}_0)$ are highly stochastically dependent over the range of x values. Therefore a formal test appears to be out of reach. [If extensive data are available, resampling methods (Hall 1985b) may be useful here.] Second, the formula for $\nu(x\mathcal{S}_0)$ depends on shape distribution only through α and β, and so nonrejection of the Boolean model hypothesis in no sense amounts to endorsement of a given model for shape distribution. Third, it must be recognized that in many circumstances where structural models are used to model natural processes, we are not totally convinced of the validity of the model. We ask that the model be a reasonable approximation, not that it mimic reality. Our lack of belief in the model may be larger than usual in the case of a Boolean model, but we may persevere with a somewhat inappropriate model because alternatives that take account of interactions between random sets can be prohibitively complex.

One may "test" for isotropy by taking the test set $\mathcal{S}_0$ to be a line segment and by estimating $\nu(x\mathcal{S}_0)$, which should be linear in x and with gradient independent of the orientation of $\mathcal{S}_0$, for different orientations. Perhaps more interestingly, a similar approach may be used to estimate the preferred alignment of sets in a Boolean model of nonisotropic convex sets. Specifically, suppose $\mathcal{S}_0$ is a line segment of unit length whose *normal* makes an angle θ to a predetermined direction. Assume the shapes generating the Boolean model are distributed as S, which is convex but not necessarily isotropic. Arguing as in the proof of Theorem 4.1 we may show that the probability that xS_0 is white equals

$$q(xS_0) = \exp[-\lambda\{\alpha + xg(\theta)\}], \qquad (5.2)$$

where $\alpha \equiv E(\|S\|_2)$ and $g(\theta)$ equals the mean thickness of S *in direction* θ. [Note particularly formula (4.11) in Chapter 4. See also Exercise 3.14 in Chapter 3, and Miles (1974a).] We may estimate $q(xS_0)$ in the manner described in (5.1), by counting. A plot of $-\log \hat{q}(xS_0)$ against x should be approximately linear with gradient $\lambda g(\theta)$, and so we may estimate $\lambda g(\theta)$ by means of simple linear regression. Arguing thus we may estimate the directions in which shapes have least and greatest mean thickness.

An alternative approach is via analysis of line transect samples. Let $\mathcal{L}$ be an infinite line with *normal* in direction θ. The Boolean model pattern divides $\mathcal{L}$ into a sequence of white (uncovered) and covered segments. The length of each covered segment has the distribution of busy period in a certain $M/G/\infty$ queue; see Chapter 2. Lengths of different white segments along $\mathcal{L}$ are independent and identically distributed exponential random variables with mean $\{\lambda g(\theta)\}^{-1}$; see Chapter 2, note formula (5.2), and observe that the length Z of an arbitrary white segment satisfies

$$P(Z > x) = q(xS_0)/P(\text{left-hand endpoint uncovered})$$
$$= \exp\{-x\lambda g(\theta)\}, \qquad x > 0.$$

Therefore $\lambda g(\theta)$ may be estimated by counting white segments on line transect samples. Note that unless the observation region is large, this procedure may be affected by a version of length-biased sampling. Long white segments are more likely than short segments to overlap the boundary of the observation region.

If data are recorded on a lattice then it can be difficult to estimate $\lambda g(\theta)$ for all values of θ. Some orientations are easier to treat than others. If the lattice is square then eight orientations are reasonably simple; these are 0, θ_0, $\pi/4$, $(\pi/2) - \theta_0$ $\pi/2$, $(\pi/2) + \theta_0$, $3\pi/4$ and $\pi - \theta_0$, where

$$\theta_0 \equiv \sin^{-1}(5^{-1/2}) \simeq 0.148\pi.$$

Figure 5.1 shows how these orientations are related to lattice structure.

The validity of specific Boolean models, in which shape distribution is fully parameterized, may be tested by Monte Carlo means. The technique usually requires that a certain characteristic of a coverage process, such as the point covariance function (see

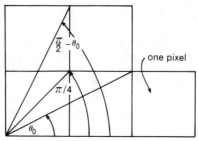

Figure 5.1. Canonical orientations in a square lattice. Here $\theta_0 \equiv \sin^{-1}(5^{-1/2})$.

Section 5.2 below), be estimated from the data and from a sequence of simulations of the claimed model. The model may be "tested" by ranking the observed characteristic together with simulated values, and thus determining how "extreme" the observed characteristic is. Functions may be ranked in terms of L^2 distance. If parameters in the model have to be estimated then the Monte Carlo test may proceed via cross-validation, dividing the observed coverage pattern into two parts and using one part to estimate parameters and the other to test goodness-of-fit. See Diggle (1981) for further discussion. Barnard (1963), Hope (1968), and Marriott (1979) have studied Monte Carlo tests in more classical circumstances.

This procedure obviously requires large amounts of numerical work. Its power hinges on the extent to which the specified characteristic reflects essential qualities of the model. For example, one can envisage situations where coverage processes with quite different physical properties have similar distributions of first contact; see Section 5.2 below.

Monte Carlo methods may also be used to estimate parameters. Again, the technique pivots on choice of a certain model characteristic. Having determined the observed value of the characteristic from data, extensive simulations may be employed to assess the relative likelihood of the observed characteristic for various parameter values. The parameter estimated (by simulation) to be most likely to have given rise to the observed value of the characteristic is deemed to be the parameter estimate [see Diggle and Gratton (1984)]. Of course, if the distribution of the characteristic could be written down in closed form then this approach could proceed after the usual fashion of maximum likelihood estimation, without any simulation. Unfortunately, likelihoods can almost never be written down for useful characteristics of coverage processes, even though they may be estimated by simulation. Most of the present chapter examines techniques for inference which circumvent the need for Monte Carlo methods.

There are several elementary coverage models derived from Boolean models, and which can be used as the basis for mathematically tractable theories of inference. In the simplest case, a Boolean model divides the plane into just two phases—covered and uncovered. However, *multiphase models* (sometimes called texture models) are also available. Examples were discussed in Chapter 1, Section 1.8. Alternatively, we might suppose two independent Boolean models C_1 and C_2 to be superimposed in the plane. A point $x \in \mathbb{R}^2$ may be classified as being from phase 1 if it is covered by C_1, from phase 2 if it is covered by C_2 but not by C_1, and from phase 3 otherwise. In this model, phase 1 "dominates" both the others. Techniques from Chapters 3 and 4 are readily used to describe properties of the model. For example, the amount of area classified as phase 2 within a region $\mathcal{R}$ equals

$$A_2 \equiv V_1 - V_{12},$$

where V_1 is the vacancy within $\mathcal{R}$ resulting from C_1 and V_{12} is the vacancy within $\mathcal{R}$ resulting from the Boolean model $C_1 \cup C_2$. The mean and variance of A_2, and a large-region central limit theorem for A_2, are easily obtained using arguments from Chapter 3. There are obvious n-phase versions of this model. Multiphase models based on random functions, with a dominant phase, may also be constructed (compare Chapter 1, Sections 1.4 and 1.8).

The *dead leaves* model is essentially a time-dependent process depositing random sets. Let C_i, $-\infty < i < \infty$, be *independent* copies of a Boolean model C, and let $Q \equiv \{\varsigma_i, -\infty < i < \infty\}$ be a Poisson process on the real line (the time axis) with intensity θ. Take Q to be independent of the C_i's, and assume the points of Q are indexed such that $\ldots < \varsigma_{-1} < \varsigma_0 < \varsigma_1 < \ldots$. Imagine depositing all the sets from C_i at time ς_i, $-\infty < i < \infty$, in such a manner that the sets from C_i hide any sets from C_j, $j < i$, which they overlap. If the shapes generating the C_i's are open and nonempty then with probability 1, at each time t each point of $\mathbb{R}^2$ is covered by sets from an infinite number of Boolean models, in an "order" determined by the time-dependent rule for depositing sets. Figure 5.2 illustrates the configuration at an arbitrary time. We may view the pattern as a time-dependent multiphase pattern, as follows. Assume that $\varsigma_{i_0} \leq t < \varsigma_{i_0+1}$. A point $x \in \mathbb{R}^2$ is in phase 0 at time t if it is covered by C_{i_0}, and in phase $i \geq 1$ if it is not covered by $C_{i_0}, C_{i_0-1}, \ldots, C_{i_0-i+1}$ but is covered by C_{i_0-i}. At any time, each point is in *some* phase $i \geq 0$. The phase

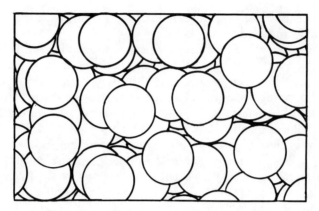

Figure 5.2. Dead leaves model. Component Boolean models are generated by fixed-radius disks.

$\pi(\mathbf{x};t)$ of $\mathbf{x} \in \mathbb{R}^k$ at time t has a geometric distribution:

$$P\{\pi(\mathbf{x};t) = n\} = p^n(1-p), \qquad n \geq 0,$$

where $p \equiv e^{-\alpha\lambda}$ is the proportion of the plane not covered by any one of the Boolean models $\mathcal{C}_i$, α is the mean area of shapes generating $\mathcal{C}_i$, and λ is the intensity of the Poisson process driving $\mathcal{C}_i$. Arguing much as in Chapter 3 we may give exact formulas for the mean and variance of the proportion of any given region that is in phase i, and prove a large-region central limit theorem for this proportion. In particular, the area at time t classified as phase i within a region $\mathcal{R}$ equals

$$A_i \equiv V_{i_0,i-1} - V_{i_0,i}$$

where V_{rs} equals the vacancy within $\mathcal{R}$ resulting from the Boolean model

$$\mathcal{C}_r \cup \mathcal{C}_{r-1} \cup \ldots \cup \mathcal{C}_{r-s}.$$

Finally, we remind the reader of the definition of a Riemann-measurable set, $\mathcal{R}$: it is a set whose indicator function,

$$i_{\mathcal{R}}(\mathbf{x}) \equiv \begin{cases} 1 & \text{if } \mathbf{x} \in \mathcal{R} \\ 0 & \text{otherwise}, \end{cases}$$

is directly Riemann integrable. A Riemann-measurable set is necessarily bounded. We shall assume that data are obtained from observation of the Boolean model within a Riemann-measurable region.

5.2. POINT COVARIANCE AND DISTRIBUTION OF FIRST CONTACT DISTANCE

Let $\mathcal{C}$ be a Boolean model in $\mathbb{R}^2$, driven by a Poisson process of intensity λ and generated by shapes distributed as S. We showed in Section 4.1 that the number of random sets in $\mathcal{C}$ that intersect a given fixed set S_0 is Poisson-distributed with mean

$$\nu(S_0) \equiv \lambda E[\|\{\mathbf{x} : (\mathbf{x} + S) \cap S_0 \neq \phi\}\|]. \tag{5.3}$$

The uncovered parts of $\mathbb{R}^2$ are sometimes referred to as the *pores*, and the sets comprising the covered parts as the *grains*. The *porosity* p of the Boolean model equals the proportion of the plane that is uncovered:

$$p \equiv P(\mathbf{0}\text{ white}) = e^{-\alpha\lambda}, \tag{5.4}$$

where $\alpha \equiv E(\|S\|)$ is the expected area of S; see formula (4.7) in Chapter 4.

Let $\mathbf{u}$ and $\mathbf{v}$ be elements of $\mathbb{R}^2$, and put $\mathbf{z} \equiv \mathbf{u} - \mathbf{v}$. The *covariance of the pores* is

$$\begin{aligned}
C_P(\mathbf{z}) &\equiv \operatorname{cov}\{I(\mathbf{u}\text{ white}), I(\mathbf{v}\text{ white})\} \\
&= P(\mathbf{u}, \mathbf{v}\text{ both white}) - P(\mathbf{u}\text{ white})P(\mathbf{v}\text{ white}) \\
&= P(\mathbf{0}, \mathbf{z}\text{ both white}) - P(\mathbf{0}\text{ white})^2 \\
&= \exp[-\nu(\{\mathbf{0}, \mathbf{z}\})] - p^2 \\
&= \exp[-\lambda E\{\|(\mathbf{z} + S) \cup S\|\}] - p^2,
\end{aligned} \tag{5.5}$$

where ν is given by (5.3). The *covariance of the grains* equals

$$\begin{aligned}
C_G(\mathbf{z}) &\equiv \operatorname{cov}\{I(\mathbf{u}\text{ not white}), I(\mathbf{v}\text{ not white})\} \\
&= \operatorname{cov}\{1 - I(\mathbf{u}\text{ not white}), 1 - I(\mathbf{v}\text{ not white})\} \\
&= \operatorname{cov}\{I(\mathbf{u}\text{ white}), I(\mathbf{v}\text{ white})\} \\
&= C_P(\mathbf{z}),
\end{aligned}$$

using the fact that covariance is unaffected by changes of location. The function C_P (or C_G) is sometimes known as *point covariance*.

Some authors reserve the name "covariance of the pores" for the function

$$\begin{aligned}
C_P^*(\mathbf{z}) &\equiv E\{I(\mathbf{u}\text{ white})\,I(\mathbf{v}\text{ white})\} \\
&= \exp[-\nu(\{\mathbf{0}, \mathbf{z}\})],
\end{aligned}$$

although this is not a statistical covariance in the strict sense. "Covariance of the grains" would then be

$$C_G^*(\mathbf{z}) \equiv E\{I(\mathbf{u} \text{ not white}) I(\mathbf{v} \text{ not white})\}$$
$$= 1 - 2p + C_P^*(\mathbf{z}),$$

and so is different from C_P^* unless $p = \frac{1}{2}$.

The *point variogram* is often defined as

$$\gamma(\mathbf{z}) \equiv E\{I(\mathbf{u} \text{ white}) I(\mathbf{v} \text{ not white})\}$$
$$= P(\mathbf{0} \text{ white}, \mathbf{z} \text{ not white})$$
$$= P(\mathbf{0} \text{ white}) - P(\mathbf{0}, \mathbf{z} \text{ both white})$$
$$= p(1 - p) - C_P(\mathbf{z}). \tag{5.6}$$

Strictly speaking, this is a semivariogram since it is one-half of the quantity

$$2\gamma(\mathbf{z}) \equiv \mathrm{var}\{I(\mathbf{u} \text{ white}) - I(\mathbf{0} \text{ not white})\}.$$

For any second-order stationary process $Z(\cdot)$, with variance σ^2 and correlation function ρ, the semivariogram is defined by $\gamma(\mathbf{z}) \equiv \frac{1}{2}\mathrm{var}\{Z(\mathbf{z}) - Z(\mathbf{0})\}$, and equals $\sigma^2\{1 - \rho(\mathbf{z})\}$.

Note that the functions C_P and γ are symmetric; for example, $C_P(-\mathbf{z}) = C_P(\mathbf{z})$ for all $\mathbf{z} \in \mathrm{IR}^2$. Furthermore, C_P and γ are both nonnegative.

We may estimate C_P and γ for various values of the arguments by sampling on a lattice. In particular, if points $\mathbf{u}_i, \mathbf{v}_i \in \mathrm{IR}^2$, $1 \leq i \leq n$, are such that $\mathbf{u}_i - \mathbf{v}_i = \mathbf{z}$ for each i, then

$$\hat{C}_P(\mathbf{z}) \equiv n^{-1} \sum_{i=1}^{n} I(\mathbf{u}_i, \mathbf{v}_i \text{ both white})$$
$$- \left[(2n)^{-1} \sum_{i=1}^{n} \{I(\mathbf{u}_i \text{ white}) + I(\mathbf{v}_i \text{ white})\}\right]^2 \tag{5.7}$$

estimates $C_P(\mathbf{z})$. The estimator is consistent if as $n \to \infty$ the points $\mathbf{u}_i$ and $\mathbf{v}_i$ range over an indefinitely large subset of IR^2. Commonly the term in p in the formula for C_P would be estimated slightly differently, by using vacancy or by averaging $I(\mathbf{u} \text{ white})$ over *all* points of a lattice.

Estimators of C_P are really only of use if we may employ them to estimate important characteristics of shape distribution and Poisson intensity. This requires C_P to have a relatively simple formula in terms of shape properties, and that usually means that shapes have to be random radius disks. (In practice it can be very difficult to discriminate between disks and nondisk but isotropic shapes.) We shall show that in this circumstance there is a one-to-one correspondence between the pair (p, C_P) and the pair (λ, F), where λ is Poisson intensity and F is the distribution function of disk radius. Thus, we can in principle consistently estimate λ and all properties of disk radius distribution by consistently estimating porosity and point covariance.

Notice that

$$\nu(\{0, \mathbf{z}\}) = 2\alpha\lambda - \lambda\rho(\mathbf{z}),$$

where

$$\rho(\mathbf{z}) \equiv E\{\|(\mathbf{z}+S) \cap S\|\} = \int_0^\infty r^2 B(|\mathbf{z}|/2r)\, dF(r) \tag{5.8}$$

and

$$B(x) \equiv \begin{cases} 4\int_x^1 (1-y^2)^{1/2}\, dy & \text{if } 0 \le x \le 1 \\ 0 & \text{otherwise} \end{cases} \tag{5.9}$$

is the area of the lens of intersection of two unit disks centered $2x$ apart. Therefore by (5.5),

$$C_P(\mathbf{z}) = p^2 \left\{ e^{\lambda\rho(\mathbf{z})} - 1 \right\},$$

and so we must prove that there is a one-to-one correspondence between pairs $(p, \lambda\rho)$ and pairs (λ, F).

Clearly $(p, \lambda\rho)$ is completely determined once (λ, F) is given. Conversely, suppose $(p, \lambda\rho)$ is given. If F has derivative (i.e., density) f and if $|\mathbf{z}| = z$ then

$$\rho(2z) = 4 \int_z^\infty r^2 f(r)\, dr \int_{z/r}^1 (1-y^2)^{1/2}\, dy,$$

where we write $\rho(x)$ for $\rho(\mathbf{x})$ if $|\mathbf{x}| = x$. Consequently,

$$\rho'(2z) = -2 \int_z^\infty f(r)(r^2 - z^2)^{1/2}\, dr$$

and

$$\rho''(2z) = z \int_z^\infty f(r)(r^2 - z^2)^{-1/2}\, dr.$$

The last integral equation is of Abel type, and may be solved to express f in terms of ρ, giving

$$f(r) = -\frac{2r}{\pi} \int_r^\infty \frac{d}{dz} \left\{ \frac{\rho''(2z)}{z} \right\} (z^2 - r^2)^{-1/2} dz$$

$$= (2r/\pi) \int_z^\infty z^{-2} \left\{ \rho''(2z) - 2z\rho'''(2z) \right\} (z^2 - r^2)^{-1/2} dz.$$

[Compare the argument leading to equation (1.48) in Chapter 1, Section 1.9. As in that work, the solution of this equation is numerically unstable.] Therefore the function ρ completely determines f, and likewise $\lambda\rho$ completely determines λf. Given λf we may find λ by the formula

$$\lambda = \int_0^\infty \lambda f(r) \, dr,$$

and thence f. More generally, an argument based on approximation to an arbitrary distribution by continuous distributions shows that no matter what the choice of λ and F, they are completely determined by $\lambda\rho$ provided that

$$\alpha \equiv \pi \int_0^\infty r^2 \, dF(r) < \infty.$$

Point covariance is not unique in the property that (together with porosity) it completely determines the distribution of a Boolean model of disks. However, this property is not commonplace. In particular it is not shared by the distribution of first contact distance, as we now show.

As before, let $\mathcal{C}$ be a Boolean model in which shapes are distributed as S and the driving Poisson process has intensity λ. Let X denote the distance from an arbitrary point in the plane (without loss of generality, the origin) to the nearest point covered by $\mathcal{C}$. We call X the *distance to the point of first contact*, and say that X has the distribution of first contact distance. Let $T(x)$ denote the closed disk of radius x centered at the origin. Then

$$P(X = 0) = P(\mathbf{0} \text{ not white}) = 1 - p,$$

and

$$P(X > x) = P\{T(x) \text{ white}\}$$
$$= \exp[-\nu\{T(x)\}],$$

where ν is defined by (5.3). Of course, these formulas are true quite generally; we are not at this point assuming that S is a disk or even convex. If S is isotropic and convex then by Theorem 4.1 of Chapter 4,

$$\nu\{T(x)\} = \lambda(\alpha + \pi x^2 + \beta x),$$

where $\alpha \equiv E(\|S\|_2)$ is the mean area and $\beta \equiv E(\|\partial S\|_1)$ the mean perimeter of S. Therefore

$$P(X > x) = \exp\left\{-\lambda(\alpha + \beta x + \pi x^2)\right\},$$

and the distribution function of first contact distance is

$$G(x) \equiv P(X \le x) = \begin{cases} 0, & x < 0 \\ 1 - \exp\{-\lambda(\alpha + \beta x + \pi x^2)\}, & x \ge 0. \end{cases}$$

This function is completely determined by three parameters, and depends on shape distribution only through mean area and mean perimeter. Therefore there are a great many quite different Boolean model distributions with identical distributions of first contact distance. Furthermore, a program for conducting inference in a Boolean model of random disks by estimating first contact distance would be doomed to failure if radius distribution depended on more than two parameters.

Diggle (1981) used point covariance and the method of minimizing integrated square error to fit a random disk model to data on heather in a forest clearing. The heather pattern within a 20 m × 10 m area was coded as a 200 × 100 binary Boolean model, in which each element represented a 10 cm square. Figure 5.3 depicts the resulting pattern. This pattern was divided into two 10 m × 10 m squares, and the model fitted separately to each square. Thus, the data for each estimation problem were digitized on a 100 × 100 square lattice. After experimentation, Diggle fitted a random disk model in which disk radius had a three-parameter Weibull distribution:

$$f(r) = ls(r - \delta)^{l-1} \exp\left\{-s(r - \delta)^l\right\}, \qquad r \ge \delta.$$

For any given value of the parameters (l, s, δ, λ), the function

$$C_P(\mathbf{z}) = e^{-2\alpha\lambda}[\exp\{\lambda\rho(\mathbf{z})\} - 1]$$

may be calculated numerically via formula (5.8). Since shapes are isotropic then $C_P(\mathbf{z})$ depends on $\mathbf{z}$ only through $z \equiv |\mathbf{z}|$, and so in a slight abuse of notation we shall write $C_P(z)$ instead of $C_P(\mathbf{z})$. An empirical version $\hat{C}_P$ of C_P may be calculated by averaging indicator

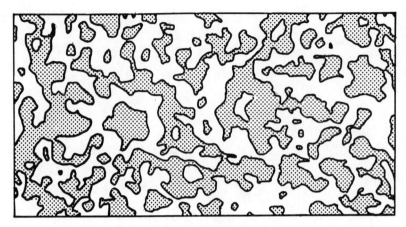

Figure 5.3. Pattern of heather in forest clearing. Data are recorded on a 200 × 100 square lattice, representing a 20 m × 10 m rectangular area. Heather occupies shaded area. [From Diggle (1981).]

functions over each 100 × 100 lattice, and parameters (l, s, δ, λ) fitted by minimizing

$$\int_0^{z_0} \left\{ \hat{C}_P(z) - C_P(z) \right\}^2 dz.$$

More generally, a weight function could be included in the integrand. There are clear limitations to the size of z_0, due to the fact that the sampling lattice is finite. Fortunately choice of z_0 turns out not to be critical, and Diggle took $z_0 = 0.1$ m (0.1 meters).

Fitted values of (l, s, δ, λ) are (0.8471, 20.6, 0.281, 2.21) and (1.011, 12.5, 0.226, 2.11) for the left- and right-hand squares, respectively. (Units are meters.) Mean disk radius under the model is

$$E(R) = \delta + \int_0^{\infty} l s r^l \exp(-s r^l)\, dr = \delta + s^{-1/l} \Gamma(1 + l^{-1}).$$

Furthermore,

$$E(R - \delta)^2 = \int_0^{\infty} l s r^{l+1} \exp(-s r^l)\, dr = s^{-2/l} \Gamma(1 + 2l^{-1}),$$

so that disk radius has variance

$$\text{var}(R) = s^{-2/l} \left\{ \Gamma(1 + 2l^{-1}) - \Gamma(1 + l^{-1})^2 \right\}$$

under the model. Substituting the parameter estimates into these equations we conclude that under the disk model, mean and standard deviation of disk radius are approximately 0.31 and 0.04 in

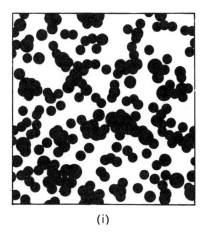 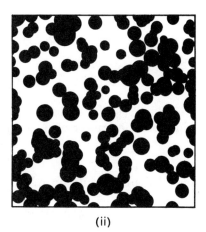

(i)   (ii)

Figure 5.4. Realizations of fitted model. Each simulation models half of the pattern in Figure 5.2. Square (i) corresponds to values (0.8471, 20.6, 0.281, 2.21) and square (ii) to values (1.011, 12.5, 0.226, 2.11) of the parameters (l, s, δ, λ). [From Diggle (1981).]

the case of the left-hand square of heather, and 0.31 and 0.08 in the case of the right-hand square. Together the data suggest an average heather bush radius of approximately 30 cm, with standard deviation less than one-third of the mean, and about two bushes per square meter. Figure 5.4 depicts typical realizations of the fitted models.

As Diggle (1981) notes, there are obvious differences between data and fitted model. In particular the fitted model has a lower frequency of small isolated patches than the data, due in part to the low value of fitted radius standard deviation relative to fitted mean. Nevertheless, a Boolean model of random disks is a plausible *approximation* on physical grounds. Some of the differences between data and fitted model would diminish if the disks were given ragged edges.

5.3. COUNTING METHODS FOR INFERENCE IN A BOOLEAN MODEL OF CONVEX SETS

Consider a Boolean model $\mathcal{C}$ in $\mathbb{R}^2$ in which the driving Poisson process has intensity λ and shapes are distributed as the random *convex isotropic* set S. Let S_0 be any fixed convex subset of $\mathbb{R}^2$. The probability that S_0 is white, or equivalently that it has no intersection with any random set from $\mathcal{C}$, may be deduced from

Theorem 4.1; it is

$$q(\mathcal{S}_0) \equiv \exp\{-\nu(\mathcal{S}_0)\}$$

where

$$\nu(\mathcal{S}_0) \equiv \lambda\left\{\alpha + \|\mathcal{S}_0\|_2 + (2\pi)^{-1}\beta\|\partial\mathcal{S}_0\|_1\right\}.$$

Here $\alpha \equiv E(\|S\|_2)$ and $\beta \equiv E(\|\partial S\|_1)$ are respectively the mean area and mean perimeter of S. If $\mathcal{S}_0$ is a point (that is, a singleton) then

$$q(\mathcal{S}_0) = e^{-\alpha\lambda},$$

and so depends only on $\alpha\lambda$. If $\mathcal{S}_0$ is a line segment of length l then

$$q(\mathcal{S}_0) = \exp\left\{-\lambda(\alpha + \pi^{-1}l\beta)\right\},$$

and so depends only on $\alpha\lambda$ and $\beta\lambda$. If $\mathcal{S}_0$ is a nondegenerate convex set then $q(\mathcal{S}_0)$ depends on each of $\alpha\lambda$, $\beta\lambda$, and λ in a nontrivial way. Given estimates of $q(\mathcal{S}_0)$ for each of these versions of $\mathcal{S}_0$ we may readily construct estimates of $\alpha\lambda$, $\beta\lambda$, and λ, and hence of α, β, and λ. The techniques developed in the present section are based on this elementary argument, and lead to estimators of the triple (α, β, λ).

Suppose the coverage pattern is observed within a region $\mathcal{R}$. Construct a regular lattice (that is, a grid) inside $\mathcal{R}$ in which each edge or bond has length l, each face is convex and has area a, and each face is bordered by m edges. For example, a square lattice has $a = l^2$ and $m = 4$. Let n_0, n_1, and n_2 denote the numbers of vertices, edges, and faces, respectively, and let N_0, N_1, and N_2 be the numbers of these that are white. Then as described above,

$$E(N_0/n_0) = e^{-\alpha\lambda} \equiv \exp(-t_0),$$
$$E(N_1/n_1) = \exp\left\{-\lambda(\alpha + \pi^{-1}l\beta)\right\} \equiv \exp(-t_1)$$

and

$$E(N_2/n_2) = \exp\left[-\lambda\left\{\alpha + a + (2\pi)^{-1}ml\beta\right\}\right] \equiv \exp(-t_2).$$

Therefore $T_i \equiv \log(n_i/N_i)$ is an estimator of t_i. If we replace t_i by T_i in the formulas above, and solve the resulting equations for α, β, and λ, we obtain the estimates

$$\hat{\lambda} \equiv a^{-1}\left\{(\tfrac{1}{2}m - 1)T_0 - \tfrac{1}{2}mT_1 + T_2\right\}, \qquad (5.10)$$

$$\hat{\alpha} \equiv T_0/\hat{\lambda} \qquad (5.11)$$

and

$$\hat{\beta} \equiv \pi(l\hat{\lambda})^{-1}(T_1 - T_0). \tag{5.12}$$

Under the assumption that $E\{(\text{rad}\,S)^2\} < \infty$, joint central limit theorems may be proved for N_0, N_1, and N_2 and so also for $\hat{\alpha}$, $\hat{\beta}$, and $\hat{\lambda}$. Techniques are similar to those used to derive Theorem 3.5 in Chapter 3. Asymptotic variances are quite complex, and provide little insight into the way choice of lattice edge width affects performance of $\hat{\alpha}, \hat{\beta}$, and $\hat{\lambda}$. In order-of-magnitude terms, $\text{var}(N_i) = O(\|\mathcal{R}\|)$ as the observation region $\mathcal{R}$ increases in a manner that preserves its shape, again provided $E\{(\text{rad}\,S)^2\} < \infty$. This is most easily seen in the case where $\text{rad}\,S$ is essentially bounded— say $\text{rad}\,S \leq r_0$ with probability 1. Then, any two vertices, edges or faces that are distant at least $2r_0$ apart at their closest separation are white or nonwhite independently of one another. This means that we may divide all faces (for example) within $\mathcal{R}$ among a fixed number c of classes, such that all faces within class i ($1 \leq i \leq c$) are independently white or nonwhite, and c does not change as $\mathcal{R}$ increases. Since the total number of faces increases linearly with $\|\mathcal{R}\|$, and since the variance of a sum of independent random variables equals the sum of the variances, then $\text{var}(N_2) = O(\|\mathcal{R}\|)$ as $\mathcal{R}$ increases. Likewise, $\text{var}(N_0)$ and $\text{var}(N_1)$ are also $O(\|\mathcal{R}\|)$, and in consequence $T_i = t_i + O_p(\|\mathcal{R}\|^{-1/2})$. We may now deduce from (5.10), (5.11), and (5.12) that $\hat{\lambda} - \lambda$, $\hat{\alpha} - \alpha$, and $\hat{\beta} - \beta$ are all $O_p(\|\mathcal{R}\|^{-1/2})$ as $\mathcal{R}$ increases. This is the analogue of $\sqrt{n}$-consistency for a traditional statistical estimator.

If lattice edge width is small then faces will tend to behave like vertices, and so will contain relatively little information. If edge width is large then relatively few faces or edges will be white, leading to lower and more variable counts and to higher estimator variance. In principle an "optimal" edge width that minimizes asymptotic variance will exist for each of $\hat{\alpha}$, $\hat{\beta}$, and $\hat{\lambda}$, but since formulas for variances are so complex we are forced to rely on more *ad hoc* recommendations. We suggest that if shapes are not too elongated, lattice edge width be chosen so that face area is approximately equal to mean shape area, α. If a prior estimate of α is not available then trial counts could be made.

To illustrate the use of the procedure we apply it to the heather data of Diggle (1981) depicted in Figure 5.3. So as to fit the counting lattice neatly inside the figure we work in units of $l = 0.98$ m instead of 1 m. The counting lattice is taken to have square faces. In order

TABLE 5.1 Summary of Counts on Five Square Lattices

Edge Width	Lattice Size	N_0	N_1	N_2	$\hat{\alpha}$	$\hat{\beta}$	$\hat{\lambda}$
$\frac{1}{4}l$	40×81	1775	2572	813	0.281	1.72	2.27
$\frac{3}{8}l$	26×54	773	920	194	0.256	1.65	2.55
$\frac{1}{2}l$	20×40	455	474	70	0.270	1.67	2.36
$\frac{5}{8}l$	16×32	287	239	19	0.261	1.66	2.57
$\frac{3}{4}l$	13×27	203	144	8	0.306	1.95	2.15

A lattice size of $r \times s$ indicates r rows of s square faces.

to assess the effect of lattice edge width on counts we try five different edge widths: $\frac{1}{4}l$, $\frac{3}{8}l$, $\frac{1}{2}l$, $\frac{5}{8}l$, and $\frac{3}{4}l$. Lattices with edge widths below $\frac{1}{4}l$ are unsatisfactory because their faces tend to behave like points, being either white or completely covered, and because of limits to resolution since the data are recorded on a square lattice with edge width approximately $\frac{1}{10}l$. Lattices with edge widths larger than $\frac{3}{4}l$ are subject to large sampling fluctuations, since only a small proportion of faces are white.

Results of the counting experiments are summarized in Table 5.1. These data suggest that there are between 2 and 2.5 heather bushes per square meter, and that mean heather bush area is approximately $\hat{\alpha} = 0.27$ m^2 and mean perimeter approximately $\hat{\beta} = 1.7$ m. If heather bushes are disk shaped, these estimates translate to a mean radius $[= (2\pi)^{-1}\beta]$ of about 27 cm and radius standard deviation $[= \{\pi^{-1}\alpha - (2\pi)^{-2}\beta^2\}^{1/2}]$ of approximately 12 cm.

Independent biological evidence (Diggle 1981, p. 532) suggests an upper bound to heather bush radius of about 50 cm. Spatial correlation analysis described in Section 5.4 below indicates that the correlation (and point covariance) between any two points in the plane vanishes at a separation of about 80 cm, suggesting that maximum shape radius is aproximately 40 cm. Taking heather plants to be disks with radius uniformly distributed between 0 and 50 cm or between 0 and 40 cm, or conducting a trial count in which $\hat{\alpha}$ is between 0.25 and 0.30 m^2, and equating mean lattice face area to mean shape area, we see that a lattice with edge width approximately 50 cm is indicated. This crude argument would lead us to select from Table 5.1 the lattice with edge width $\frac{1}{2}l$. Figure 5.5 illustrates the lattice in this special case.

Each of these conclusions is based on the assumption that sets generating the Boolean model are convex and isotropic. Evidence concerning this hypothesis may be obtained by "testing," as outlined in Section 5.1. According to the theory there, the function $-\log q(xS_0)$ [where S_0 is a square and $q(xS_0)$ denotes the probability that xS_0 is white] should be quadratic in x. We may estimate $q(xS_0)$ by counting [see formula (5.1)]. Figure 5.6, based on data kindly supplied by Dr. G. M. Laslett, summarizes the results of such counting experiments for the heather data. It also illustrates estimates of the functions $-\log q(xS_0)$ in the case where xS_0 is a horizontal line segment or a vertical line segment. If shapes generating the Boolean model were truly convex and isotropic then these last two functions should coincide.

Similar evidence of anisotropy may be obtained from an analysis along lines suggested by Renshaw and Ford (1983). The question at issue is not the existence of anisotropy, but whether or not its presence has an important effect on techniques that assume isotropy. Figure 5.7, discussed in the next section, suggests that it may not.

5.4. SPATIAL CORRELATION ANALYSIS

The point covariance function C_P introduced in Section 5.2 [formula (5.5)] may be viewed as a version of the autocorrelation function common in time series analysis and its applications. Point covariance is readily estimated by counting on a lattice [formula (5.7)], and a plot of the estimate provides useful information about the way in which strength of spatial interaction varies with distance and direction. Figure 5.7 depicts the estimated point covariance function for the heather data of Diggle (1981); the raw data were presented in Figure 5.3.

If the coverage process is a Boolean model then it follows from the definition of C_P that $C_P(0) = p(1-p)$, where p equals porosity. If shapes generating the Boolean model are distributed as the convex set S (not necessarily isotropic) then the function

$$e(\mathbf{z}) \equiv E\{\|(\mathbf{z}+S) \cup S\|\}$$

is radially increasing, in the sense that $e(c_1\mathbf{z}) \le e(c_2\mathbf{z})$ whenever $0 < c_1 \le c_2 < \infty$ and $\mathbf{z} \in \mathbb{R}^2$. If S is isotropic (but not necessarily convex) then e is radially symmetric, in the sense that $e(\mathbf{z})$ depends

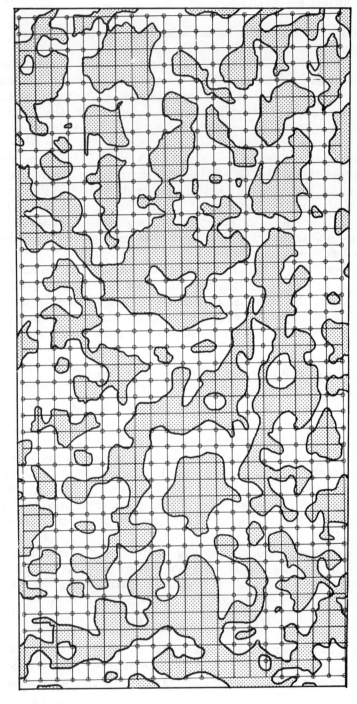

Figure 5.5. Heather data from Figure 5.3 superimposed on 40 × 20 square lattice. White vertices are encircled. [From Hall (1985h).]

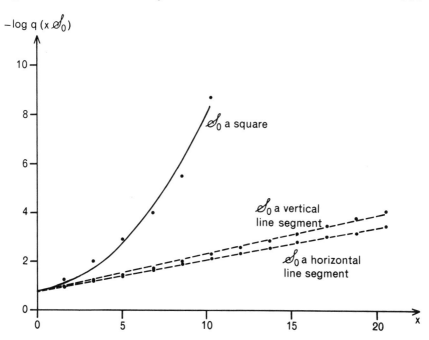

Figure 5.6. Function $-\log q(xS_0)$, in the case where S_0 is a square or line segment. Edge width of square S_0 and lengths of line segments are all approximately 1.7 m. Scale on horizontal axis is in units of meters. (Data supplied by Dr. G. M. Laslett.)

on $\mathbf{z}$ only through $|\mathbf{z}|$. Therefore the point covariance function,

$$C_P(\mathbf{z}) = \exp\{-\lambda e(\mathbf{z})\} - p^2,$$

is radially decreasing in a Boolean model of convex sets and radially symmetric in a Boolean model of isotropic sets. Of course, $C_P(\mathbf{z}) \geq 0$ and $C_P(\mathbf{z}) \to 0$ as $|\mathbf{z}| \to \infty$.

The estimates plotted in Figure 5.7 suggest that to a reasonable approximation, the "true" point covariance function of the process generating the heather data is both radially decreasing and radially symmetric. Estimated point covariance is virtually 0 at a separation of 100 cm, suggesting that 50 cm is an upper bound to shape radius. In this section we show how to use spatial correlation analysis to obtain more concise information about maximum shape radius.

Consider a Boolean model $\mathcal{C}$ in $\mathbb{R}^2$, where the driving Poisson process has intensity λ and each shape is distributed as S. Suppose the resulting spatial pattern is sampled at vertices of a regular lattice distributed across the observation region $\mathcal{R}$. The observation at

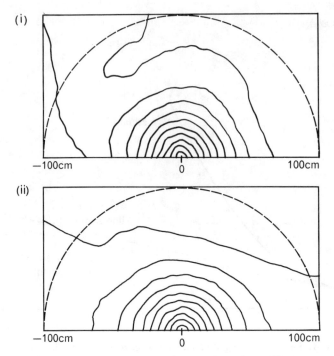

Figure 5.7. Estimated point covariance, C_P, for data of Figure 5.3. Part (i) corresponds to the left-hand half of Figure 5.3, and part (ii) to the right-hand half. The origin is at the bottom center of each diagram, and $C_P(0) \doteq 0.25$ in each case. Each contour represents 10% of $C_P(0)$, or approximately 0.025, and contour values *decrease* away from the origin. The broken arcs define a range of 100 cm away from the origin. [After Diggle (1981).]

each vertex is either white W or black B, corresponding to uncovered or covered, respectively. Certain vertices separated by a fixed distance $b > 0$ are considered to be bonded to one another; that is, they are joined by an edge, in graph-theoretic notation (Bollabás 1979, Chapter 1). We ask that the resulting graph $\mathcal{G}$ have the following properties:

(i) no vertices are distant $< b$ apart,

(ii) each pair of bonded vertices is distant precisely b apart,

(iii) $\mathcal{G}$ is regular of degree ν; that is, each vertex of $\mathcal{G}$

has precisely ν edges radiating from it.   (5.13)

Property (iii) requires us to interpret properties (i) and (ii) in a torroidal fashion on the perimeter of $\mathcal{G}$. This convention permits considerable simplification of notation and reduction in weight of algebra. Some of our theoretical results are of an asymptotic nature, and hold as the number of vertices in the graph increases. They continue to be valid if the torus convention is disregarded, but are much more tedious to state and derive without that convention. There is no need to persevere with the convention in applications, and our application at the end of this section to Diggle's (1981) heather data gives details of test statistics that avoid the convention.

An edge joining a black to a white vertex is called a BW edge. Similarly we define BB and WW edges. Let m and n be the total numbers of edges and vertices, respectively, let N equal the total number of black vertices, and let M, M', and M'' be the total numbers of BW, BB, and WW edges, respectively. Then

$$M' = \tfrac{1}{2}(N\nu - M) \quad \text{and} \quad M'' = m - \tfrac{1}{2}N\nu - \tfrac{1}{2}M.$$

Therefore the variables $M - E(M \mid N)$, $M' - E(M' \mid N)$, and $M'' - E(M'' \mid N)$ are simply constant multiples of one another. Since the probability that any given vertex is black is unknown, inference would usually be based on properties of the conditional distribution of counts, given N. Consequently, inferences based on M, M', or M'' would produce *identical* results. The results would be asymptotically equivalent if the torus convention were ignored.

Conditional on N, the probability that a given edge is BW equals

$$2 \frac{N}{n} \frac{n - N}{n - 1},$$

under the hypothesis that vertex colors are independent and identically distributed. Let E_0 denote expectation under this hypothesis. We base inference on

$$M - E_0(M \mid N) = M - \{\nu N(n - N)/(n - 1)\}.$$

It may be proved that

$$
\begin{aligned}
&E_0(M^2 \mid N) - \{E_0(M \mid N)\}^2 \\
&= \frac{2\nu(n - \nu - 1)N(n - N)}{(n - 1)(n - 2)(n - 3)} \left\{ \frac{N(n - N)}{n - 1} - 1 \right\},
\end{aligned}
$$

and so the statistic

$$U \equiv \left\{ \frac{M(n - 1)}{\nu N(n - N)} - 1 \right\} \left[\frac{2(n - \nu - 1)}{\nu(n - 2)(n - 3)} \left\{ 1 - \frac{n - 1}{N(n - N)} \right\} \right]^{-1/2}$$

$$(5.14)$$

satisfies $E_0(U \mid N) = 0$ and $E_0(U^2 \mid N) = 1$.

Define $R \equiv \text{rad}(S)$ and $r_0 \equiv \text{ess sup} R$, and assume $0 < r_0 < \infty$. If the length b of the edges of the graph is no less than $2r_0$ then all vertices are colored independently. However, if $b < 2r_0$ and S is a random isotropic open set then colors of bonded vertices are correlated. In many cases this is true without the conditions of isotropy and openness, but to avoid having to discuss exceptions it is convenient to make these assumptions. When colors at bonded vertices are correlated there is a lower proportion of BW edges than would occur under the hypothesis of independence, and then U has a predisposition toward negative values. An asymptotic (that is, large-region), α-level test of $H_0 : r_0 \leq \frac{1}{2}b$ against $H_1 : r_0 > \frac{1}{2}b$ may be conducted by comparing the statistic U with the lower α-level critical point of the standard normal distribution, and rejecting H_0 if U is less than that point. It may be proved (see Theorem 5.1 below) that as the observation region $\mathcal{R}$ increases, U becomes asymptotically normal $N(0,1)$ if $r_0 \leq \frac{1}{2}b$, and that $U \to -\infty$ in probability if $r_0 > \frac{1}{2}b$.

To obtain a little detail about the power of this test we study a sequence of "local alternatives." Consider the Boolean model $\mathcal{C}_\delta$ in which the driving Poisson process has intensity λ and shapes are distributed as $(1+\delta)S$, rather than S, for some $\delta \geq 0$. Let $\mathbf{u}$ and $\mathbf{v}$ be coordinates of two vertices P and Q, and put $\mathbf{z} \equiv \mathbf{u} - \mathbf{v}$. Write $\alpha \equiv E(\|S\|)$ for the mean area of S. Then

$$P(P, Q \text{ have different colors}) = 2P(P \text{ white}, Q \text{ black})$$
$$= 2\gamma_\delta(\mathbf{z})$$
$$= 2p_\delta(1 - p_\delta) - 2C_{P,\delta}(\mathbf{z}),$$

where $p_\delta \equiv \exp\{-\alpha\lambda(1+\delta)^2\}$ denotes porosity in $\mathcal{C}_\delta$, and γ_δ and $C_{P,\delta}$ are, respectively, the point variogram and point covariance of $\mathcal{C}_\delta$ [see (5.5) and (5.6)]. If we continue to define $r_0 \equiv \text{ess sup rad} S$ then the essential supremum of radius of shapes generating $\mathcal{C}_\delta$ is $(1+\delta)r_0$. The theorem below describes asymptotic properties of U in the case where the observation region $\mathcal{R}$ is a rectangle, and lattice edge width b is taken to be $2r_0$. Recall that m equals the number of edges of the graph $\mathcal{G}$ within the observation region $\mathcal{R}$. We keep the degree and edge width of $\mathcal{G}$, and the ratio of the side lengths of $\mathcal{R}$, fixed as $\mathcal{R}$ increases. Recall that m denotes the total number of edges in the lattice, and diverges to $+\infty$ as $\mathcal{R}$ increases.

Theorem 5.1. *Suppose the Boolean model $\mathcal{C}_\delta$ is observed within the rectangle $\mathcal{R}$, and the resulting spatial pattern is sampled at ver-*

tices of a regular graph $\mathcal{G}$ satisfying condition (5.13) with $b = 2r_0$. Assume S is isotropic, let $\mathbf{z}$ be any vector of length $2r_0$, and write p_δ and $C_{P,\delta}$ for porosity and point covariance of C_δ, respectively. The statistic U defined at (5.14) admits the expansion

$$U = U' - m^{1/2} C_{P,\delta}(\mathbf{z}) \{p_0(1 - p_0)\}^{-1} (1 + \Delta), \qquad (5.15)$$

where as $\mathcal{R}$ increases and $\delta \to 0$, U' is asymptotically normal $N(0,1)$ and $\Delta \to 0$ in probability. If $b \geq 2r_0$ and $\delta = 0$ then U is asymptotically normal $N(0,1)$ as $\mathcal{R}$ increases.

A proof of Theorem 5.1 is deferred until the end of the section.
To explain the behavior of $C_{P,\delta}(\mathbf{z})$, observe from (5.5) that as $\delta \to 0$,

$$C_{P,\delta}(\mathbf{z}) \to C_{P,0}(\mathbf{z}) \equiv \exp[-\lambda E\{\|(\mathbf{z} + S) \cup S\|\}] - \exp\{-2\lambda E(\|S\|)\}.$$

The right-hand side here is strictly positive if $b < 2r_0$, and 0 otherwise.

To explore the implications of this result, suppose S is a disk of fixed radius r_0. Write η for the area of the lens of intersection of two disks with radii $(1 + \delta)r_0$, centered $2r_0$ apart. A little trigonometry and algebra show that

$$\eta \sim 2^{7/2} 3^{-1} r_0^2 \delta^{3/2}$$

as $\delta \to 0$, and so

$$C_{P,\delta}(\mathbf{z}) = p_\delta^2 (e^{\lambda \eta} - 1) \sim \text{const.} \, \delta^{3/2}$$

as $\delta \to 0$. Substituting into (5.15) and noting that m increases in proportion to $\mathcal{R}$, we see that

$$U \simeq U' - \text{const.} \|\mathcal{R}\|^{1/2} \delta^{3/2}$$

as $\mathcal{R}$ increases and $\delta \to 0$. Therefore the greatest detectable discrepancy δ from the null hypothesis is of order $\delta \simeq \|\mathcal{R}\|^{-1/3}$ for large regions $\mathcal{R}$.

Another implication of (5.15) is that power is maximized by maximizing total edge count, m. In this sense the optimal graph is based on the triangular lattice, and has degree $\nu = 6$ (see Exercise 5.6).

The test described above is asymptotic in character, since it is based on the approximate normality of U. Exact critical points conditional on N may be derived by simulation, but more simply, an exact test (admittedly with low power) may be conducted as follows. Given a regular graph $\mathcal{G}$ satisfying (5.13), construct a simple

closed curve passing through as many vertices as possible by deleting a subset of the edges of $\mathcal{G}$. We shall assume that the curve passes
through *all* vertices; this construction is usually possible. Break the
curve at an arbitrary vertex, to form a linear sequence $\mathcal{L}$ of edges.
Under the null hypothesis that $r_0 \leq \frac{1}{2}b$, and conditional on the total
number N of black vertices, relative positions of black and white
vertices along $\mathcal{L}$ are those of order statistics from two independent
and identically distributed samples, one of size N and the other of
size $n - N$. The number of BW edges in $\mathcal{L}$ equals the number of
runs in the combined sample, minus one. Therefore an edge count
test based on $\mathcal{L}$ is equivalent to a Wald-Wolfowitz runs test, for
which exact critical points are readily available (Siegel 1956, pp.
136–145, 252–253). If the graph $\mathcal{G}$ is of degree ν then asymptotic
efficiency of the reduced test relative to the full test may be shown
to equal $2/\nu$.

In applications of these testing procedures we would usually conduct a sequence of tests using increasing values of lattice edge width
b, with the aim of determining the edge width where the tests start
to accept the null hypothesis. Formal theoretical justification for
this procedure may be developed quite easily (see Exercise 5.7). To
illustrate the technique we apply it to Diggle's (1981) heather data,
depicted in Figure 5.3.

We employ the equilateral triangular lattice throughout. In applications it is usually inconvenient to use the torus convention.
However, dropping the convention requires introduction of a more
complicated version of the test statistic U, as follows. Suppose the
lattice is constructed as in Figure 5.8, with r rows each of s vertices. Let M equal the number of BW edges in the resulting graph,
and N the number of black vertices. There are $n = rs$ vertices and
$m = 3rs - 2(r + s) + 1$ edges overall. Under the null hypothesis that
maximum shape radius r_0 is less than half edge width, the mean of
M given N equals $2mN(n - N)/\{n(n - 1)\}$. Define

$$W \equiv \frac{4N(n - N)}{n(n - 1)(n - 2)(n - 3)}\left[N(n - N)\left\{\frac{2m^2(2n - 3)}{n(n - 1)} - (m + D)\right\}\right.$$
$$\left. - (n - 1)(m^2 - m - D) + \tfrac{1}{4}(n - 2)(n - 3)(2m + D)\right],$$

where

$$D \equiv 2(15rs - 17r - 18s + 18),$$

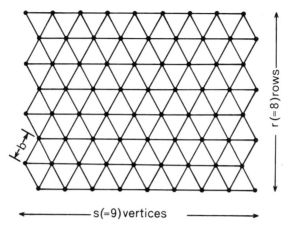

Figure 5.8. Triangular lattice of $r(=8)$ rows of $s(=9)$ vertices. Vertices and edges of the resulting graph are indicated by dots and lines, respectively. The graph does not satisfy point (iii) of (5.13), but is useful in applications.

and put

$$U \equiv W^{-1/2}\left\{M - \frac{2mN(n-N)}{n(n-1)}\right\}. \qquad (5.16)$$

It may be proved that under the null hypothesis, U has zero mean and unit variance (both conditional on N).

Values of this statistic were computed in the case of Diggle's (1981) heather data, for lattice edge widths 40 cm, 60 cm, 80 cm, 100 cm, 120 cm, and 140 cm. To check on the effect which placement of the lattice had on U, two counts were made for each lattice edge width, the lattice for the second count having its edges parallel to those of the first but its vertices at the centroids of the triangular faces of the first. The results are summarized in Table 5.2. They suggest a maximum shape diameter of approximately 80 cm, corresponding to a maximum radius of about 40 cm. The analyses in Sections 5.2 and 5.3 indicate mean shape radius to be approximately 30 cm.

Proof of Theorem 5.1. As $\mathcal{R}$ increases and $\delta \to 0$, $N/\{n(1-p_\delta)\}$ $\to 1$ in probability and $p_\delta \to p_0$. Therefore

$$U = \{M(n-1) - \nu N(n-N)\}(2\nu)^{-1/2}n^{-3/2}$$
$$\times \{p(1-p)\}^{-1}\{1 + o_p(1)\}, \qquad (5.17)$$

TABLE 5.2 Summary of Counts on Eight Triangular Lattices

Edge Width b (in cm)	Lattice Size	Values of U, defined at (5.16)
40	27×47	$-19, -15$
60	18×31	$-4.7, -3.0$
80	15×25	$-2.4, -1.5$
100	11×19	$-0.2, +0.3$
120	16×10	$-1.3, 0.0$
140	14×8	$+0.3, +1.2$

A lattice size of $r \times s$ indicates r rows of s vertices, as shown in Figure 5.8. Raw data are depicted in Figure 5.3.

where we write p for p_δ. Let E_1 denote expectation under the local alternative hypothesis, and put $C \equiv C_{P,\delta}(\mathbf{z})$. The first step is to prove that

$$E_1\{M(n-1) - \nu N(n-N)\}n^{-3/2} = -\nu n^{1/2}C + o(n^{1/2}C) \quad (5.18)$$

as $\mathcal{R}$ increases and $\delta \to 0$.

Number the vertices from 1 to n. Let $\mathcal{N}_i$ be the set of vertices bonded to vertex i, and $\mathcal{M}_i$ the set of vertices that are distant precisely b units from vertex i but are not bonded to vertex i. Write m_i for the number of elements of $\mathcal{M}_i$. Then $0 \le m_i \le 4$. Let I_i be the indicator of the event that vertex i is black. Then

$$N(n-N) = \sum_i \sum_j I_i(1-I_j) = M + \sum_i \sum_{j \notin \mathcal{N}_i} I_i(1-I_j),$$

whence

$$M(n-1) - \nu N(n-N) = (n-\nu-1)M - \nu \sum_i \sum_{j \notin \mathcal{N}_i} I_i(1-I_j). \quad (5.19)$$

If $j \notin \mathcal{N}_i$ and δ is small then $E_1\{I_i(1-I_j)\}$ equals 0 if $j = i$, $p(1-p)$ if $j \notin \mathcal{M}_i$ and $j \ne i$, and $p(1-p) - C$ if $j \in \mathcal{M}_i$. Consequently,

$$E_1\left\{\sum_i \sum_{j \notin \mathcal{N}_i} I_i(1-I_j)\right\} = n(n-\nu-1)p(1-p) - C\sum_{i=1}^n m_i. \quad (5.20)$$

Additionally,

$$E_1(M) = m\{2p(1-p) - 2C\} = n\nu p(1-p) - n\nu C. \quad (5.21)$$

[Strictly speaking, the last formula assumes that the Boolean model is viewed using the torus convention on the rectangle $\mathcal{R}$; see Chapter 1, Section 1.5. However the error committed by ignoring this convention equals $o(nC)$, and so is negligible.] Result (5.18) follows from (5.19)–(5.21).

Theorem 5.1 is a consequence of (5.17), (5.18), and

$$\{Z - E_1(Z)\}\,(2\nu)^{-1/2}n^{-3/2}\,\{p(1-p)\}^{-1} \to N(0,1) \quad \text{in distribution,} \tag{5.22}$$

where $Z \equiv M(n-1) - \nu N(n-N)$. We conclude by deriving (5.22). For *any* real q,

$$Z = -(n-1)\sum_i \sum_{j\in\mathcal{N}_i}(I_i-q)(I_j-q)$$

$$+\nu\left\{\sum_i(I_i-q)\right\}^2 - \nu\sum_i(I_i-q)^2.$$

We take $q \equiv 1-p = 1-p_\delta$. If δ is sufficiently small then each $I_i - q$ is correlated with no more than six variables $I_j - q$, and so

$$E_1\left\{\sum_i(I_i-q)\right\}^2 + E_1\left\{\sum_i(I_i-q)^2\right\} = O(n)$$

as $n \to \infty$. Therefore

$$Z - E_1(Z) = -(n-1)\sum_i\sum_{j\in\mathcal{N}_i}X_{ij} + O_p(n),$$

where $X_{ij} \equiv (I_i-q)(I_j-q) - E_1\{(I_i-q)(I_j-q)\}$. In consequence, result (5.22) will follow if we prove that with

$$X \equiv \sum_i\sum_{j\in\mathcal{N}_i}X_{ij},$$

we have

$$X(2\nu n)^{-1/2}\,\{p(1-p)\}^{-1} \to N(0,1) \quad \text{in distribution} \tag{5.23}$$

as $n \to \infty$.

Suppose the rectangular region $\mathcal{R}$ has dimensions $a_1 l \times a_2 l$, where a_1 and a_2 are fixed and $l \to \infty$ as $\mathcal{R}$ increases. Let c be a large positive constant, put $d \equiv 2r_0 + 1$, and divide $\mathcal{R}$ as nearly as possible into a lattice of squares of side length $c+d$. Inside each square, inscribe a concentric square of side length c. This results in a sequence of squares $\mathcal{D}_1, \ldots, \mathcal{D}_s$ say, all subsets of $\mathcal{R}$ and of side length c, and each separated from its neighbors by distance d. Let $\mathcal{I}_{kk}$ denote the set of all indices i such that vertex i and all the vertices in $\mathcal{N}_i$ lie within $\mathcal{D}_k$. Define

$$Y_k \equiv \sum_{i \in \mathcal{I}_{kk}} \sum_{j \in \mathcal{N}_i} X_{ij} \quad \text{and} \quad Y \equiv \sum_{k=1}^{s} Y_k.$$

Then

$$\lim_{c \to \infty} \limsup_{n \to \infty} n^{-1} E_1 \left\{ (X - Y)^2 \right\} = 0.$$

Furthermore, for small values of δ the variables Y_k are stochastically independent. An application of Lindeberg's central limit theorem (Appendix II) shows that their sum is asymptotically normally distributed with zero mean and variance $E_1(Y^2)$. Furthermore,

$$\left| E_1(X^2) - E_1(Y^2) \right| \le 4 \left[E_1 \left\{ (X - Y)^2 \right\} \left\{ E_1(X^2) + E_1(X - Y)^2 \right\} \right]^{1/2},$$

and so

$$\lim_{c \to \infty} \limsup_{n \to \infty} \left| n^{-1} E_1(X^2) - n^{-1} E_1(Y^2) \right| = 0,$$

provided that for some $\sigma^2 \ge 0$,

$$n^{-1} E_1(X^2) \to \sigma^2 < \infty. \tag{5.24}$$

Therefore the proof of (5.23) will be complete if we establish (5.24) with $\sigma^2 = 2\nu p_0(1 - p_0)$. This demonstration requires only a little algebra.

Proof of a central limit theorem for U under the assumption that $\delta = 0$ and $b > 2r_0$ follows the same lines, but is even easier. $\square$

5.5. INFERENCE IN A BOOLEAN MODEL OF NORMAL-RADIUS DISKS

Consider a Boolean model of disks in the plane, and suppose disk radius is approximately normally distributed with mean μ and variance σ^2. We must assume that σ is considerably less than μ, for otherwise we may have a problem with disks of negative radius! The condition $\sigma/\mu < \frac{1}{3}$ is therefore imposed. Dupač (1980) suggests a technique for estimating μ, σ^2, and Poisson intensity λ, based on vacancy and properties of circular clumps. A circular clump is a clump of disks whose boundary is a perfect circle, and so is comprised of a primary disk possibly containing smaller disks. The boundary of the primary disk forms the boundary of the clump. To describe Dupač's estimators, we begin by deriving the density of radius distribution for circular clumps.

Let R have the distribution of disk radius, with density function f. The probability that no random disks from the Boolean model intersect the *boundary* of a given disk of radius r equals $e^{-\lambda\eta}$ where

$$\eta \equiv \pi E \left[(r+R)^2 - \{\max(r-R,0)\}^2\right]$$
$$= \pi E \left[(r+R)^2 - (r-R)^2 + \{\min(r-R,0)\}^2\right]$$
$$= \pi \left\{4rE(R) + \int_r^\infty (x-r)^2 f(x)\,dx\right\}.$$

Therefore the density of the radius of circular clumps is proportional to

$$f(r)\exp\left[-\lambda\pi\left\{4rE(R) + \int_r^\infty (x-r)^2 f(x)\,dx\right\}\right]. \qquad (5.25)$$

Let ϕ and Φ denote the standard normal density and distribution functions, respectively. Put

$$s \equiv (r-\mu)/\sigma \quad \text{and} \quad \psi(s) \equiv \{1-\Phi(s)\}(1+s^2) - s\phi(s).$$

When f is the normal density with mean μ and variance σ^2,

$$\int_r^\infty (x-r)^2 f(x)\,dx = \int_{(r-\mu)/\sigma}^\infty (\sigma x + \mu - r)^2 \phi(x)\,dx$$
$$= \sigma^2 \int_s^\infty (x^2 - 2sx + s^2)\phi(x)\,dx$$
$$= \sigma^2 \psi(s).$$

Therefore the function at (5.25) is proportional to

$$\exp\left[-\lambda\pi\left\{4\mu\sigma s + \sigma^2\psi(s)\right\} - \tfrac{1}{2}s^2\right]. \qquad (5.26)$$

For part of its range the function $\psi(x)$ may be approximated by $\tfrac{1}{2}(x-1)^2$. In particular,

$$\psi(x) = \tfrac{1}{2}(x-1)^2 + 0.202x + O(x^3)$$

TABLE 5.3 Percentage Differences Between Approximate and Exact Values of μ'

	σ/μ	
$\lambda\pi\mu^2$	$\frac{1}{3}$	$\frac{1}{6}$
$\frac{1}{3}$	-0.6	-0.02
$\frac{1}{6}$	-0.3	-0.01

Tabulated are values of $100(\mu'_{approx}/\mu'_{exact} - 1)$, where μ'_{approx} equals the right-hand side of equation (5.27). (Dupač 1980.)

as $x \to 0$. Replacing $\psi(s)$ by $\frac{1}{2}(s-1)^2$ in (5.26), we obtain the function

$$\exp\left[-\lambda\pi\left\{4\mu\sigma s + \tfrac{1}{2}\sigma^2(s-1)^2\right\} - \tfrac{1}{2}s^2\right] = \text{const.} \exp\left[-\tfrac{1}{2}(\lambda\pi + \sigma^{-2})\right.$$
$$\left. \times \left\{r - (\lambda\pi + \sigma^{-2})^{-1}(\mu\sigma^{-2} - \lambda\pi(3\mu - \sigma))\right\}^2\right],$$

where the constant does not depend on r. Let μ' and σ'^2 denote, respectively, the mean and variance of the radius of circular clumps. We conclude from the foregoing that the distribution of radius of circular clumps is approximately normal with mean

$$\mu' \simeq (\lambda\pi + \sigma^{-2})^{-1}\left\{\mu\sigma^{-2} - \lambda\pi(3\mu - \sigma)\right\} \tag{5.27}$$

and variance

$$\sigma'^2 \simeq (\lambda\pi + \sigma^{-2})^{-1}. \tag{5.28}$$

These approximations are quite good if we insist that $\sigma/\mu < \frac{1}{3}$ and $\lambda\pi\mu^2 < \frac{1}{3}$ (see Table 5.3). Given the former condition, the latter is almost the same as $\alpha\lambda < \frac{1}{3}$, where α is mean disk area, that is, almost the same as $p = e^{-\alpha\lambda} > 0.7$, where p is porosity. Therefore the condition $\lambda\pi\mu^2 < \frac{1}{3}$ ensures a relatively small area covered by disks, and also a relatively large number of circular clumps.

Let $\hat{\mu}'$ and $\hat{\sigma}'^2$ denote sample-based estimates of μ' and σ'^2, and V equal vacancy within an observation region $\mathcal{R}$. Take $\hat{p}$ equal to estimated porosity: $\hat{p} \equiv \|\mathcal{R}\|^{-1}V$. Notice that

$$p = \exp\left\{-\lambda\pi(\mu^2 + \sigma^2)\right\}. \tag{5.29}$$

Replacing μ', σ'^2, and p in (5.27), (5.28), and (5.29) by the estimates $\hat{\mu}'$, $\hat{\sigma}'^2$, and $\hat{p}$, we obtain three equations that may be solved simultaneously by iteration to give estimates $\hat{\mu}$, $\hat{\sigma}^2$, and $\hat{\lambda}$ of μ, σ^2, and λ, respectively:

$$\hat{\mu}' = (\hat{\lambda}\pi + \hat{\sigma}^{-2})\left\{\hat{\mu}\hat{\sigma}^{-2} - \hat{\lambda}\pi(3\hat{\mu} - \hat{\sigma})\right\},$$

$$\hat{\sigma}'^2 = (\hat{\lambda}\pi + \hat{\sigma}^{-2})^{-1}, \qquad \hat{p} = \exp\left\{-\hat{\lambda}\pi(\hat{\mu}^2 + \hat{\sigma}^2)\right\}.$$

5.6. CURVATURE-BASED INFERENCE IN A BOOLEAN MODEL OF DISKS

Let $\mathcal{C} \equiv \{\boldsymbol{\xi}_i + S_i, i \geq 1\}$ be a Boolean model in which the driving Poisson process $\mathcal{P} \equiv \{\boldsymbol{\xi}_i, i \geq 1\}$ has intensity λ, and in which shapes $\{S_i, i \geq 1\}$ are distributed as S, where S is a (closed) disk centered at the origin and of random radius R. In theory, the radius and visible curvature of each piece of each disk protruding from a clump anywhere in the observation region may be determined precisely. The present section develops a theory of inference that makes use of all these possible measurements. The work is of technical interest because it comes close to a unified theory of inference in random-disk Boolean models. It permits the use of analogues of much of the methodology from classical statistical theory—maximum likelihood estimation, inference based on the method of moments, likelihood ratio testing, nonparametric density estimation, etc. However, practical applications are very limited, owing to the difficulty of determining disk radius from small portions of disks protruding from clumps. Among other things, the technique obviously relies heavily on the assumption that random sets are exactly disk-shaped. A further drawback is that individual measurements of curvature are difficult to make digitally on image analysis equipment. This disadvantage is shared by some curvature-based estimators occasionally used in stereology.

We begin by introducing notation. Let $\partial(\boldsymbol{\xi}_i + S_i)$ denote the boundary of the random disk $\boldsymbol{\xi}_i + S_i$, and suppose the Boolean model pattern is observed within a bounded region $\mathcal{R}$. Write R_i for the radius of S_i. The total visible curvature within $\mathcal{R}$ of $\partial(\boldsymbol{\xi}_i + S_i)$ is the total angle subtended at the center of $\boldsymbol{\xi}_i + S_i$ by that part of the (possibly broken) curve $\mathcal{R} \cap \partial(\boldsymbol{\xi}_i + S_i)$ which is not covered by other sets. Call it Φ_i. We are interested in quantities such as

$$T \equiv \sum_{i=1}^{\infty} \Phi_i A(R_i), \tag{5.30}$$

where A is a given function. If $E(R^2) < \infty$ then with probability 1 the series contains only a finite number of nonzero terms. Therefore T is well-defined.

The statistic T is observable (provided the model is exactly true, as outlined two paragraphs above). This statistic is easier to calculate than the technical definition might suggest. Note that formula (5.30) is additive, and so individual observations of visible curvature may be added in one-by-one, by working methodically around the boundary of each clump. The total contribution to T from $\xi_i + S_i$, perhaps due to several different protrusions to the boundary of a clump, does not have to be treated at once. For example, Figure 5.9 depicts a portion of the disk $\xi_i + S_i$ protruding from a clump. Let L and W be, respectively, the length and width of the protrusion (see Figure 5.9 for definitions). The radius of S_i and contribution to uncovered curvature of that piece of protrusion are given by

$$R_i = \begin{cases} (W/2) + (L^2/8W) & \text{if } 2W < L \\ L/2 & \text{otherwise} \end{cases}$$

$$\Theta = \begin{cases} 2\sin^{-1}(L/2R_i) & \text{if } 2W < L \\ 2\pi - 2\cos^{-1}\{(W - \frac{1}{2}L)/R_i\} & \text{otherwise,} \end{cases} \tag{5.31}$$

respectively. The statistic T is obtained by adding $\Theta A(R_i)$ over all protrusions contained within $\mathcal{R}$. If a protrusion is intersected by the boundary of $\mathcal{R}$, then instead of L and W we use the length and width of that part of the protrusion that lies inside $\mathcal{R}$.

Curvature-based inference is founded on the fact that as the observation region $\mathcal{R}$ increases, the ratio

$$\hat{\mu} \equiv \left\{ \sum_{i=1}^{\infty} \Phi_i A(R_i) \right\} \Big/ \left(\sum_{i=1}^{\infty} \Phi_i \right) \tag{5.32}$$

converges to $\mu \equiv E\{A(R)\}$. Thus, weighting the observed values $A(R_i)$ by their associated visible curvatures permits us to recover properties of the true distribution of radius. In fact, we may treat each R_i as though it were independent of all others, and weight it as though it were observed a total number of times proportional to its associated uncovered curvature. Arguing in this manner we may view inference in a Boolean model of random disks from the point of view of inference based on a sequence of independent observations. Notice that we do not require any special correction for edge effects.

A simple way of describing large-region properties of $\hat{\mu}$ is to give formulas for the mean and variance of statistics such as T, defined

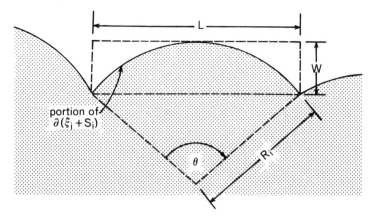

Figure 5.9. Calculation of radius and visible curvature. Measurements L and W are, respectively, length and width of a protrusion from a clump. Radius R_i and contribution Θ to visible curvature for this protrusion are given by formula (5.31).

at (5.30). This we do next. Let $\alpha \equiv E(\|S\|) = \pi E(R^2)$, and assume the observation region $\mathcal{R}$ is Riemann measurable.

Theorem 5.2. *If $E(R^2) < \infty$ and $E\{|A(R)|\} < \infty$ then*

$$E(T) = 2\pi\lambda\|\mathcal{R}\|e^{-\alpha\lambda}E\{A(R)\}. \tag{5.33}$$

If $E(R^4) < \infty$ and $E\{A(R)^2\} < \infty$ then

$$\mathrm{var}(T) = O(\|\mathcal{R}\|), \tag{5.34}$$

assuming $\mathcal{R}$ increases as a scale multiple of a fixed Riemann-measurable set.

A proof of Theorem 5.2 is deferred until the end of the section. Result (5.33) is an easy corollary of the marked point process argument discussed in Section 4.1, as our proof will show. An exact formula for $\mathrm{var}(T)$ may be written down using properties of second moment measures of marked point processes. However, the formula is quite complex, and does not concern us here. It is not difficult to determine an asymptotic expression for $\mathrm{var}(T)$, of the form $\mathrm{var}(T) \sim \mathrm{const.}\,\|\mathcal{R}\|$ as $\mathcal{R}$ increases (see Theorem 5.3 below).

If we apply Theorem 5.2 first to the numerator and then to the denominator of the ratio defining $\hat{\mu}$ [formula (5.32)], we conclude that if $E(R^4) < \infty$ and $E\{A(R)^2\} < \infty$ then

$$\hat{\mu} = \frac{2\pi\lambda\|\mathcal{R}\|e^{-\alpha\lambda}E\{A(R)\} + O_p(\|\mathcal{R}\|^{1/2})}{2\pi\lambda\|\mathcal{R}\|e^{-\alpha\lambda} + O_p(\|\mathcal{R}\|^{1/2})}$$

$$= E\{A(R)\} + O_p(\|\mathcal{R}\|^{-1/2})$$

as $\mathcal{R}$ increases. Therefore $\hat{\mu}$ converges to μ at the rate $\|\mathcal{R}\|^{-1/2}$, which is the analogue of $\sqrt{n}$-consistency in classical statistical theory. Thus we may consistently estimate moments of arbitrary functions of radius. In particular, the mean $\mu \equiv E(R)$ and variance $\sigma^2 \equiv \text{var}(R)$ of radius are consistently estimated by

$$\hat{\mu} \equiv \left(\sum_{i=1}^{\infty}\Phi_i R_i\right)\bigg/\left(\sum_{i=1}^{\infty}\Phi_i\right) \quad \text{and}$$

$$\hat{\sigma}^2 \equiv \left(\sum_{i=1}^{\infty}\Phi_i R_i^2\right)\bigg/\left(\sum_{i=1}^{\infty}\Phi_i\right) - \hat{\mu}^2, \tag{5.35}$$

respectively.

Suppose radius distribution has density $f(r\,|\,\boldsymbol{\theta})$, where $\boldsymbol{\theta}$ is a finite vector of unknown parameters. We may develop an analogue of the likelihood principle for estimating $\boldsymbol{\theta}$, by arguing as follows. Remember that our philosophy is to regard radius values as though they were independent, and to weight each value as though it were observed a number of times proportional to its associated visible curvature within the observation region $\mathcal{R}$. This leads us to consider the pseudo likelihood,

$$L \equiv \prod_{i=1}^{\infty} f(R_i\,|\,\boldsymbol{\theta})^{\Phi_i}, \tag{5.36}$$

which is observable. Of course only a finite number of the variables Φ_i are nonzero, and so all but a finite number of terms in the infinite product at (5.36) are equal to unity. Write $\boldsymbol{\theta} = (\theta_1, \theta_2, \dots)$, and define

$$A_j(r\,|\,\boldsymbol{\theta}) \equiv \{f(r\,|\,\boldsymbol{\theta})\}^{-1} \frac{\partial}{\partial\theta_j}f(r\,|\,\boldsymbol{\theta}). \tag{5.37}$$

The "likelihood equations" defining the "maximum likelihood estimator" $\hat{\boldsymbol{\theta}} = (\hat{\theta}_1, \hat{\theta}_2, \dots)$ are obtained by differentiating $\log L$ with respect to θ_j for $j \geq 1$, and equating to zero. They have the form

$$\sum_{i=1}^{\infty} \Phi_i A_j(R_i\,|\,\hat{\boldsymbol{\theta}}) = 0, \qquad j \geq 1. \tag{5.38}$$

For example if f is the $N(\mu,\sigma^2)$ density and $\boldsymbol{\theta} = (\theta_1,\theta_2) = (\mu,\sigma^2)$, as in Section 5.5, then the solutions of the "likelihood equations" are just the estimators $\hat{\mu}$ and $\hat{\sigma}^2$ defined at (5.35).

To ascertain properties of $\hat{\boldsymbol{\theta}}$ we need only mimic arguments used in classical likelihood theory; for the latter, see Kendall and Stuart (1979, Chapter 18). Define

$$A_{jk}(r\,|\,\boldsymbol{\theta}) \equiv \frac{\partial}{\partial\theta_k}A_j(r\,|\,\boldsymbol{\theta}),$$

and for a given value $\boldsymbol{\theta}_0$ of $\boldsymbol{\theta}$ put

$$T_j \equiv \sum_{i=1}^{\infty}\Phi_i A_j(R_i\,|\,\boldsymbol{\theta}_0) \quad \text{and} \quad T_{jk} \equiv \sum_{i=1}^{\infty}\Phi_i A_{jk}(R_i\,|\,\boldsymbol{\theta}_0).$$

Suppose the true value of $\boldsymbol{\theta}$ is $\boldsymbol{\theta}_0 \equiv (\theta_{01},\theta_{02},\dots)^T$. Assuming $E(R^4) < \infty$ and the same regularity conditions on f needed in classical likelihood theory (Wald 1943; Le Cam 1970; Kendall and Stuart 1979, p. 56ff), the jth likelihood equation at (5.38) may be expressed in a short Taylor series as

$$
\begin{aligned}
O &= \sum_{i=1}^{\infty}\Phi_i A_j(R_i\,|\,\hat{\boldsymbol{\theta}}) \\
&= \sum_{i=1}^{\infty}\Phi_i A_j(R_i\,|\,\boldsymbol{\theta}_0) + \sum_{k}(\hat{\theta}_k - \theta_{0k})\sum_{i=1}^{\infty}\Phi_i A_{jk}(R_i\,|\,\boldsymbol{\theta}_0) \\
&\quad + \text{smaller order terms} \\
&= T_j + \sum_{k}(\hat{\theta}_k - \theta_{0k})T_{jk} + \text{smaller order terms},
\end{aligned}
\tag{5.39}
$$

say. Both T_j and T_{jk} are versions of the statistic T defined at (5.30). Note particularly that

$$E(T_j) = 2\pi\lambda\|\mathcal{R}\|e^{-\alpha\lambda}E\{A_j(R\,|\,\boldsymbol{\theta}_0)\} = 0,$$

so that $T_j = O_p(\|\mathcal{R}\|^{1/2})$; and

$$E(T_{jk}) = 2\pi\lambda\|\mathcal{R}\|e^{-\alpha\lambda}\sigma_{jk} \quad \text{where} \quad \sigma_{jk} \equiv E\{A_{jk}(R\,|\,\boldsymbol{\theta}_0)\},$$

so that $T_{jk} = 2\pi\lambda\|\mathcal{R}\|e^{-\alpha\lambda}\sigma_{jk} + O_p(\|\mathcal{R}\|^{1/2})$. Writing $\mathbf{T} \equiv (T_1,T_2,\dots)^T$ and $\boldsymbol{\Sigma} \equiv (\sigma_{ij})$, where superscript T denotes transpose, we conclude from (5.39) that

$$\hat{\boldsymbol{\theta}} - \boldsymbol{\theta}_0 = -(2\pi\lambda\|\mathcal{R}\|e^{-\alpha\lambda}\boldsymbol{\Sigma})^{-1}\mathbf{T} + o_p(\|\mathcal{R}\|^{-1/2}) \tag{5.40}$$

as the observation region $\mathcal{R}$ increases.

One consequence of (5.40) is that $\hat{\boldsymbol{\theta}} - \boldsymbol{\theta}_0 = O_p(\|\mathcal{R}\|^{-1/2})$ as $\mathcal{R}$ increases. The theorem below gives a large-region expression for the variance of statistics such as T defined at (5.30). From that and an analogue for covariance we may deduce a large-region formula for the variance of $\mathbf{T}$. However, that expression is rather complex.

Given a planar curve Γ, let $\mathbf{t} = \mathbf{t}(\mathbf{x})$ denote a unit vector in the direction of the tangent to Γ at $\mathbf{x} \in \Gamma$. Define

$$e(\mathbf{x}) \equiv E\{\|(\mathbf{x}+S) \cup S\|\} = 2\alpha - E\left\{ R^2 B(|\mathbf{x}|/2R) \right\},$$

where $B(x)$ denotes the area of the lens of intersection of two unit disks centered $2x$ apart; see (5.9). Put

$$\tau_1 \equiv E\left[A^2(R) \int_{\partial S} |dt(\mathbf{x}_1)| \int_{\partial S} \exp\left\{ -\lambda e(\mathbf{x}_1 - \mathbf{x}_2) \right\} |dt(\mathbf{x}_2)| \right]$$

and

$$\tau_2 \equiv 2\pi E\{A(R)\} E\left(A(R) \int_{\mathbb{R}^2} d\mathbf{x} \int_{\partial(\mathbf{x}+S)} [\exp\{-\lambda e(\mathbf{y})\} \right.$$
$$\left. - e^{-2\alpha\lambda}] |dt(\mathbf{y})| \right),$$

remembering that S is a disk centered at the origin and of radius R. Define T as at (5.30).

Theorem 5.3. *If $E(R^4) < \infty$ and $E\{A(R)^2\} < \infty$ then*

$$\mathrm{var}(T) = (\lambda\tau_1 + \lambda^2\tau_2)\|\mathcal{R}\| + o(\|\mathcal{R}\|), \qquad (5.41)$$

as $\mathcal{R}$ increases as a scale multiple of a fixed Riemann-measurable set. Both τ_1 and τ_2 are finite.

Notice that τ_2 vanishes if $E\{A(R)\} = 0$. To see the implications of this property, let $\hat{\mu}$ be the statistic defined at (5.32) and take $\mu \equiv E\{A(R)\}$. Then

$$\hat{\mu} - \mu = \left[\sum_i \Phi_i \{A(R_i) - \mu\} \right] \Big/ \left(\sum_i \Phi_i \right)$$
$$= (2\pi\lambda\|\mathcal{R}\|e^{-\alpha\lambda})^{-1} \sum_i \Phi_i \{A(R_i) - \mu\} + O_p(\|\mathcal{R}\|^{-1}).$$

$$(5.42)$$

The first term on the right-hand side is the dominant one, being of order $\|\mathcal{R}\|^{-1/2}$. Furthermore, $A(R) - \mu$ has zero mean, and so its corresponding value of τ_2 vanishes. Our proof of Theorem 5.3 will show that the term $\lambda\tau_1$ in formula (5.41) derives from the sum of the variances of the terms in the series T, while $\lambda^2\tau_2$ comes from the sum of the covariances. When $\tau_2 = 0$, the summands are asymptotically uncorrelated, a description that also applies to the series on the right-hand side of (5.42). In this sense, the individual terms that contribute to the estimator $\hat{\mu}$ are asymptotically uncorrelated, and behave rather like independent random variables.

So far we have avoided mentioning distributional properties of estimators, preferring instead to describe asymptotic properties of the first two moments. However, large-region central limit theorems are readily derived. The proofs are tedious rather than difficult, and closely parallel the argument used to establish Theorem 3.5 in Chapter 3. Assuming the conditions of Theorem 5.3, the statistic T defined at (5.30) is asymptotically normally distributed with mean $E(T)$ and variance var(T).

We conclude this section with proofs of Theorems 5.2 and 5.3.

Proof of Theorems 5.2 and 5.3. Recall that $\mathcal{C} = \{\xi_j + S_j, j \geq 1\}$ is the Boolean model generated by shapes distributed as S, and $\mathcal{P} = \{\xi_j, j \geq 1\}$ is the Poisson process driving $\mathcal{C}$. To prove (5.33) we assign marks to points ξ_i of $\mathcal{P}$, and use marked point process techniques from Section 4.1 of Chapter 4, as follows. Let $\mathcal{C}_i \equiv \{\xi_j + S_j : j \geq 1, j \neq i\}$ represent the sets in $\mathcal{C}$ excluding $\xi_i + S_i$, and let $\xi_i + \Gamma_i$ be the set of all points in $\partial(\xi_i + S_i)$ not covered by any sets from $\mathcal{C}_i$. Then Γ_i is a broken curve comprised of a collection of segments of a circle centered at the origin. Assign the point ξ_i the mark (Γ_i, R_i). Given a curve Γ such as Γ_i, define $f(\mathbf{x}, \Gamma, r)$ to be the total curvature of $(\mathbf{x} + \Gamma) \cap \mathcal{R}$, multiplied by $A(r)$. In this notation,

$$T = \sum_{i=1}^{\infty} f(\xi_i, \Gamma_i, R_i).$$

Take the generic random disk S, centered at the origin and with radius R, to be independent of $\mathcal{C}$, and let Γ be the set of all points in ∂S not covered by $\mathcal{C}$. By the Campbell theorem [formulas (4.4) and (4.8), Chapter 4] applied separately to the positive and negative parts of f, we obtain

$$E(T) = \lambda \int_{\mathbb{R}^2} E\{f(\mathbf{x}, \Gamma, R)\}\, d\mathbf{x}.$$

Writing $t(y)$ for the unit tangent to $(x + \Gamma) \cap \mathcal{R}$ at a point y, and changing an order of integration, we obtain

$$
\begin{aligned}
A(R)^{-1} \int_{\mathbb{R}^2} f(x, \Gamma, R) \, dx &= \int_{\mathbb{R}^2} dx \int_{(x+\Gamma) \cap \mathcal{R}} |dt(y)| \\
&= \int_{\mathbb{R}^2} dx \int_{\Gamma \cap (-x+\mathcal{R})} |dt(y)| \\
&= \int_{\Gamma} |dt(y)| \int_{(-y+\mathcal{R})} dx \\
&= \|\mathcal{R}\| \int_{\Gamma} |dt(y)|.
\end{aligned}
$$

Multiplying by $A(R)$ and taking expectations we conclude that

$$
E(T) = \lambda \|\mathcal{R}\| E \{A(R) \times (\text{total curvature of } \Gamma)\}. \tag{5.43}
$$

(See Chapter 4, Section 4.3 for a definition of curvature in terms of an integral.) Conditional on R, ∂S has total curvature 2π, and each element of ∂S is uncovered with probability $e^{-\alpha \lambda}$. Therefore conditional on R, the expected total curvature of Γ equals $2\pi e^{-\alpha \lambda}$. Result (5.33) now follows from (5.43).

The first step in proving (5.34) and Theorem 5.3 is to derive an exact expression for $\text{var}(T)$. As mentioned earlier, $\text{var}(T)$ may be written down immediately from formulas for second moment measures of marked point processes, but since those results would not be of use to us in the future we do not develop them here. Instead we use the following argument. We claim that

$$
E(T^2) = \lambda \tau_3 + \lambda^2 \tau_4 \tag{5.44}
$$

where

$$
\begin{aligned}
\tau_3 \equiv E \Big[A^2(R) \int_{\mathbb{R}^2} dx \int_{\mathcal{R} \cap \partial(x+S)} |dt(y_1)| \\
\times \int_{\mathcal{R} \cap \partial(x+S)} \exp \{-\lambda e(y_1 - y_2)\} |dt(y_2)| \Big]
\end{aligned}
$$

and

$$
\begin{aligned}
\tau_4 \equiv E \Big[A(R_1) A(R_2) \int_{\mathbb{R}^2} dx_1 \int_{\mathbb{R}^2} dx_2 \\
\times \int_{\mathcal{R} \cap \partial(x_1+S_1)} |dt(y_1)| \int_{\mathcal{R} \cap \partial(x_2+S_2)} \exp \{-\lambda e(y_1 - y_2)\} |dt(y_2)| \Big].
\end{aligned}
$$

To prove (5.44), assume temporarily that disk radius is essentially bounded by r. Let $\mathcal{A}$ be a bounded set sufficiently large to contain

$$\mathcal{R}^r \equiv \left\{ \mathbf{y} \in \mathbb{R}^k : |\mathbf{x} - \mathbf{y}| < r \text{ for some } \mathbf{x} \in \mathcal{R} \right\}. \tag{5.45}$$

(Remember that a Riemann-measurable set $\mathcal{R}$ is necessarily bounded.) Any random set in the Boolean model and intersecting $\mathcal{R}$ must be centered within $\mathcal{A}$. Let M be a Poisson-distributed random variable with mean $\lambda\|\mathcal{A}\|$. Conditional on M, let $\boldsymbol{\xi}_{(1)}, \ldots, \boldsymbol{\xi}_{(M)}$ be independent and uniformly distributed over $\mathcal{A}$, and let $S_{(1)}, \ldots, S_{(M)}$ be independent copies of S, independent also of the $\boldsymbol{\xi}_{(i)}$'s. Then T has the same distribution as

$$T' \equiv \sum_{i=1}^{M} \Phi_{(i)} A(R_{(i)}), \tag{5.46}$$

where $R_{(i)}$ and $\Phi_{(i)}$ are, respectively, the radius of $S_{(i)}$ and the total visible curvature of $\partial_i \equiv \mathcal{R} \cap \partial(\boldsymbol{\xi}_{(i)} + S_{(i)})$. [Here, a point of $\partial(\boldsymbol{\xi}_{(i)} + S_{(i)})$ is said to be visible if it is covered by none of the sets $\boldsymbol{\xi}_{(j)} + S_{(j)}$, $j \geq 1$ and $j \neq i$.] Notice that

$$E(\Phi_{(1)}^2 \mid \boldsymbol{\xi}_{(1)}, S_{(1)}; M \geq 1)$$
$$= \int_{\partial_1} \int_{\partial_1} P(\mathbf{y}_1, \mathbf{y}_2 \text{ both not covered} \mid \boldsymbol{\xi}_{(1)}, S_{(1)})$$
$$\times |dt(\mathbf{y}_1)||dt(\mathbf{y}_2)|$$
$$= \int_{\partial_1} \int_{\partial_1} \exp\{-\lambda e(\mathbf{y}_1 - \mathbf{y}_2)\} |dt(\mathbf{y}_1)||dt(\mathbf{y}_2)|,$$

and similarly,

$$E(\Phi_{(1)}\Phi_{(2)} \mid \boldsymbol{\xi}_{(1)}, \boldsymbol{\xi}_{(2)}, S_{(1)}, S_{(2)}; M \geq 2)$$
$$= \int_{\partial_1} \int_{\partial_2} \exp\{-\lambda e(\mathbf{y}_1 - \mathbf{y}_2)\} |dt(\mathbf{y}_1)||dt(\mathbf{y}_2)|.$$

In consequence,

$$E\left\{ \Phi_{(1)}^2 A(R_{(1)})^2 \mid M \geq 1 \right\} = \|\mathcal{A}\|^{-1} \tau_3 \tag{5.47}$$

and

$$E\{\Phi_{(1)}\Phi_{(2)} A(R_{(1)})A(R_{(2)}) \mid M \geq 2\} = \|\mathcal{A}\|^{-2} \tau_4.$$

Squaring both sides of (5.46) and taking expectations we find that

$$E(T^2) = E(T'^2)$$
$$= E(M)E\left\{\Phi_{(1)}^2 A(R_{(1)})^2 \mid M \geq 1\right\}$$
$$+ E\{M(M-1)\} E\left\{\Phi_{(1)}\Phi_{(2)}A(R_{(1)})A(R_{(2)}) \mid M \geq 2\right\}.$$

Furthermore, $E(M) = \text{var}(M) = \lambda\|\mathcal{A}\|$. Result (5.44) follows on combining the formulas from (5.47) down.

A truncation and approximation argument shows that (5.44) holds without the assumption that $\text{rad}\, S$ is essentially bounded. Thus we conclude from (5.33) and (5.44) that

$$\text{var}(T) = \lambda\tau_3 + \lambda^2\tau_5,$$

where τ_3 is as before and

$$\tau_5 \equiv E\left(A(R_1)A(R_2)\int_{\mathbb{R}^2} d\mathbf{x}_1 \int_{\mathbb{R}^2} d\mathbf{x}_2 \int_{\mathcal{R}\cap\partial(x_1+S_1)} |dt(\mathbf{y}_1)| \int_{\mathcal{R}\cap\partial(x_2+S_2)}\right.$$
$$\left.\times [\exp\{-\lambda e(\mathbf{y}_1 - \mathbf{y}_2)\} - e^{-2\alpha\lambda}]|dt(\mathbf{y}_2)|\right).$$

We shall prove that as $\mathcal{R}$ increases,

$$\tau_3 = \|\mathcal{R}\|\tau_1 + o(\|\mathcal{R}\|) \tag{5.48}$$

and

$$\tau_5 = \|\mathcal{R}\|\tau_2 + o(\|\mathcal{R}\|). \tag{5.49}$$

Observe that with $f(\mathbf{x}) \equiv \exp\{-\lambda e(\mathbf{x})\}$ we have

$$\tau_3 - \|\mathcal{R}\|\tau_1$$
$$= E\left\{A^2(R)\int_{\tilde{\mathcal{R}}} d\mathbf{x}\int_{\mathcal{R}\cap\partial(x+S)}|dt(\mathbf{y}_1)|\int_{\mathcal{R}\cap\partial(x+S)}f(\mathbf{y}_1-\mathbf{y}_2)|dt(\mathbf{y}_2)|\right\}$$
$$- E\left\{A^2(R)\int_{\mathcal{R}} d\mathbf{x}\int_{\tilde{\mathcal{R}}\cap\partial(x+S)}|dt(\mathbf{y}_1)|\int_{\mathcal{R}\cap\partial(x+S)}f(\mathbf{y}_1-\mathbf{y}_2)|dt(\mathbf{y}_2)|\right\}$$
$$- E\left\{A^2(R)\int_{\mathcal{R}} d\mathbf{x}\int_{\partial(x+S)}|dt(\mathbf{y}_1)|\int_{\tilde{\mathcal{R}}\cap\partial(x+S)}f(\mathbf{y}_1-\mathbf{y}_2)|dt(\mathbf{y}_2)|\right\}. \tag{5.50}$$

The first term on the right-hand side is dominated by

$$t \equiv E\left(A^2(R)\int_{\tilde{\mathcal{R}}}[c\{\mathcal{R}\cap\partial(\mathbf{x}+S)\}]^2 d\mathbf{x}\right),$$

where $c(\Gamma)$ denotes the total curvature of a broken planar curve Γ. If $\mathcal{R} = r\mathcal{R}_0$ for a fixed set $\mathcal{R}_0$ and an increasing scale factor r, then

$$t = r^2 E\left\{A^2(R)J(R)\right\}$$

where

$$J(R) \equiv \int_{\dot{\mathcal{R}}_0} \left[c\left\{\mathcal{R}_0 \cap \partial(\mathbf{x} + r^{-1}S)\right\}\right]^2 d\mathbf{x}.$$

It is easy to see that $J(R) \to 0$ almost surely as $r \to \infty$. Since $\mathcal{R}_0$ is bounded then

$$c\left\{\mathcal{R}_0 \cap \partial(\mathbf{x} + r^{-1}S)\right\} \le C^{1/2}\min(1, rR^{-1})$$

for a constant $C > 0$, and so $J(R)$ is dominated by

$$C\min(1, r^2 R^{-2}) \int_{\dot{\mathcal{R}}_0} I\left\{\mathcal{R}_0 \cap (\mathbf{x} + r^{-1}S) \ne \phi\right\} d\mathbf{x}$$

$$\le C\min(1, r^2 R^{-2}) \|\mathcal{R}_0^{r^{-1}R}\|$$

$$\le C(\|\mathcal{R}_0^1\| + \sup_{x>1} x^{-2}\|\mathcal{R}_0^x\|) < \infty,$$

where $\mathcal{R}_0^z$ is as defined at (5.45). It now follows from the dominated convergence theorem that $t = o(\|\mathcal{R}\|)$ as $\mathcal{R}$ increases.

The second term on the right-hand side of (5.50) is dominated by

$$4\pi^2 E\left[A^2(R)\int_{\mathcal{R}} I\{\mathcal{R} \cap (\mathbf{x} + S) \ne \phi\}\, d\mathbf{x}\right]$$

$$= 4\pi^2 r^2 E\left[A^2(R)\int_{\mathcal{R}_0} I\left\{\mathcal{R}_0 \cap (\mathbf{x} + r^{-1}S) \ne \phi\right\} d\mathbf{x}\right]$$

$$= o(\|\mathcal{R}\|),$$

again using the dominated convergence theorem. Similarly, the third term equals $o(\|\mathcal{R}\|)$. Result (5.48) follows on combining the estimates from (5.50) down.

We may write

$$\|\mathcal{R}\|\tau_2 = E\left[A(R_1)A(R_2)\int_{\mathcal{R}} d\mathbf{x}_1 \int_{\mathbb{R}^2} d\mathbf{x}_2 \int_{\partial(\mathbf{x}_1 + S_1)} |dt(\mathbf{y}_1)|\right.$$

$$\left. \times \int_{\partial(\mathbf{x}_2 + S_2)} \left\{f(\mathbf{y}_1 - \mathbf{y}_2) - e^{-2\alpha\lambda}\right\} |dt(\mathbf{y}_2)|\right],$$

where S_1 and S_2 are independent copies of S and $R_i \equiv \mathrm{rad}(S_i)$. The argument that earlier gave (5.50) now yields

$$|\tau_5 - \|\mathcal{R}\|\tau_2| \le 2E(J_1) + (2\pi)^2 E(J_2) + (2\pi)^2 E(J_3), \qquad (5.51)$$

where

$$J_1 \equiv I(R_1 \le R_2)|A(R_1)A(R_2)| \int_{\check{\mathcal{R}}} c\{\mathcal{R} \cap \partial(\mathbf{x}_1 + S_1)\}\, d\mathbf{x}_1$$
$$\times \int_{\mathbb{R}^2} c\{\mathcal{R} \cap \partial(\mathbf{x}_2 + S_2)\}\, F(\mathbf{x}_1,\mathbf{x}_2)\, d\mathbf{x}_2,$$
$$J_2 \equiv |A(R_1)A(R_2)| \int_{\mathcal{R}} d\mathbf{x}_1 \int_{\check{\mathcal{R}}} F(\mathbf{x}_1,\mathbf{x}_2)\, d\mathbf{x}_2,$$
$$J_3 \equiv |A(R_1)A(R_2)| \int_{\mathcal{R}} d\mathbf{x}_1 \int_{\mathcal{R}} F(\mathbf{x}_1,\mathbf{x}_2)$$
$$\times I\{\tilde{\mathcal{R}} \cap (\mathbf{x}_1 + S_1) \ne \phi \text{ or } \tilde{\mathcal{R}} \cap (\mathbf{x}_2 + S_2) \ne \phi\}\, d\mathbf{x}_2,$$

and the *random* function F is given by

$$F(\mathbf{x}_1,\mathbf{x}_2) \equiv \sup_{\mathbf{y}_i \in \mathbf{x}_i + S_i;\, i=1,2} \left\{ f(\mathbf{y}_1 - \mathbf{y}_2) - e^{-2\alpha\lambda} \right\}.$$

Note that $c\{\mathcal{R} \cap \partial(\mathbf{x} + S_i)\} \le \mathrm{const.}\, \min(1, rR_i^{-1})$, and also

$$F(\mathbf{x}_1,\mathbf{x}_2) \le g(|\mathbf{x}_1 - \mathbf{x}_2| - R_1 - R_2)$$

where $g(z) \equiv \lambda\pi E\{r^2 I(R > \tfrac{1}{2}z)\}$. Therefore

$$\int_{\mathbb{R}^2} F(\mathbf{x}_1,\mathbf{x}_2)\, d\mathbf{x}_2$$
$$\le g(0) \int_{|\mathbf{y}| \le R_1 + R_2} d\mathbf{y} + \int_{|\mathbf{y}| > R_1 + R_2} g(|\mathbf{y}| - R_1 - R_2)\, d\mathbf{y}$$
$$= g(0)\pi(R_1 + R_2)^2 + 2\pi \int_0^\infty g(z)(z + R_1 + R_2)\, dz$$
$$\le \mathrm{const.}\,(1 + R_1 + R_2)^2, \qquad (5.52)$$

since $zg(z)$ has a finite integral by virtue of the fact that $E(R^4) < \infty$.

Consequently if $R_1 \le R_2$,

$$\int_{\tilde{R}} c\{\mathcal{R} \cap \partial(\mathbf{x}_1 + S_1)\} d\mathbf{x}_1 \int_{\mathbb{R}^2} c\{\mathcal{R} \cap \partial(\mathbf{x}_2 + S_2)\} F(\mathbf{x}_1, \mathbf{x}_2) d\mathbf{x}_2$$

$$\le \text{const.} \min(1, r R_1^{-1}) \min(1, r R_2^{-1})(1 + R_1 + R_2)^2 r^2$$

$$\times \int_{\tilde{R}_0} I\{\mathcal{R}_0 \cap \partial(\mathbf{x} + r^{-1} S_1) \ne \phi\} d\mathbf{x}$$

$$\le \text{const.} \left\{ \min(1, r R_1^{-1}) \right\}^2 (1 + R_1 + R_2)^2 r^2 \|\mathcal{R}_0^{r^{-1} R_1}\|$$

$$\le \text{const.} (1 + R_1 + R_2)^2 r^2. \tag{5.53}$$

From (5.52) and (5.53) we conclude that for $i = 1, 2$, and 3, $\|\mathcal{R}\|^{-1} J_i \le K_i$ where K_i does not depend on $\mathcal{R}$ and $E(K_i) < \infty$. It is easy to see that each $\|\mathcal{R}\|^{-1} J_i \to 0$ almost surely as $\mathcal{R}$ increases, and so it follows from (5.51) and the dominated convergence theorem that $|\tau_5 - \|\mathcal{R}\| \tau_2| = o(\|\mathcal{R}\|)$ as $\mathcal{R}$ increases. This proves (5.49), and from that result and (5.48) follows Theorem 5.3. Part (5.34) of Theorem 5.2 is immediate. $\square$

EXERCISES

5.1 Prove formula (5.2).

5.2 Consider a Boolean model $\mathcal{C}$ of random-radius spheres in $\mathbb{R}^k$. Assume sphere radius has distribution function F and the driving Poisson process has intensity λ. Let p be porosity and C_P be pore covariance. Show that there is a one-to-one correspondence between pairs (p, C_P) and pairs (λ, F). [Hint: The cases of odd k and even k are slightly different.]

5.3 Let $\mathcal{C}$ be as in Exercise 5.2, and let G be the distribution function of first contact distance. What is the greatest number of parameters of sphere radius distribution that can be determined from G?

Exercises 5.4–5.7 refer to the lattice-based spatial correlation analysis discussed in Section 5.4. All use graphs satisfying the torus convention, part (iii) of (5.13).

5.4 Show that if ν, m, N, M, M', and M'' are, respectively, graph order, number of edges, number of black vertices, number of BW edges, number of BB edges, and number of WW edges, then

$$M' = \tfrac{1}{2}(N\nu - M) \quad \text{and} \quad M'' = m - \tfrac{1}{2}N\nu - \tfrac{1}{2}M.$$

5.5 Let E_0 denote expectation under the null hypothesis that vertex colors (B, W) are independent and identically distributed. Show that

$$E_0(M \mid N) = M - \nu N(n - N)/(n - 1)$$

and

$$\begin{aligned}
E_0(M^2 \mid N) = {} & \frac{\nu^2 N(n - N)}{n - 1} \\
& + \frac{\nu N(N - 1)(n - N)(n - N - 1)}{(n - 1)(n - 2)(n - 3)} \{(n - 4)\nu + 2\},
\end{aligned}$$

and thence that

$$\mathrm{var}_0(M \mid N) = \frac{2\nu(n - \nu - 1)N(n - N)}{(n - 1)(n - 2)(n - 3)} \left\{ \frac{N(n - N)}{n - 1} - 1 \right\}.$$

5.6 Consider two tests of the type described in Section 5.4, of the hypothesis $H_0 : r_0 \le \tfrac{1}{2}b$ against $H_1 : r_0 > \tfrac{1}{2}b$. Suppose the first test uses a graph of degree ν_1, and the second a graph of degree ν_2. Choose rectangular regions $\mathcal{R}_1$ and $\mathcal{R}_2$, with side lengths in common proportion, such that power against a local alternative $\mathcal{C}_\delta$ (see Theorem 5.1) is the same for each test. Define the ratio $\|\mathcal{R}_2\|/\|\mathcal{R}_1\|$ to be the efficiency of test 1 relative to test 2. Arguing in this way, show that a test based on the square lattice graph ($\nu = 4$) is 57.7% efficient (asymptotically, as regions increase and $\delta \to 0$) relative to a test based on the triangular lattice graph, and a test based on the hexagonal lattice graph, ($\nu = 3$) is 33% efficient (asymptotically) relative to a test based on the triangular lattice graph. [The definition of efficiency given here is a natural extension of the classical one; for the latter, see for example Cox and Hinkley (1974, pp. 337–338).]

5.7 Consider a Boolean model $\mathcal{C}$ in which the driving Poisson process has intensity λ and shapes are distributed as the random open isotropic set S. Suppose the Boolean model pattern is observed within a region $\mathcal{R}$. Consistent estimation of $r_0 \equiv \mathrm{ess\ sup\,rad}\,S$ may be achieved by using a nested sequence of hypothesis tests, as follows. Put $b_0 \equiv 2r_0$, and suppose we know that $b' < b_0 < \infty$ where b' is any fixed positive constant. Construct the statistic $U = U(b)$, defined at (5.14), for a sequence of regular graphs on $\mathcal{R}$, the kth having edge width $b_k \equiv b' + \epsilon k$ where $k \geq 0$. Here ϵ is a small positive constant, which of course must be allowed to decrease as the amount of information (i.e., the size of $\mathcal{R}$) increases. Let $-z < 0$ denote the critical point for the test, and put

$$\hat{b} \equiv \begin{cases} \inf\{b_k : U(b_k) > -z\} & \text{if set nonempty} \\ b' & \text{otherwise.} \end{cases}$$

Assume for definiteness that $\mathcal{R}$ is a rectangle of dimensions $a_1 l \times a_2 l$, where a_1 and a_2 are held fixed as l increases. Prove that if

$$\epsilon = \epsilon(\mathcal{R}) \to 0 \quad \text{and} \quad \|\mathcal{R}\|^{-1} \log \epsilon^{-1} \to 0,$$

and

$$z = z(\mathcal{R}) \to \infty \quad \text{and} \quad \|\mathcal{R}\|^{-1} z^2 \to 0$$

as $\mathcal{R}$ increases, then $\hat{b} \to b_0$ in probability as $\mathcal{R}$ increases.

Of course, the size of $\mathcal{R}$ imposes a practical upper bound to the number of values b_k for which we may construct the statistic $U(b_k)$. However, no formal condition is needed to accommodate this fact. Neither do we make any assumptions about graphs with different edge lengths b_k being aligned with one another in some way, or even being of the same degree.

[Hint: Let $\mathcal{B} \equiv \{b_k : U(b_k) > -z\}$, and fix $\eta \in (0, b_0 - b')$. We must show that

$$P(b < b_0 + \eta, \text{ some } b \in \mathcal{B}) \to 1$$

and

$$P(b > b_0 - \eta, \text{ all } b \in \mathcal{B}) \to 1. \tag{5.A}$$

The first result follows from the fact that with $b'' \equiv \inf\{b_k : b_k > b_0\}$, $U(b'')$ is asymptotically normal $N(0,1)$. To establish (5.A), let I_i be the indicator of the event that vertex i is black, and put $q \equiv 1 - p$ where p equals porosity. By grouping vertex indices i into groups within which I_i are independent, and using Bernstein's inequality (e.g., Hoeffding 1963), show that for each $\xi > 0$,

$$P\left\{ \sup_{b' \leq b_k \leq b_0 - \eta} \left| \sum_i (I_i - q) \right| > \xi n \right\} \to 0.$$

Prove from this that with $Z(b) \equiv M(n-1) - \nu N(n-N)$ and

$$Z_1(b) \equiv \sum_i \sum_{j \in \mathcal{N}_i} (I_i - q)(I_j - q),$$

we have

$$\sup_{b' \leq b_k \leq b_0 - \eta} |Z(b_k) + n Z_1(b_k)| = o_p(n^2).$$

Using a similar argument show that

$$\sup_{b' \leq b_k \leq b_0 - \eta} |Z_1(b_k) - E\{Z_1(b_k)\}| = o_p(n).$$

Finally, prove that for large regions,

$$\inf_{b' \leq b_k \leq b_0 - \eta} E\{Z_1(b_k)\} \geq \varsigma n$$

for some $\varsigma > 0$. Combine these estimates.]

Exercises 5.8 and 5.9 refer to the problem of curvature-based inference in a Boolean model of random disks, discussed in Section 5.6.

5.8 Describe how to use curvature-based methods to conduct likelihood ratio tests of hypotheses concerning Boolean models of random disks.

5.9 A kernel density estimator based on a random sample $X_1, \ldots, X_n$ from a distribution with unknown density f may be constructed as follows. Let K be a bounded, symmetric function vanishing outside a compact set and satisfying

$$\int_{-\infty}^{\infty} K(z)\, dz = 1.$$

The random function

$$\hat{f}(x \mid h) \equiv (nh)^{-1} \sum_{i=1}^{n} K\{(x - X_i)/h\}$$

consistently estimates f, provided f is bounded and continuous and $h = h(n) \to 0$ and $nh \to \infty$ as n increases. If f has two or more bounded derivatives then asymptotic formulas for the variance and bias of f may be developed; see for example Rosenblatt (1971). Write down an analogue of $\hat{f}$ for estimating the unknown density f of radius distribution, when the data have the form (R_i, Φ_i) discussed in Section 5.6.

Show that under the assumption $E(R^4) < \infty$, your estimator admits the expansion

$$\hat{f}(x \mid h) = f(x) + b(x \mid h) + \mu_2^{-1} \left[h^{-1} \sum_i \Phi_i K \left\{ (x - R_i)/h \right\} - \mu_1 \right]$$

$$+ O_p(\|\mathcal{R}\|^{-1/2}), \tag{5.B}$$

where

$$b(x \mid h) \equiv \int_{-\infty}^{\infty} K(z) \left\{ f(x - hz) - f(x) \right\} dz$$

is the classical expression for bias of a density estimator,

$$\mu_1 \equiv 2\pi\lambda \|\mathcal{R}\| e^{-\alpha\lambda} \left\{ f(x) + b(x \mid h) \right\} \quad \text{and}$$

$$\mu_2 \equiv 2\pi\lambda \|\mathcal{R}\| e^{-\alpha\lambda}.$$

Write down a formula for the large-region variance of the second-last term in (5.B).

NOTES

Section 5.1
Serra (1982, pp. 495–496 and 501) treats tests of the Boolean model hypothesis, and Diggle and Gratton (1984) examine Monte Carlo inference. Serra (1982, Chapters XIII and XIV) discusses some of the available non-Boolean models, including multiphase texture models, graded texture models, and dead leaves models. See also Davy (1982). Ripley (1981) and Diggle (1983) discuss general models for spatial point processes, from which general coverage process models may be constructed. (See also Chapter 3, Section 3.8.) Stoyan, Kendall, and Mecke (1987, Chapters 3, 6, 9, and 11) provide an excellent account of the use of a wide range of coverage process models.

Our assessment of the merits and disadvantages of Boolean schemes as models for real physical phenomena is not dissimilar to that of Stoyan, Kendall, and Mecke (1987). Comment those authors (p. 69): "In description of given samples and situations in nature the Boolean model has a three-fold value. First, a relatively parsimonious description is available .... Second, it is at least plausible that the model assumptions may aid understanding of the process of formation of the structure (though ... this is by no means invariably the case ...). Third, the formulas ... derived for the Boolean model may be used for estimation of quantities not available for direct measurements."

Section 5.2

A discussion of elementary properties of point covariance and first contact distance may be found in Serra (1982, pp. 487–489), albeit with slightly different definitions of pore and grain covariances. The numerical work in Section 5.2 is taken from Diggle (1981). Diggle conducts a Monte Carlo test of goodness of fit of the Boolean model, using distribution of first contact distance as the discriminating characteristic. We have not discussed those results here. In view of the properties of first contact distance derived in Section 5.2, it would seem that first contact distance is not the best discriminator of properties of a Boolean model of random disks. Point covariance is a more powerful device.

Section 5.3

The work in Section 5.3 is taken from Hall (1985h), and bears a filial resemblance to much older lattice-based counting methods. Serra's (1982) monograph provides an extraordinarily valuable account of practicable techniques of this kind. See also Mead (1974) and Kellerer (1985).

Section 5.4

The spatial correlation analysis described in Section 5.4 may be viewed as an application of techniques for assessing association in statistical maps. The reference most relevant to our work is Moran (1948); see also Moran (1947, 1950a), Krishna Iyer (1949), and Geary (1954). Cliff and Ord (1973, 1981) give a contemporary account, and discuss more general counting statistics that employ all possible pairs of sites and use general weight functions. The fact that tests based on BW, BB, and WW counts are equivalent (under the torus convention) or asymptotically equivalent (generally) in our context contrasts markedly with properties of edge count statistics in other cases. There, statistics based on BW counts can strongly outperform those based on BB or WW counts; see Cliff and Ord (1975; 1981, Chapter 6). All these statistics are related to coefficients of autocorrelation and cross-correlation, of which Davis (1973, p. 232ff) gives a very accessible account in an applied context.

The techniques studied in Section 5.4 are readily extended to cases where a Boolean model confers on $\mathbb{R}^2$ a range of colors richer than simply (B,W). For example, in some situations each point has a color dictated by the *number* of random sets that cover it. There, edge count statistics should be replaced by more general coefficients of spatial correlation, such as those proposed by Moran (1950a) and Geary (1954). For example, Moran's coefficient calculated for a νth degree graph has the form

$$\left\{ \sum_i \sum_{j \in \mathcal{N}_i} (X_i - \overline{X})(X_j - \overline{X}) \right\} \left\{ \nu \sum_i (X_i - \overline{X})^2 \right\}^{-1},$$

where X_i denotes the color of vertex i and $\mathcal{N}_i$ is the set of vertices bonded to vertex i.

Section 5.5
The work in Section 5.5 for Boolean models of normal-radius disks is taken from Dupač (1980). Dupač also discusses techniques for estimating variances of the estimates. Note however that the estimates of μ', σ'^2, and λ will not be consistent, owing to the approximations inherent in their definitions.

Section 5.6
An approach to curvature-based inference in a Boolean model of random disks appears in Mullins (1976, 1978). However, the time-dependent character of Mullins' context is quite different from that here, and does not permit the basic equations developed in Theorems 5.2 and 5.3. Nevertheless it is of interest to note that in Mullins' lunar crater applications it is feasible to measure radius and visible curvature of portions of disks. See also Cross and Fisher (1968). One may readily generalize the curvature-based approach, for example by supposing that disks can be of various types and that types can be identified from portions of disks protruding from clumps. In this case disk radius is just one of several characteristics describing type. If one of these characteristics is time then one may develop a very general time-dependent model, with a rich potential for inference. Unfortunately the disadvantages and limitations discussed in the first paragraph of Section 5.6 persist.

APPENDIX I

Direct Radon-Nikodym Theorem

Consider a simple point process Q on the positive half-line, in which successive points occur at locations $T_1, T_2, \ldots$ with $0 \equiv T_0 < T_1 < T_2 < \ldots$. Let

$$N_t \equiv \sup \{i \geq 0 : T_i \leq t\}$$

denote the number of points occurring in time interval $[0, t]$. Suppose the random variables N_t are defined on the probability space $(\Omega, \mathcal{F}, P_0)$, and write $\mathcal{F}_t$ for the σ-field generated by $\{N_s, s \leq t\}$. We are concerned only with values of $t \leq 1$. Assume N_t has a $(P_0, \mathcal{F}_t)$-predictable intensity λ_t, given by

$$\lambda_t \, dt = E(dN_t \mid \mathcal{F}_{t-}) = P_0(dN_t = 1 \mid \mathcal{F}_{t-}).$$

See Brémaud (1981, pp. 8, 27, and 31ff) for a formal definition of a predictable stochastic process and a predictable intensity. Essentially, an $\mathcal{F}_t$-predictable process is one that is left-continuous and adapted to $\mathcal{F}_t$. Let μ_t be any other $\mathcal{F}_t$-predictable stochastic process. Define the measure P on $(\Omega, \mathcal{F})$ by

$$\frac{dP}{dP_0} = \mathcal{L}_0 \, ,$$

where

$$\mathcal{L}_0 \equiv \left\{ \prod_{i=1}^{\infty} \mu_{T_i} I(T_i \leq 1) \right\} \exp \left\{ \int_0^1 (1 - \mu_t) \lambda_{0t} \, dt \right\}.$$

The direct Radon-Nikodym theorem follows.

Theorem A.1. *If P is a proper probability measure on $(\Omega, \mathcal{F})$ then the point process N_t, $0 \le t \le 1$, defined on $(\Omega, \mathcal{F}, P)$, has predictable intensity $\lambda_t = \lambda_{0t} \mu_t$.*

See Brémaud (1981, pp. 166–167) for discussion and a proof.

We shall apply this result in the situation of Section 2.6. Suppose a renewal process is operating on the positive half-line. Write $N = N_1$ for the number of renewals in $[0,1]$ and T_i for the occurrence time of the ith renewal, and let $h(\cdot \,|\, \boldsymbol{\theta}_0)$ be a left-continuous version of the density of renewal life. Then

$$\lambda_{0t} = \frac{h(t - T_t \,|\, \boldsymbol{\theta}_0)}{1 - H(t - T_t \,|\, \boldsymbol{\theta}_0)},$$

where

$$H(t \,|\, \boldsymbol{\theta}) \equiv \int_0^t h(u \,|\, \boldsymbol{\theta}) \, du$$

and T_t is the occurrence time of the last point to occur *before* time t. If the process N_t $(0 \le t \le 1)$, defined on the probability space $(\Omega, \mathcal{F}, P)$, is to have intensity

$$\lambda_t \equiv \frac{h(t - T_t \,|\, \boldsymbol{\theta})}{1 - H(t - T_t \,|\, \boldsymbol{\theta})},$$

then by the theorem, we require

$$\mu_t = \frac{\lambda_t}{\lambda_{0t}} = \frac{h(t - T_t \,|\, \boldsymbol{\theta})}{h(t - T_t \,|\, \boldsymbol{\theta}_0)} \frac{1 - H(t - T_t \,|\, \boldsymbol{\theta}_0)}{1 - H(t - T_t \,|\, \boldsymbol{\theta})}.$$

In this case,

$$\int_0^1 (1 - \mu_t) \lambda_{0t} \, dt = \int_0^1 (\lambda_{0t} - \lambda_t) \, dt$$

$$= \sum_{i=1}^{\infty} \left(\int_{T_{i-1}}^{T_i} \lambda_{0t} \, dt - \int_{T_{i-1}}^{T_i} \lambda_t \, dt \right) I(T_i \le 1)$$

$$+ \left(\int_{T_N}^1 \lambda_{0t} \, dt - \int_{T_N}^1 \lambda_t \, dt \right)$$

$$= \sum_{i=1}^{\infty} \log[\{1 - H(T_i - T_{i-1} \,|\, \boldsymbol{\theta})\} / \{1 - H(T_i - T_{i-1} \,|\, \boldsymbol{\theta}_0)\}] I(T_i \le 1)$$

$$+ \log[\{1 - H(1 - T_N \,|\, \boldsymbol{\theta})\} / \{1 - H(1 - T_N \,|\, \boldsymbol{\theta}_0)\}],$$

and so

$$\mathcal{L}_0 = \left\{ \prod_{i=1}^{\infty} \frac{h(T_i - T_{i-1} \mid \boldsymbol{\theta})}{h(T_i - T_{i-1} \mid \boldsymbol{\theta}_0)} I(T_i \le 1) \right\} \frac{1 - H(1 - T_N \mid \boldsymbol{\theta})}{1 - H(1 - T_N \mid \boldsymbol{\theta}_0)}.$$

The likelihood ratio is given by

$$\mathcal{L} \equiv \frac{dP}{dP_0} = \mathcal{L}_0 = \left\{ \prod_{i=1}^{N} \frac{h(X_i \mid \boldsymbol{\theta})}{h(X_i \mid \boldsymbol{\theta}_0)} \right\} \frac{1 - H(1 - T_N \mid \boldsymbol{\theta})}{1 - H(1 - T_N \mid \boldsymbol{\theta}_0)},$$

where we have written $X_i = T_i - T_{i-1}$. This agrees with the formula in Section 2.6; note that we have taken $t = 1$.

APPENDIX II

Central Limit Theorem, Poisson Limit Theorem, Ergodic Theorem, and Law of Large Numbers

(i) CENTRAL LIMIT THEOREM

Let $X_1, X_2, \ldots$ be independent and identically distributed random variables, with mean $\mu \equiv E(X_1)$ and finite variance $\sigma^2 \equiv \text{var}(X_1)$. Write $N(0, \sigma^2)$ for the normal distribution with 0 mean and variance σ^2. Then

$$n^{-1/2} \sum_{i=1}^{n} (X_i - \mu) \to N(0, \sigma^2) \qquad (A.1)$$

in distribution as $n \to \infty$. See, for example, Billingsley (1979, Theorem 27.1, p. 308) or Chung (1974, Theorem 6.4.3, p. 169).

Sometimes we wish to use this result when the number of terms in the series is *random*. Replace n by $N = N(n)$, a random variable taking positive integer values. Assume that as $n \to \infty$, $N(n)/n \to c$ in probability, where c is a positive constant. Then

$$N^{-1/2} \sum_{i=1}^{N} (X_i - \mu) \to N(0, \sigma^2)$$

in distribution as $n \to \infty$ (see, e.g., Chung 1974, Theorem 7.3.2, p. 216), or equivalently,

$$(nc)^{-1/2} \sum_{i=1}^{N} (X_i - \mu) \to N(0, \sigma^2).$$

The index n is only a "dummy variable" in these results. If t is a continuous parameter and $N = N(t)$ satisfies $N(t)/t \to c$ in probability as $t \to \infty$, then

$$(tc)^{-1/2} \sum_{i=1}^{N} (X_i - \mu) \to N(0, \sigma^2)$$

in distribution as $t \to \infty$.

Result (A.1) is the classic form of the central limit theorem for a sum of independent random variables. But in some situations where we wish to apply the central limit theorem, summands are not identically distributed. Lyapounov's central limit theorem can be useful in this context. Let $\{m_n\}$ be a sequence of positive integers diverging to $+\infty$, and suppose $\{X_{ni}, 1 \leq i \leq m_n, n \geq 1\}$ is an array of random variables such that each "row" $\{X_{ni}, 1 \leq i \leq m_n\}$ of the array is a collection of independent variables. Assume each $E(X_{ni}) = 0$ and $E(X_{ni}^2) < \infty$, and put

$$S_n \equiv \sum_{i=1}^{m_n} X_{ni}$$

and $s_n^2 \equiv E(S_n^2)$. *Lyapounov's central limit theorem* declares that if

$$s_n^{-3} \sum_{i=1}^{m_n} E|X_{ni}|^3 \to 0 \qquad (A.2)$$

as $n \to \infty$, then

$$s_n^{-1} S_n \to N(0, 1)$$

in distribution as $n \to \infty$ (Billingsley 1979, p. 312; Chung 1974, p. 200). Condition (A.2) is one form of *Lyapounov's condition*.

Again, n is only a dummy variable. The theorem continues to hold if sequences are indexed by a continuous parameter t (diverging to $+\infty$) rather than the discrete parameter n.

Lyapounov's central limit theorem is a consequence of *Lindeberg's central limit theorem*, which declares that for the triangular array defined above [with each $E(X_{ni}) = 0$ and $E(X_{ni}^2) < \infty$], *Lindeberg's condition*,

$$\text{for all } \epsilon > 0, \quad s_n^{-2} \sum_{i=1}^{m_n} E\left\{X_{ni}^2 I(|X_{ni}| > \epsilon s_n)\right\} \to 0,$$

is sufficient for

$$s_n^{-1} S_n \to N(0, 1)$$

in distribution (Billingsley 1979, p. 310; Chung 1974, p. 205).

The *Berry-Esseen theorem* (sometimes called "Esseen's inequality") provides a rate of convergence in Lyapounov's central limit theorem. It asserts that

$$\sup_{-\infty < x < \infty} |P(s_n^{-1} S_n \le x) - \Phi(x)| \le C_0 s_n^{-3} \sum_{i=1}^{m_n} E|X_{ni}|^3,$$

where C_0 is an absolute constant and Φ is the standard normal distribution function [see Petrov (1975, Theorem 3, p. 111)]. It is known that we may take $C_0 = 1$.

(ii) POISSON LIMIT THEOREM

The normal distribution is one of a class of infinitely divisible limits possible for distributions of sums of independent random variables. Another limit, the Poisson, is appropriate when data represent occurrences of independent, rare events. Assume that each row $\{X_{ni}, 1 \le i \le m_n\}$ in the array $\{X_{ni}, 1 \le i \le m_n, n \ge 1\}$ is a sequence of independent, nonnegative, *integer-valued* random variables. In order that the sum

$$S_n \equiv \sum_{i=1}^{m_n} X_{ni}$$

converge in distribution to a Poisson variable with mean $\gamma \ge 0$, meaning that

$$P(S_n = m) \to \gamma^m e^{-\gamma}/m!, \qquad m \ge 0,$$

as $n \to \infty$, it is sufficient that

$$\max_{1 \le i \le m_n} P(X_{ni} \ge 1) \to 0,$$

$$\sum_{i=1}^{m_n} P(X_{ni} = 1) \to \gamma$$

and

$$\sum_{i=1}^{m_n} P(X_{ni} \ge 2) \to 0$$

[see Gnedenko and Kolmogorov (1968, p. 132)].

(iii) ERGODIC THEOREM

Let $\mathcal{I}_k$ denote the sequence of k-vectors all of whose elements are integers. Given two members $\mathbf{i}^{(j)} = (i_1^{(j)}, \ldots, i_k^{(j)})$ $(j = 1, 2)$ of $\mathcal{I}_k$, write $\mathbf{i}^{(1)} \leq \mathbf{i}^{(2)}$ to denote that $i_j^{(1)} \leq i_j^{(2)}$ for $1 \leq j \leq k$. Let $\{X_\mathbf{i}, \mathbf{i} \in \mathcal{I}_k\}$ be a sequence of real-valued random variables indexed by elements of $\mathcal{I}_k$. We call the sequence *stationary* if for each finite subset of $\mathcal{T}$ of $\mathcal{I}_k$, and each element $\mathbf{i}$ of $\mathcal{I}_k$, the sequences $\{X_\mathbf{j}, \mathbf{j} \in \mathcal{T}\}$ and $\{X_\mathbf{j}, \mathbf{j} \in \mathbf{i} + \mathcal{T}\}$ have the same joint distribution. Of course, this implies that the $X_\mathbf{i}$'s are identically distributed.

The ergodic theorem follows. Write $\mathbf{0} \equiv (0, \ldots, 0)$, $\mathbf{1} \equiv (1, \ldots, 1)$ and $\mathbf{n} \equiv (n_1, \ldots, n_k)$.

Theorem A.2. *Assume $\{X_\mathbf{i}, \mathbf{i} \in \mathcal{I}_k\}$ is stationary, and $E|X_\mathbf{0}|^p < \infty$ for some $p > 1$. Then the sequence of normalized partial sums*

$$(n_1 \ldots n_k)^{-1} \sum_{\mathbf{0} \leq \mathbf{i} \leq \mathbf{n}-\mathbf{1}} X_\mathbf{i}, \qquad \mathbf{n} \geq \mathbf{0}, \tag{A.3}$$

converges with probability 1 as $n_1, \ldots, n_k \to \infty$. The convergence is also in L^p norm.

Dunford and Schwartz (1958, Chapter VIII, Section 6) state and prove this and more general results. Breiman (1968, Chapter 6, Section 5) gives an excellent account of the case $k = 1$.

(iv) LAWS OF LARGE NUMBERS

In the case $k = 1$ the condition $p > 1$ in Theorem A.2 may be relaxed to $p \geq 1$, but $p = 1$ fails to be sufficient when $k > 1$. This is most easily seen by considering the case where the $X_\mathbf{i}$'s are stochastically independent and identically distributed. In this situation the tail σ-field of the sequence of sums defined at (A.3) is trivial (i.e., contains only sets of measure 0 or 1), and so any limit which the sequence has must be almost surely constant. In fact the limit must be $E(X_\mathbf{0})$, as follows from L^p convergence. It may be proved that the condition

$$E[|X_\mathbf{0}| \{\log(1 + |X_\mathbf{0}|)\}^{k-1}] < \infty \tag{A.4}$$

is necessary and sufficient for almost sure convergence of the series in (A.3) when the summands are independent [see Smythe (1973)].

Condition (A.4) is known to be sufficient in the general stationary case (Dunford 1951; Zygmund 1951). The result

$$(n_1 \ldots n_k)^{-1} \sum_{0 \leq i \leq n-1} X_i \to E(X_0)$$

almost surely is known as the *strong law of large numbers*. Its weaker form, in which convergence is only in probability, is called the *weak law of large numbers*.

APPENDIX III

Shepp's Coverage Theorem

The Cantor ternary set is the set that remains after removing the open middle one-third— i.e., the set $\left(\frac{1}{3}, \frac{2}{3}\right)$—from the interval $[0, 1]$; removing the open middle one-third from each of the remaining intervals $[0, \frac{1}{3}]$ and $[\frac{2}{3}, 1]$; and so on. The Cantor set has an uncountable infinity of elements, yet has measure 0. A "random Cantor set" could be defined as follows. Let $\mathcal{P}_1, \mathcal{P}_2, \ldots$ be independent Poisson processes on the real line $\mathbb{R}$, each of intensity $\lambda > 0$, and let $1 \geq t_1 \geq t_2 \geq \ldots$ be a sequence of positive numbers converging to 0. The random Cantor set $\hat{C}$ is that subset of $\mathbb{R}$ that remains after removing from the real line all sets of points of the form $\xi + (0, t_j)$, for $\xi \in \mathcal{P}_j$ and $j \geq 1$.

The real number x is in one of the sets $\xi + (0, t_j)$, with $\xi \in \mathcal{P}_j$, if and only if some point of $\mathcal{P}_j$ lies within $(x - t_j, x)$. The probability of this event is $\exp(-\lambda t_j)$. Therefore the chance that x is not covered by any set of the form $\xi + (0, t_j)$ for some $\xi \in \mathcal{P}_j$ and $j \geq 1$ equals

$$\prod_{j=1}^{\infty} \{1 - \exp(-\lambda t_j)\},$$

which is 0 if and only if

$$\sum_{j=1}^{\infty} t_j = \infty. \tag{A.5}$$

It follows that $\hat{C}$ is of measure 0 with probability 1 if and only if (A.5) holds.

Shepp (1972b) showed that $\hat{C}$ is empty with probability 0 or 1 according as the series

$$\sum_{n=1}^{\infty} n^{-2} \exp\{\lambda(t_1 + \cdots + t_n)\} \qquad \text{(A.6)}$$

converges or diverges. The condition that this series be infinite is strictly stronger than (A.5). (Actually, Shepp's full result applies to interval processes where interval lengths have a general distribution, not just the discrete distribution assumed above.) Note that

$$P(\hat{C} = \phi) = 1$$

if and only if

$$P\{\hat{C} \cap (a,b) = \phi\} = 1$$

for any interval (a,b) with $-\infty \leq a < b \leq \infty$.

Shepp's theorem verifies a conjecture of Mandelbrot, who introduced and studied these interval processes (Mandelbrot 1972a). Problems of this type were first examined by Dvoretzky (1956), who treated the distribution of random arcs placed randomly and uniformly around the circumference of a circle. Suppose the circumference is of unit length, and the arcs have lengths $t_1 \geq t_2 \geq \ldots$, with $t_1 < 1$. Dvoretzky noted that the uncovered length of circumference has measure 0 with probability 1 if (A.5) holds, but that that condition is not sufficient for almost sure complete coverage. Shepp (1972a) proved that complete coverage occurs with probability 1 if and only if the series at (A.6) (with $\lambda = 1$) diverges. See also Flatto and Konheim (1962), Billard (1965), Kahane (1968), and Mandelbrot (1972b).

APPENDIX IV

Mean Content and Mean Square Content of Cells Formed by Poisson Field of Random Planes

In Section 3.5 we defined a Poisson field of $(k-1)$-dimensional planes in $\mathbb{R}^k$, of intensity μ, to be a stochastic sequence of planes whose normals to the origin have independent, uniformly distributed orientations, and whose successive distances from the origin are points of a Poisson process of intensity μ on $(0,\infty)$. It is easily shown that this definition is independent of the choice of origin. Let $\mathcal{V}^\circ$ denote a typical (convex) cell formed by members of the field.

(i) DERIVATION OF FORMULA FOR $E(\|\mathcal{V}^\circ\|)$

We shall prove that

$$E(\|\mathcal{V}^\circ\|) = \left\{ \pi^{1/2} \Gamma(k+1) \Big/ \Gamma\left(\frac{k+1}{2}\right) \right\} \left\{ 2\Gamma\left(\frac{k+1}{2}\right) \Big/ \mu \Gamma\left(\frac{k}{2}\right) \right\}^k .$$

$$(A.7)$$

Note that this is quite different from the expected content of the cell containing a specified point, such as the origin.

Our proof is based on that of Goudsmit (1945), who treated the case $k = 2$. Consider a $(k+1)$-dimensional sphere of radius r, whose north pole is uppermost (see Figure A.1 for the case $k = 2$). Place n "great circles" on the sphere, in such a manner that the "upward"-pointing normals to the k-dimensional planes of these "circles" are distributed independently and uniformly on the surface of the upper

353

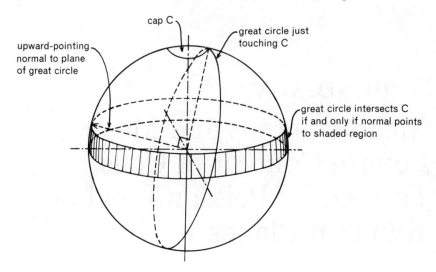

Figure A.1. Scheme for derivation of $E(\|\mathcal{V}^{\circ}\|)$, in the case $k = 2$.

hemisphere. As r increases, any bounded region of the surface will flatten out into a subset of $\mathbb{R}^{k}$, and the inscribed "circles" will become $(k - 1)$-dimensional planes distributed in $\mathbb{R}^{k}$ according to a Poisson field. We must let n and r increase together in such a way that the resulting Poisson field has intensity μ.

Let s_k denote the surface content of a unit-radius k-dimensional sphere. Place a circular cap C of unit radius over the north pole of the sphere. The probability that the cap is intersected by the ith great circle is asymptotic to

$$\frac{(\text{radius of } C) \times (\text{perimeter of great circle})}{\text{surface content of upper hemisphere}} = \frac{s_k r^{k-1}}{\frac{1}{2} s_{k+1} r^{k}}$$
$$= (2s_k/s_{k+1}) r^{-1},$$

as $r \to \infty$. Therefore the expected number of great circles that intersect the cap is approximately $(2s_k/s_{k+1})(n/r)$. Now, the expected number of $(k - 1)$-dimensional planes from a Poisson field of intensity μ that intersect a unit sphere in $\mathbb{R}^{k}$ equals μ. Thus, we should choose n and r such that $(2s_k/s_{k+1})(n/r) \to \mu$, i.e., $n/r \to \mu s_{k+1}/2s_k$.

The expected content of an arbitrary region $\mathcal{V}^{\circ}$ formed by a Poisson field of planes is the limit of the expected content of the regions formed by intersecting great circles:

$$E(\|\mathcal{V}^\circ\|) = \lim_{n/r \to \mu s_{k+1}/2s_k}$$

$$\times \frac{\text{total content of } (k+1)\text{-dimensional sphere surface}}{\text{total number of regions}}.$$

The total content of the sphere surface equals $s_{k+1}r^k$, while the total number $C(n, k+1)$ of regions formed by the pattern of great circles may be found from *Schläfli's formula*:

$$C(n, k+1) = 2\sum_{i=0}^{k} \binom{n-1}{i} \sim 2n^k/k!$$

as $n \to \infty$. See below for a derivation of Schläfli's formula. The asymptotic relation follows from Sterling's formula and the fact that the series is dominated by the $i = k$ term.

Consequently,

$$E(\|\mathcal{V}^\circ\|) = \lim_{n/r \to \mu s_{k+1}/2s_k} (s_{k+1}r^k)/(2n^k/k!)$$

$$= k!\tfrac{1}{2}s_{k+1}(2s_k/\mu s_{k+1})^k$$

$$= \left\{ k!\,\pi^{1/2} \Big/ \Gamma\Big(\frac{k+1}{2}\Big) \right\} \left\{ 2\Gamma\Big(\frac{k+1}{2}\Big) \Big/ \mu\Gamma\Big(\frac{k}{2}\Big) \right\}^k, \quad (A.8)$$

using the fact that $s_k = 2\pi^{k/2}/\Gamma(k/2)$. This completes our derivation of (A.7).

(ii) DERIVATION OF SCHLÄFLI'S FORMULA

Schläfli's formula,

$$C(n, k) = 2\sum_{i=0}^{k-1} \binom{n-1}{i}, \qquad n \geq 1 \quad \text{and} \quad k \geq 2, \qquad (A.9)$$

is an expression for the number $C(n, k)$ of regions formed by the pattern of n "great circles" on the surface of a k-dimensional sphere, given that every k-element subset of the normals to the great circle planes is linearly independent. To derive the formula, observe that the $(n+1)$th great circle is itself the surface of a $(k-1)$-dimensional sphere, and the intersections it makes with the other n great circles on the k-dimensional sphere, are themselves great circles on the surface of a $(k-1)$-dimensional sphere. Therefore the number of intersections that the $(n+1)$th great circle on the k-dimensional sphere makes with the other n great circles is $C(n, k-1)$. Each intersection corresponds to the formation of a new region on the surface of the k-dimensional sphere. Therefore if a new great circle is added to n great circles already existing on the surface of a k-dimensional sphere, another $C(n, k-1)$ regions are formed. That is,

$$C(n+1, k) = C(n, k) + C(n, k-1). \qquad (A.10)$$

Clearly $C(1, k) = 2$, and $C(n, 2) = 2n$. Formula (A.9) may now be deduced from (A.10) by iteration. [Repeated application of (A.10) gives a formula for general $C(n, k)$ in terms of $C(1, j)$ and $C(m, 2)$ for $j \leq k$ and $m \leq n$.]

The techniques used above may also be employed to prove the following lemma, which is needed in part (iii) below.

Lemma A.3. *Let $\mathcal{F}$ be a Poisson field of $(k-1)$-dimensional planes in $\mathbb{R}^k$ with intensity μ, and let Π be any fixed m-dimensional plane in $\mathbb{R}^k$, where $1 \leq m \leq k-1$. (We may regard Π as a copy of $\mathbb{R}^m$.) The set of intersections of $\mathcal{F}$ with Π is a Poisson field of $(m-1)$-dimensional planes in Π, whose intensity is given by*

$$\mu^* = \Gamma\left(\frac{k}{2}\right) \Gamma\left(\frac{m+1}{2}\right) \mu \Big/ \Gamma\left(\frac{k+1}{2}\right) \Gamma\left(\frac{m}{2}\right).$$

Proof. The Poisson-ness of the new process Π is clear, given the homogeneity of the original Poisson field in $\mathbb{R}^k$. We derive only the formula for μ^*, and this we do by going back to the "spherical" model for $\mathcal{F}$, in which $(k-1)$-dimensional planes are thought of as n "great circles" on a $(k+1)$-dimensional sphere of radius r. We consider first the case $m = k - 1$.

The number of intersections of any one of the n great circles with the other $n-1$ great circles is given by Schläfli's formula:

$$C(n-1, k) = 2 \sum_{i=0}^{k-1} \binom{n-2}{i} \sim 2n^{k-1}/(k-1)!$$

as $n \to \infty$. The "circumference" of each great circle is $s_k r^{k-1}$. Therefore the number of intersections per unit "length" of circumference is asymptotic to

$$\left\{ 2n^{k-1}/(k-1)! \right\} / s_k r^{k-1} = \left\{ 2/(k-1)! s_k \right\} (n/r)^{k-1}. \qquad (A.11)$$

As noted earlier in this appendix, if we let $n \to \infty$ and $r \to \infty$ in such a way that $n/r \to \mu s_{k+1}/2s_k$, then the great circles become $(k-1)$-dimensional planes in a Poisson field of intensity μ in $\mathbb{R}^k$. We may now deduce from (A.11) that if Π is a fixed $(k-1)$-dimensional plane in $\mathbb{R}^k$, the mean $(k-1)$-dimensional content of an arbitrary cell formed by intersections with Π of a Poisson field of $(k-1)$-dimensional planes with intensity μ in $\mathbb{R}^k$ equals

$$\left[\left\{ 2/(k-1)! s_k \right\} (\mu s_{k+1}/2s_k)^{k-1} \right]^{-1} = \left\{ (k-1)! s_k/2 \right\} (2s_k/\mu s_{k+1})^{k-1}. \qquad (A.12)$$

Each intersection with Π results in a $(k-2)$-dimensional plane contained within Π, and the sum total of these is a Poisson field of $(k-2)$-dimensional planes within Π. We derive the intensity, ν, of this field by comparing the expected cell content at (A.12) with that derived at (A.8). Note that in the latter formula we must replace k by $k-1$ and μ by ν. Thus,

$$\left\{ (k-1)! s_k/2 \right\} (2s_k/\mu s_{k+1})^{k-1} = \left\{ (k-1)! s_k/2 \right\} (2s_{k-1}/\nu s_k)^{k-1},$$

whence

$$\nu = (s_{k-1} s_{k+1}/s_k^2)\mu. \qquad (A.13)$$

This proves Lemma A.3 in the case $m = k-1$. To deduce the lemma for general m, $1 \le m \le k-1$, we note that by (A.13) and induction over k, the intensity in that case is given by

$$(s_{k-1} s_{k+1}/s_k^2)(s_{k-2} s_k/s_{k-1}^2) \ldots (s_m s_{m+2}/s_{m+1}^2)\mu$$
$$= (s_{k+1} s_m/s_k s_{m+1})\mu.$$

Lemma A.3 follows on noting that $s_k = 2\pi^{k/2}/\Gamma(k/2)$. $\square$

(iii)   DERIVATION OF FORMULA FOR $E(\|\mathcal{V}^\circ\|^2)$

We prove that

$$E(\|\mathcal{V}^\circ\|^2) = \left\{ 2\pi^{1/2}\Gamma(k)\Gamma(k+1) \Big/ \Gamma\left(\frac{k}{2}\right)\Gamma\left(\frac{k+1}{2}\right) \right\}$$
$$\times \left\{ 2\pi^{1/2}\Gamma\left(\frac{k+1}{2}\right) \Big/ \mu\Gamma\left(\frac{k}{2}\right) \right\}^{2k}. \qquad (A.14)$$

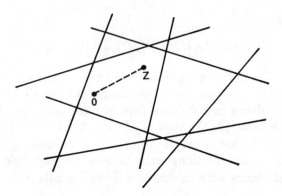

Figure A.2. Scheme for derivation of $E(\|\mathcal{V}^\circ\|^2)$, in the case $k = 2$.

The derivation consists of obtaining two different formulas for the probability that two points, **0** and **Z**, lie within the same convex cell. Figure A.2 illustrates the case $k = 2$.

Let h denote the probability density of $\|\mathcal{V}^\circ\|$, and let $\mathcal{R}$ be a large k-dimensional rectangular prism in $\mathbb{R}^k$ containing the origin **0**. Write $N(\mathcal{R})$ for the total number of convex cells in $\mathcal{R}$ formed by the Poisson field of planes. Then $N(\mathcal{R})h(x)\,dx$ is approximately the total number of cells whose k-dimensional content lies between x and $x + dx$, and $x \times N(\mathcal{R})h(x)\,dx$ is approximately the total content of all these cells. Therefore the probability that the origin lies within a cell whose content is between x and $x + dx$ is very nearly

$$xN(\mathcal{R})h(x)\,dx/\|\mathcal{R}\|.$$

Place a point **Z** at random into $\mathcal{R}$. Given that **0** lies within a cell whose content equals x, the probability that **Z** lies in the same cell as **0** equals $x/\|\mathcal{R}\|$. Integrating over all possible values of x, we see that the probability that **0** and **Z** lie within the same cell, no matter what its size, is asymptotic to

$$
\begin{aligned}
p &\equiv \int_0^\infty (x/\|\mathcal{R}\|)xN(\mathcal{R})h(x)\,dx/\|\mathcal{R}\| \\
&= \left\{ N(\mathcal{R})/\|\mathcal{R}\|^2 \right\} \int_0^\infty x^2 h(x)\,dx \\
&= \left\{ N(\mathcal{R})/\|\mathcal{R}\|^2 \right\} E(\|\mathcal{V}^\circ\|^2), \quad\quad\quad\text{(A.15)}
\end{aligned}
$$

as $\|\mathcal{R}\| \to \infty$.

Taking $m = 1$ in Lemma A.3 we see that the Poisson field of intercepts on a given straight line resulting from $(k-1)$-dimensional planes in $\mathbb{R}^k$ has intensity

$$\left\{\Gamma\!\left(\frac{k}{2}\right)\Big/\Gamma\!\left(\frac{k+1}{2}\right)\pi^{1/2}\right\}\mu.$$

That is, the Poisson process formed by the intercepts on an arbitrary straight line has intensity

$$\kappa \equiv \left\{\Gamma\!\left(\frac{k}{2}\right)\Big/2\pi^{1/2}\Gamma\!\left(\frac{k+1}{2}\right)\right\}\mu.$$

Place a point $\mathbf{Z}$ at random into our rectangular prism $\mathcal{R}$, and consider the straight line segment $\mathcal{L}$ joining $\mathbf{Z}$ to the origin. The probability that $\mathbf{Z}$ is distant between t and $t+dt$ from $\mathbf{0}$ equals $s_k t^{k-1}\,dt/\|\mathcal{R}\|$. Given that $|\mathbf{Z}| = t$, the probability that $\mathbf{0}$ and $\mathbf{Z}$ lie within the same cell equals the probability that the line $\mathcal{L}$ suffers no intersection by the $(k-1)$-dimensional planes of the Poisson field in $\mathbb{R}^k$. This probability is

$$e^{-\kappa t} = \exp\left[-\left\{\Gamma\!\left(\frac{k}{2}\right)\Big/2\pi^{1/2}\Gamma\!\left(\frac{k+1}{2}\right)\right\}\mu t\right].$$

Therefore

$$p = \int_0^\infty \exp\left[-\left\{\Gamma\!\left(\frac{k}{2}\right)\Big/2\pi^{1/2}\Gamma\!\left(\frac{k+1}{2}\right)\right\}\mu t\right]s_k t^{k-1}\,dt\Big/\|\mathcal{R}\|$$

$$= (k-1)!\left\{2\pi^{1/2}\Gamma\!\left(\frac{k+1}{2}\right)\Big/\mu\Gamma\!\left(\frac{k}{2}\right)\right\}^k s_k\Big/\|\mathcal{R}\|. \qquad (A.16)$$

Combining (A.15) and (A.16), and noting that

$$\|\mathcal{R}\|/N(\mathcal{R}) \to E(\|\mathcal{V}^\circ\|)$$

$$= \left\{\pi^{1/2}\Gamma(k+1)\Big/\Gamma\!\left(\frac{k+1}{2}\right)\right\}\left\{2\Gamma\!\left(\frac{k+1}{2}\right)\Big/\mu\Gamma\!\left(\frac{k}{2}\right)\right\}^k$$

as $\|\mathcal{R}\| \to \infty$ [see part (i) of this appendix], we find that

$$E(\|\mathcal{V}^\circ\|^2) = \Gamma(k)\pi^{k/2}s_k\left\{\pi^{1/2}\Gamma(k+1)\Big/\Gamma\!\left(\frac{k+1}{2}\right)\right\}$$

$$\times \left\{2\Gamma\!\left(\frac{k+1}{2}\right)\Big/\mu\Gamma\!\left(\frac{k}{2}\right)\right\}^{2k}.$$

Result (A.14) follows immediately on noting that $s_k = 2\pi^{k/2}/\Gamma(k/2)$.

APPENDIX V
Multitype Branching Processes

We give here a very quick development of elementary theory for multitype branching processes with a countable number of types. Excellent accounts of the finite type case are given by Athreya and Ney (1972, Chapter V) and Karlin and Taylor (1975, p. 411f).

Consider a population of individuals each of which is classified as being of one or other of at most a countable infinity of types. We represent the types by integers $0, 1, 2, \ldots$. Assume the individuals reproduce according to the following rules. Given that an individual of type i is present in the nth generation, that individual spawns $N \geq 0$ children in the $(n+1)$th generation, where N has distribution function F. Conditional on N, the types of the N progeny are independent and identically distributed with distribution function G_i. Thus, both F and G_i, $i \geq 0$, are proper distribution functions with support $\{0, 1, 2, \ldots\}$. Each individual $\mathcal{I}$ in the nth generation reproduces independently of all other individuals in that generation, and according to a distribution function that depends on the previous history of the process only through the type of $\mathcal{I}$.

Let Z_{ni} denote the number of individuals of type i in the nth generation. Put $\mathbf{Z}_n \equiv (Z_{n0}, Z_{n1}, \ldots)$. The last property announced in the previous paragraph indicates that $\mathbf{Z}_n$ is a Markov chain (Karlin and Taylor 1975, Chapter 2). To determine transition probabilities for this chain, let $p^{(i)}(j_0, j_1, \ldots)$ denote the chance that a type i parent produces j_0 individuals of type 0, j_1 individuals of type 1, and so on. Then

$$\sum_{j_0 \geq 0, j_1 \geq 0, \dots} p^{(i)}(j_0, j_1, \dots) = 1.$$

Write $\mathbf{j} \equiv (j_0, j_1, \dots)$ and $\mathbf{s} \equiv (s_0, s_1, \dots)$, and let $\mathbf{j} \geq \mathbf{0}$ mean that $j_0 \geq 0$, $j_1 \geq 1, \dots$. Assuming temporarily that there is only a finite number of types, put

$$\mathbf{s^j} \equiv s_0^{j_0} s_1^{j_1} \dots,$$

$$f^{(i)}(\mathbf{s}) \equiv \sum_{\mathbf{j} \geq \mathbf{0}} p^{(i)}(\mathbf{j}) \mathbf{s^j},$$

$$\mathbf{f}(\mathbf{s}) \equiv (f^{(0)}(\mathbf{s}), f^{(1)}(\mathbf{s}), \dots)$$

and

$$P(\mathbf{i}, \mathbf{j}) = \text{coefficient of } \mathbf{s^j} \text{ in } \{\mathbf{f}(\mathbf{s})\}^{\mathbf{i}}$$
$$= \text{coefficient of } \mathbf{s^j} \text{ in } f^{(0)}(\mathbf{s})^{i_0} f^{(1)}(\mathbf{s})^{i_1} \dots. \quad \text{(A.17)}$$

Then the Markov chain $\{Z_n, n \geq 0\}$ has transition probability function $P(\mathbf{i}, \mathbf{j})$:

$$P(\mathbf{i}, \mathbf{j}) = P(Z_{n+1} = \mathbf{j} \mid Z_n = \mathbf{i}). \quad \text{(A.18)}$$

The transition probability function in the case of a countable infinity of types may be obtained informally, by treating (A.17) as a relation among vectors $(i_0, i_1, \dots)$ rather than among real numbers $s_0^{i_0} s_1^{i_1} \dots$; or formally, by introducing a new type -1, defining an individual to have type -1 if its true type is $\geq m$, following the arguments leading to (A.18), and then letting $m \to \infty$.

The probability that a given type i individual in the nth generation produces precisely k individuals of type j in the $(n+1)$th generation equals

$$p^{(i,j)}(k) \equiv \sum_{k_l \geq 0: l \neq j} p^{(1)}(k_0, \dots, k_{j-1}, k, k_{j+1}, \dots).$$

The expected number of type j individuals born to a single type i individual is

$$\mu_{ij} \equiv \sum_{k=1}^{\infty} k p^{(i,j)}(k)$$

$$= \frac{\partial}{\partial s_j} f^{(i)}(\mathbf{s})|_{\mathbf{s}=1},$$

where $1 \equiv (1,1,\dots)$, the latter formula holding in the case of a finite number of types. Let $\mathbf{M}$ denote the square matrix (μ_{ij}). The numbers (finite or infinite) of rows and columns of $\mathbf{M}$ are the same as the number of types. Let $\mathbf{i} = (i_0, i_1, \dots)$ be a vector of nonnegative integers. It follows easily from the definition of the process $\{\mathbf{Z}_n, n \geq 0\}$ that

$$E(\mathbf{Z}_{n+1} \mid \mathbf{Z}_n = \mathbf{i}) = \mathbf{iM},$$

and thence by induction that

$$E(\mathbf{Z}_n \mid \mathbf{Z}_0 = \mathbf{i}) = \mathbf{iM}^n, \qquad n \geq 1.$$

APPENDIX VI
Lattice Percolation

We give here a brief sketch of some of the main features of lattice percolation. The interested reader should consult Essam (1972), Smythe and Wierman (1978), or Kesten (1982) for further details.

The notion of percolation was introduced by Broadbent (1954), Broadbent and Hammersley (1957), and Hammersley (1957) to model the movement of fluid through a random porous medium; for example, of water through a sponge. Consider the medium to be composed of an array of "pipes" laid out in a regular lattice, such as a square lattice in two dimensions or a cubic lattice in three dimensions. We might model the randomness of the medium by asking that each pipe be open with probability p and closed with probability $1 - p$, independently of all other pipes. Given that the fluid enters the medium at some point P, is there a nonzero probability that it will penetrate arbitrarily far into the medium? In other words, is there a positive probability of an infinite connected sequence of open pipes starting from P?

When abstracted into mathematical formalism this problem becomes one of *bond* (or edge) *percolation*. Suppose each bond in a regular lattice is deleted with probability $1 - p$ and retained with probability p, independently of all other bonds. Then we ask, do there exist values of $p < 1$ for which any given bond has positive probability of being linked to an infinite connected cluster (that is, an infinite clump) of other retained bonds? The answer is generally "yes," and the value of p beyond which infinite clumps start to form is termed the *critical probability*.

In a slightly different problem, one might have all the pipes in the medium open but have each junction of pipes open with probabilty p and closed with probability $1 - p$, independently of all other junctions. Fluid can pass through any pipe linking two open junctions, but cannot cross closed junctions. The mathematical formulation of this situation is a problem of *site percolation*. Open junctions (sites) are said to be *occupied*. There is generally a critical probability beyond which infinite clumps (of occupied sites) have positive probability of occurring.

Let N equal the total number of bonds linked to a given arbitrary site P by passable paths, in a bond or a site percolation problem. We say that "percolation occurs" if $P(N = \infty) > 0$. At least two types of critical probability may be distinguished:

$$p_H \equiv \sup \{0 \leq p \leq 1 : P(N = \infty) = 0\} \qquad (A.19)$$

and

$$p_T \equiv \sup \{0 \leq p \leq 1 : E(N) < \infty\}.$$

Clearly $0 \leq p_T \leq p_H \leq 1$. In most cases of importance, $0 < p_H = p_T < 1$, although the identity $p_H = p_T$ and the common value can be very difficult to establish rigorously. Examples of critical probabilities for regular two-dimensional lattices include:

bond percolation on the triangular lattice:

$$p_H = p_T = 2\sin(\pi/18) \doteq 0.35;$$

site percolation on the triangular lattice:

$$p_H = p_T = 0.5;$$

bond percolation on the square lattice:

$$p_H = p_T = 0.5;$$

site percolation on the square lattice:

$$p_H = p_T \doteq 0.59 \text{ (by simulation)};$$

bond percolation on the hexagonal lattice:

$$p_H = p_T = 1 - 2\sin(\pi/18) \doteq 0.65;$$

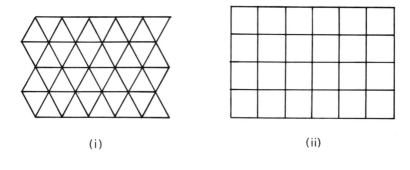

(i)   (ii)

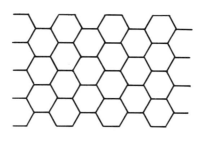

(iii)

Figure A.3. (i) Triangular, (ii) square, and (iii) hexagonal lattices in the plane.

site percolation on the hexagonal lattice:

$$p_H = p_T \doteq 0.70 \text{ (by simulation)}.$$

Figure A.3 illustrates triangular, square, and hexagonal lattices.

Relatively few explicit values of critical probabilities are known for percolation in higher dimensions. However, various inequalities may be established. If $p_H^{(k)}$ is the critical probability defined at (A.19) for site percolation on a k-dimensional cubic lattice, then $p_H^{(k)} > p_H^{(2)}$ if $k \geq 3$. Higuchi (1982) has shown by rigorous argument that $p_H^{(2)} > 0.5$, and simulations suggest $p_H^{(2)} \doteq 0.59$. A counting argument may be used to prove that $p_H^{(k)} < 1$ for each k. This result was employed in the proof of Theorem 4.13 in Chapter 4.

If $p > p_H$ in a site or bond percolation problem on a regular lattice, then with probability 1 an infinite cluster occurs somewhere in the lattice.

References

Abend, K., Harley, T. J., and Kanal, L. N. (1965). Classification of binary random patterns. *IEEE Trans. Inform. Theory* **11**, 538-544.

Abramowitz, M., and Stegun, I. A. (1965). *Handbook of Mathematical Functions*. New York: Dover.

Adams, D. J., and Matheson, A. J. (1972). Computation of dense random packings of hard spheres. *J. Chem. Phys.* **56**, 1989-1994.

Ahuja, N. (1981a). Mosaic models for images, I. Geometric properties of components in cell-structure mosaics. *Inform. Sci.* **23**, 69-104.

Ahuja, N. (1981b). Mosaic models for images, II. Geometric properties of components in coverage mosaics. *Inform. Sci.* **23**, 159-200.

Ahuja, N. (1981c). Mosaic models for images, III. Spatial correlation in mosaics. *Inform. Sci.* **24**, 43-69.

Ahuja, N., and Dubitzki, T. (1980). Some experiments with mosaic models for images. *IEEE Trans. Systems, Man, Cybernet.* **SMC-10**, 744-749.

Ahuja, N., and Rosenfeld, A. (1981). Mosaic models for textures. *IEEE Trans. Pattern Anal., Machine Intelligence* **PAM-3**, 1-11.

Ahuja, N., and Rosenfeld, A. (1982). Image models. In: *Handbook of Statistics*, **2**, Classification, Pattern Recognition and Reduction of Dimensionality, Ed. P. R. Krishnaiah and L. N. Kanal, Chapter 17, pp. 383-397. Amsterdam: North-Holland.

Ailam, G. (1966). Moments of coverage and coverage spaces. *J. Appl. Prob.* **3**, 550-555.

369

Ailam, G. (1968). On probability properties of measures of random sets and the asymptotic behavior of empirical distribution functions. *J. Appl. Prob.* **5**, 196-202.

Ailam, G. (1970). The asymptotic distribution of the measure of random sets with application to the classical occupancy problem and suggestions for curve fitting. *Ann. Math. Statist.* **41**, 427-439.

Akeda, Y., and Hori, M. (1976). On random sequential packing in two and three dimensions. *Biometrika* **63**, 361-366.

Aldous, D. J. (1984). *Probability Approximations via the Poisson Clumping Heuristic.* Unpublished Lecture Notes.

Ambartzumian, R. V. (1965). The Poisson superposition of clusters. (Russian.) *Dokl. Akad. Nauk SSSR* **41**, 73-80.

Ambartzumian, R. V. (1966). On the Poisson clustering process. *Bull. Int. Statist. Inst.* **41**, 388-390.

Ambartzumian, R. V. (1971). Probability distributions in the geometry of clusters. *Studia Sci. Math. Hungar.* **6**, 235-241.

Ambartzumian, R. V. (1982). *Combinatorial Integral Geometry, with Applications to Mathematical Stereology.* New York: Wiley.

Ambegaokar, V., Halperin, B. I., and Langer, J. S. (1971). Hopping conductivity in disordered systems. *Phys. Rev. B* **4**, 2612-2620.

Armitage, P. (1949). An overlap problem arising in particle counting. *Biometrika* **36**, 257-266.

Athreya, K. B., and Ney, P. E. (1972). *Branching Processes.* New York: Springer.

Baddeley, A. (1977). A fourth note on recent research in geometrical probability. *Adv. Appl. Prob.* **9**, 824-860.

Baddeley, A. (1980a). A limit theorem for statistics of spatial data. *Adv. Appl. Prob.* **12**, 447-461.

Baddeley, A. (1980b). Absolute curvatures in integral geometry. *Math. Proc. Cambridge Phil. Soc.* **88**, 45-58.

Baddeley, A. (1982). Stochastic geometry: An introduction and reading-list. *Int. Statist. Rev.* **50**, 179-193.

Bánkövi, G. (1962). On gaps generated by a random space-filling procedure. *Publ. Math. Inst. Hungar. Acad. Sci.* **7**, 395-407.

Barnard, G. A. (1963). Contribution to discussion. *J. Roy. Statist. Soc. Ser. B* **25**, 294.

Bartlett, M. S. (1964). A note on spatial pattern. *Biometrics* **20**, 891-892.

Bartlett, M. S. (1974). The statistical analysis of spatial pattern. *Adv. Appl. Prob.* **6**, 336-358.

Bartlett, M. S. (1975). *The Statistical Analysis of Spatial Pattern.* London: Chapman and Hall.

Baticle, M. (1933a, 1933b, 1935). Le problème de la répartition. *Compt. Rend.* **196**, 1945-1946; **197**, 632-634; **201**, 862-864.

Bernal, J. D. (1959). A geometrical approach to the structure of liquids. *Nature (London)* **183**, 141-147.

Bernal, J. D. (1960). Geometry of the structure of monatomic liquids. *Nature (London)* **185**, 68-70.

Bernal, J. D., and Mason, J. (1960). Co-ordination of randomly packed spheres. *Nature (London)* **188**, 910-911.

Berryman, J. G. (1983). Random close packing of hard spheres and disks. *Phys. Rev. A* **27**, 1053-1061.

Billard, P. (1965). Séries de Fourier aléatoirement bornées, continues, uniformément convergentes. *Ann. Sci. Ecole Norm. Sup.* **83**, 131-179.

Billingsley, P. (1968). *Convergence of Probability Measures.* New York: Wiley.

Billingsley, P. (1979). *Probability and Measure.* New York: Wiley.

Blaisdell, B. E., and Solomon, H. (1970). On random sequential packing in the plane and a conjecture of Palásti. *J. Appl. Prob.* **7**, 667-698.

Blaisdell, B. E., and Solomon, H. (1982). Random sequential packing in Euclidean spaces of dimensions three and four and a conjecture of Pálasti. *J. Appl. Prob.* **19**, 382-390.

Blaschke, W. (1949). *Vorlesungen über Integralgeometrie.* New York: Chelsea.

Blödner, R., Mühlig, P., and Nagel, W. (1984). The comparison by simulation of solutions of Wicksell's corpuscle problem. *J. Microsc.* **135**, 61-74.

Bollabás, B. (1979). *Graph Theory: An Introductory Course.* New York: Springer.

Bonnesen, T., and Fenchel, W. (1948). *Theorie der Konvexen Körper.* New York: Chelsea.

Breiman, L. (1968). *Probability.* Reading, Mass: Addison-Wesley.

Brémaud, P. (1981) *Point Processes and Queues.* New York: Springer.

Broadbent, S. R. (1954). Contribution to discussion. *J. Roy. Statist. Soc. Ser. B* **16**, 68.

Broadbent, S. R., and Hammersley, J. M. (1957). Percolation processes, I. Crystals and mazes. *Proc. Cambridge Phil. Soc.* **53**, 629-641.

Bronowski, J., and Neyman, J. (1945). The variance of the measure of a two-dimensional random set. *Ann. Math. Statist.* **16**, 330-341.

Chernoff, H., and Daly, J. F. (1957). The distribution of shadows. *J. Math. Mech.* **6**, 567-584.

Choquet, G. (1955). Theory of capacities. *Ann. Inst. Fourier* **5**, 131-295.

Chung, K. L. (1974). *A Course in Probability Theory,* 2nd Ed. New York: Academic Press.

Cliff, A. D., and Ord, J. K. (1973). *Spatial Autocorrelation.* London: Pion.

Cliff, A. D., and Ord, J. K. (1975). The choice of test for spatial autocorrelation. In: *Display and Analysis of Spatial Data,* Eds. J. C. Davis and M. J. McCullagh, pp. 54-57. New York: Wiley.

Cliff, A. D., and Ord, J. K. (1981). *Spatial Processes.* London: Pion.

Coleman, R. (1972). Sampling procedures for the lengths of random straight lines. *Biometrika* **59**, 415-426.

Coleman, R. (1979). *An Introduction to Mathematical Stereology.* Aarhus: Department of Theoretical Statistics, University of Aarhus.

Coleman, R. (1981). The stereological determination of particle size distributions: Theoretical considerations. *Proc. Third Eur. Symp. Stereol.,* Ljubljana, pp. 67-77.

Coniglio, A., Stanley, H. E., and Klein, W. (1979). Site-bond correlated-percolation problem: Statistical mechanical model of polymer gelation. *Phys. Rev. Lett.* **42**, 518-522.

Conolly, B. W. (1971). The busy period for the infinite capacity service system $M/G/\infty$. In: *Studi di Probabilità, Statistica e Ricerca Operativa in Onore di G. Pompilj,* pp. 521-529. Gubbio: Edizroni Oderisi.

Cooke, P. J. (1974). Bounds for coverage probabilities with applications to sequential coverage problems. *J. Appl. Prob.* **11**, 281-293.

Cooper, D. W., Feldman, H. A., and Chase, G. R. (1978). Fiber counting: A source of error corrected. *Am. Indust. Hyg. Assoc. J.* **39**, 362-367.

Copson, E. T. (1968). *Metric Spaces.* London: Cambridge University Press.

Cowan, R. (1978). The use of the ergodic theorems in random geometry. *Suppl. Adv. Appl. Prob.* **10**, 47-57.

Cowan, R. (1979). Homogeneous line-segment processes. *Math. Proc. Cambridge Phil. Soc.* **86**, 481-489.

Cowan, R. (1984). A model for random packing of disks in the neighborhood of one disk. *SIAM J. Appl. Math.* **44**, 839-853.

Cox, D. R. (1962). *Renewal Theory.* London: Methuen.

Cox, D. R. (1969). Some sampling problems in technology. In: *New Developments in Survey Sampling,* Eds. N. L. Johnson and H. Smith, pp. 506-527. New York: Wiley-Interscience.

Cox, D. R., and Hinkley, D. V. (1974). *Theoretical Statistics.* London: Chapman and Hall.

Cox, D. R., and Isham, V. (1980). *Point Processes.* London: Chapman and Hall.

Cox, D. R., and Lewis, P. A. W. (1966). *The Statistical Analysis of Series of Events.* London: Methuen.

Cox, D. R., and Miller, H. D. (1965). *The Theory of Stochastic Processes.* London: Methuen.

Cressie, N. (1976). On the logarithms of high-order spacings. *Biometrika* **63**, 343-355.

Cressie, N. (1978). A strong limit theorem for random sets. *Suppl. Adv. Appl. Prob.* **10**, 36-46.

Cressie, N. (1979). An optimal statistic based on higher order gaps. *Biometrika* **66**, 619-627.

Cross, C. A., and Fisher, D. L. (1968). The computer simulation of lunar craters. *Monthly Not. Roy. Astr. Soc.* **139**, 261-272.

Cruz-Orive, L. M. (1983). Distribution-free estimation of sphere size distribution from slabs showing overprojection and truncation, with a review of previous methods. *J. Microsc.* **131**, 265-290.

Daley, D. J. (1971). The definition of a multi-dimensional generalization of shot noise. *J. Appl. Prob.* **8**, 128-135.

Daley, D. J., and Milne, R. K. (1973). The theory of point processes: A bibliography. *Int. Statist. Rev.* **41**, 183-201.

Daley, D. J., and Vere-Jones, D. (1988). *Introduction to the Theory of Point Processes.* New York: Academic Press.

Dalton, N. W., Domb, C., and Sykes, M. F. (1964). Dependence of critical concentration of a dilute ferromagnet on the range of interaction. *Proc. Phys. Soc.* **83**, 496-498.

Daniels, H. E. (1952). The covering circle of a sample from a circular normal distribution. *Biometrika* **39**, 137-143.

Darling, D. A. (1953). On a class of problems related to the random division of an interval. *Ann. Math. Statist.* **24**, 239-253.

Davis, H. T. (1941). *The Analysis of Economic Time Series.* Bloomington, Indiana: Principia.

Davis, J. C. (1973). *Statistics and Data Analysis in Geology.* New York: Wiley.

Davy, P. J. (1978). Aspects of random set theory. *Suppl. Adv. Appl. Prob.* **10**, 36-46.

Davy, P. J. (1982). Coverage. In: *Encyclopedia of Statistical Sciences*, Eds. S. Kotz and N. L. Johnson, **2**, pp. 212-214. New York: Wiley.

Davy, P. J., and Miles, R. E. (1977). Sampling theory for opaque spatial specimens. *J. Roy. Statist. Soc. Ser. B* **39**, 56-65.

De Bruin, N. G. (1965). Asymptotic distribution of lattice points in a rectangle. *SIAM Rev.* **7**, 274-275.

Deheuvels, P. (1983). Strong bounds for multidimensional spacings. *Z. Wahrscheinlichkeitstheor. verw. Geb.* **64**, 411-424.

De Hoff, R. T. (1978). Stereological uses of the area tangent count. *Lect. Notes Biomath.* **23**, 99-113.

De Hoff, R. T., and Rhines, F. N. (1968). *Quantitative Microscopy.* New York: McGraw-Hill.

Del Pino, G. E. (1979). On the asymptotic distribution of k-spacings with applications to goodness-of-fit tests. *Ann. Statist.* **7**, 1058-1065.

Descloux, A. (1976). Some properties of the variance of the switch-count load. *Bell Syst. Tech. J.* **55**, 59-88.

Dickson, L. E. (1952). *History of the Theory of Numbers*, Vol. II. New York: Chelsea.

Diggle, P. J. (1981). Binary mosaics and the spatial pattern of heather. *Biometrics* **37**, 531-539.

Diggle, P. J. (1983). *Statistical Analysis of Spatial Point Patterns*. London: Academic Press.

Diggle, P. J., Besag, J., and Gleaves, J. T. (1976). The statistical analysis of spatial point patterns by means of distance methods. *Biometrics* **32**, 659-667.

Diggle, P. J., and Gratton, R. J. (1984). Monte Carlo methods of inference for implicit statistical models. (With discussion.) *J. Roy. Statist. Soc. Ser. B* **46**, 193-227.

Dishington, R. J. (1956). *The Expected Coverage of a Small Circular Target by a Number of Circular Bombs*. Rand Publication No. RM-413. Santa Monica, California: Rand.

Dolby, J. L., and Solomon, H. (1975). Information density phenomena and random loose packing. *J. Appl. Prob.* **12**, 364-370.

Domb, C. (1947). The problem of random intervals on a line. *Proc. Cambridge Phil. Soc.* **43**, 329-341.

Domb, C. (1972). A note on the series expansion method for clustering problems. *Biometrika* **59**, 209-211.

Domb, C. and Sykes, M. F. (1961). Cluster size in random mixtures and percolation processes. *Phys. Rev.* **122**, 77-78.

Doob, J. L., (1953). *Stochastic Processes*. New York: Wiley.

Downton, F. (1961). A note on vacancies on a line. *J. Roy. Statist. Soc. Ser. B* **23**, 207-214.

Duer, W. C., Greenstein, J. R., Oglesby, G. B., and Millero, F. J. (1977). Qualitative observations concerning packing densities for liquids, solutions and random assemblies of spheres. *J. Chem. Educ.* **54**, 139-143.

Dunford, N. (1951). An individual ergodic theorem for non-commutative transforms. *Acta Sci. Math. Szeged* **14**, 1-4.

Dunford, N., and Schwartz, J. T. (1958). *Linear Operators. Part I: General Theory*. New York: Interscience.

Dupač, V. (1980). Parameter estimation in the Poisson field of disks. *Biometrika* **67**, 187-190.

Dvoretzky, A. (1956). On covering a circle by randomly placed arcs. *Proc. Natl. Acad. Sci. U.S.A.* **42**, 199-203.

Dvoretzky, A., and Robbins, H. (1964). On the "parking" problem. *Publ. Math. Inst. Hungar. Acad. Sci. Ser. A* **9**, 209-226.

Dvurečenskij, A. (1984). The busy period of order n in the GI/D/∞ queue. *J. Appl. Prob.* **21**, 207-212.

Eberhardt, L. L. (1978). Transect methods for population studies. *J. Wild. Management* **42**, 1-31.

Eberl, W., and Hafner, R. (1971). Die asymptotische Verteilung von Koinzidenzen. *Z. Wahrlichkeitstheor. verw. Geb.* **18**, 322-332.

Eckler, A. R. (1969). A survey of coverage problems associated with point and area targets. *Technometrics* **11**, 561-589.

Eckler, A. R., and Burr, S. A. (1972). *Mathematical Models of Target Coverage and Missile Allocation.* Alexandria, Virginia: Military Operations Research Society.

Eddy, W. F. (1980). The distribution of the convex hull of a Gaussian sample. *J. Appl. Prob.* **17**, 686-695.

Eddy, W. F., and Gale, J. D. (1981). The convex hull of a spherically symmetric sample. *Adv. Appl. Prob.* **13**, 751-763.

Edens, E. (1975). Random covering of a circle. *Indag. Math.* **37**, 373-384.

Efron, B. (1965). The convex hull of a random set of points. *Biometrika* **52**, 331-343.

Elias, H. (1967). *Stereology.* Berlin: Springer.

Elliott, R. J., Heap, B. R., Morgan, D. J., and Rushbrooke, G. S. (1960). Equivalence of the critical concentrations in the ising and Heisenberg models for ferromagnetism. *Phys. Rev. Lett.* **5**, 366-367.

Essam, J. W. (1972). Percolation and cluster size. In: *Phase Transitions and Critical Phenomena* , Eds. C. Domb and M. S. Green. Chapter 6, **2**, pp. 197-270. London: Academic Press.

Fava, N. A., and Santaló, L. A. (1978). Plate and line segment processes. *J. Appl. Prob.* **15**, 494-501.

Federer, H. (1969). *Geometric Measure Theory.* Berlin: Springer.

Fejes Tóth, L. (1953). *Lagerungen in der Ebene auf der Kugel und in Raum.* Berlin: Springer.

Feller, W. (1971). *Introduction to Probability Theory and Its Applications,* Vol. 2. New York: Wiley.

Feron, R. (1972). Charactéristiques de position et de dispersion des aléatoires et des multialéatoires. *Pub. Economét.* **5**, 99-124.

Feron, R. (1976a). Ensembles aléatoires flous. *Compt. Rend. Acad. Sci. Ser. A* **282**, 903-906.

Feron, R. (1976b). Ensembles flous, ensembles aléatoires flous et économie floue. *Pub. Economét.* **9**, 51-66.

Finney, J. L. (1970). Random packings and structure of simple liquids, I. Geometry of random close packing. *Proc. Roy. Soc. London Ser. A* **319**, 479–493.

Finney, J. L., and Gotoh, K. (1975). Reply to Gamba (1975). *Nature (London)* **256**, 522.

Fisher, R. A. (1929). Tests of significance in harmonic analysis. *Proc. Roy. Soc. London Ser. A* **125**, 54–59.

Fisher, R. A. (1940). On the similarity of the distributions found for the test of significance in harmonic analysis, and in Stevens' problem in geometrical probability. *Ann. Eugen.* **10**, 14–17.

Flatto, L. (1973). A limit theorem for random coverings of a circle. *Israel J. Math.* **15**, 167–184.

Flatto, L., and Konheim, A. G. (1962). The random division of an interval and the random covering of a circle. *SIAM Rev.* **4**, 211–222.

Flatto, L., and Newman, D. J. (1977). Random coverings. *Acta Math.* **138**, 241–264.

Fortet, R., and Kambouzia, M. (1975). Ensembles aléatoires, répartitions ponctuelles aléatoires, problèmes de recouvrement. *Ann. Inst. Henri Poincaré* **B11**, 299–316.

Franken, P., König, D., Arndt, U., and Schmidt, V. (1981). *Queues and Point Processes.* Berlin: Akademie-Verlag.

Fremlin, D. H. (1976). The clustering problem: Some Monte Carlo results. *J. Phys.* **37**, 813–817.

Fullman, R. L. (1953). Measurement of particle size in opaque bodies. *J. Metals* **5**, 447–452.

Gács, P., and Szász, D. (1975). On a problem of Cox concerning point processes in $\mathbb{R}^k$ of "controlled variability." *Ann. Prob.* **3**, 597–607.

Gamba, A. (1975). Random packing of equal spheres. *Nature (London)* **256**, 521–522.

Garwood, F. (1947). The variance of the overlap of geometrical figures with reference to a bombing problem. *Biometrika* **34**, 1–17.

Garwood, F. (1960). An application of the theory of probability to vehicular-controlled traffic. *J. Roy. Statist. Soc. Suppl.* **7**, 65–77.

Gawlinski, E. T., and Redner, S. (1983). Monte-Carlo renormalisation group for continuum percolation with excluded-volume interactions. *J. Phys. A* **16**, 1063–1071.

Gawlinski, E. T., and Stanley, H. E. (1981). Continuum percolation in two dimensions: Monte Carlo tests of scaling and universality for non-interacting disks. *J. Phys. A* **14**, L291–L299.

Gayda, J. P., and Ottavi, H. (1974). Clusters of atoms coupled by long range interactions. *J. Physi.* **35**, 393–399.

Geary, R. C. (1954). The contiguity ratio and statistical mapping. *Incorp. Statistician* **5**, 115-145.

Gelbaum, B. R., and Olmstead, J. M. H. (1964). *Counterexamples in Analysis*. San Fransisco: Holden-Day.

Gelfond, A. O., and Linnik, Yu. V. (1965). *Elementary Methods in Analytic Number Theory*. Chicago: Rand McNally.

Gilbert, E. N. (1961). Random plane networks. *J. Soc. Indust. Appl. Math.* **9**, 533-543.

Gilbert, E. N. (1964). Randomly packed and solidly packed spheres. *Can. J. Math.* **16**, 286-298.

Gilbert, E. N. (1965). The probability of covering a sphere with N circular caps. *Biometrika* **52**, 323-330.

Girling, A. J. (1982). Approximate variances associated with random configurations of hard spheres. *J. Appl. Prob.* **19**, 588-596.

Glaz, J., and Naus, J. (1979). Multiple coverage of the line. *Ann. Prob.* **7**, 900-906.

Gnedenko, B. V., and Kolmogorov, A. N. (1968). *Limit Distributions for Sums of Independent Random Variables*. Reading, Mass: Addison-Wesley.

Gnedenko, B. V., and König, D. (1984). *Handbuch der Bedienungstheorie, Vol. II*. Berlin: Akademie-Verlag.

Gotoh, K., and Finney, J. L. (1974). Statistical geometrical approach to random packing density of equal spheres. *Nature (London)* **252**, 202-205.

Goudsmit, S. (1945). Random distribution of lines in a plane. *Rev. Mod. Phys.* **17**, 321-322.

Grassman, W. K. (1981). The optimal estimation of the expected number in a $M/D/\infty$ queueing system. *Operations Res.* **29**, 1208-1211.

Green, P. J., and Sibson, R. (1978). Computing Dirichlet tessellations in the plane. *Comp. J.* **21**, 168-173.

Greenberg, I. (1980). The moments of coverage of a linear set. *J. Appl. Prob.* **17**, 865-868.

Grimmett, G. R. (1976). On the number of clusters in the percolation model. *J. London Math. Soc.* **13**, 346-350.

Groves, A. D., and Smith, E. S. (1957). Salvo hit probabilities for offset circular targets. *Operations Res.* **5**, 222-228.

Guenther, W. C., and Terragno, P. J. (1964). A review of the literature on a class of coverage problems. *Ann. Math. Statist.* **35**, 232-260.

Gundersen, H. J. G. (1977). Notes on the estimation of the numerical density of arbitrary profiles: The edge effect. *J. Microsc.* **111**, 219-223.

Gundersen, H. J. G. (1978). Estimators of the number of objects per area unbiased by edge effects. *Microsc. Acta* **81**, 107-117.

Haan, S. W., and Zwanzig, R. (1977). Series expansions in a continuum percolation problem. *J. Phys. A* **10**, 1547-1555.

Hadwiger, H., and Giger, H. (1968). Über Treffzahlwahrscheinlichkeiten im Eikörperfeld. *Z. Wahrscheinlichkeitstheor. verw. Geb.* **10**, 329-334.

Hafner, R. (1972a). The asymptotic distribution of random clumps. *Computing* **10**, 335-351.

Hafner, R. (1972b). Die asymptotische Verteilung von mehrfachen Koinzidenzen. *Z. Wahrscheinlichkeitstheor. verw. Geb.* **21**, 96-108.

Hall, P. (1984a). Random, nonuniform distribution of line segments on a circle. *Stoch. Proc. Appl.* **18**, 239-261.

Hall, P. (1984b). Mean and variance of vacancy for distribution of k-dimensional spheres within k-dimensional space. *J. Appl. Prob.* **21**, 738-752.

Hall, P. (1985a). Three limit theorems for vacancy in multivariate coverage problems. *J. Multivar. Anal.* **16**, 211-236.

Hall, P. (1985b). Resampling a coverage pattern. *Stoch. Proc. Appl.* **20**, 231-246.

Hall, P. (1985c). Heavy traffic approximations for busy period in an $M/G/\infty$ queue. *Stoch. Proc. Appl.* **19**, 259-269.

Hall, P. (1985d). On inference in a one-dimensional mosaic and an $M/G/\infty$ queue. *Adv. Appl. Prob.* **17**, 210-229.

Hall, P. (1985e). Distribution of size, structure and number of vacant regions in a high-intensity mosaic. *Z. Wahscheinlichkeitstheor. verw. Geb.* **70**, 237-261.

Hall, P. (1985f). Correcting segment counts for edge effects when estimating intensity. *Biometrika* **72**, 459-463.

Hall, P. (1985g). On continuum percolation. *Ann. Prob.* **13**, 1250-1266.

Hall, P. (1985h). Counting methods for inference in binary mosaics. *Biometrics* **41**, 1049-1052.

Hall, P. (1986). Clump counts in a mosaic. *Ann. Prob.* **14**, 424-458.

Hall, P. (1987). On the processing of a motion-blurred image. *SIAM J. Appl. Math.* **47**, 441-453.

Hall, P., and Titterington, D. M. (1986). On some smoothing techniques used in image restoration. *J. Roy. Statist. Soc. Ser. B* **48**, 330-343.

Halmos, P. R. (1950). *Measure Theory.* Princeton: Van Nostrand.

Hammersley, J. M. (1957). Percolation processes, II. The connective constant. *Proc. Cambridge Phil. Soc.* **53**, 642-645.

Hammersley, J. M., Lewis, J. W. E., and Rowlinson, J. S. (1975). Relationships between the multinomial and Poisson models of stochastic processes, and between the canonical and grand canonical ensembles in statistical mechanics, with illustrations and Monte Carlo methods for the penetrable sphere model of liquid-vapour equilibrium. *Sankhyā Ser. A* **37**, 457-491.

Hanisch, K.-H. (1981). On classes of random sets and point processes. *Serdica* **7**, 160-166.

Hanisch, K.-H., and Stoyan, D. (1979). Formulas for the second-order analysis of marked point processes. *Math. Oper. Statist. Ser. Statist.* **10**, 555-560.

Hardy, G. H. (1916). The coverage order of the arithmetical functions $P(x)$ and $\Delta(x)$. *Proc. London Math. Soc.* **15**, 192-213.

Hasofer, A. M. (1963). On the reliability of the point-counter method in petrography. *Aust. J. Appl. Sci.* **14**, 168-179.

Higuchi, Y. (1982). Coexistence of the infinite (*) clusters:—a remark on the square lattice site percolation. *Z. Wahrscheinlichkeitstheor. verw. Geb.* **61**, 75-81.

Hlawka, E. (1950). Über Integrale auf konvexen Körpern, I. *Monatsh. Math.* **54**, 1-36.

Hoeffding, W. (1963). Probability inequalities for sums of bounded random variables. *J. Am. Statist. Assoc.* **58**, 13-30.

Hoffman-Jørgensen, J. (1973). Coverings of metric spaces with randomly placed balls. *Math. Scand.* **32**, 169-186.

Holcomb, D. F., Iwasawa, M., and Roberts, F. D. K. (1972). Clustering of randomly placed spheres. *Biometrika* **59**, 207-209.

Holcomb, D. F., and Rehr, J. J. (1969). Percolation in heavily doped semiconductors. *Phys. Rev.* **183**, 773-776.

Holst, L. (1972). Asymptotic normality and efficiency for certain goodness-of-fit tests. *Biometrika* **59**, 137-145.

Holst, L. (1979). Asymptotic normality of sum-functions of spacings. *Ann. Prob.* **7**, 1066-1072.

Holst, L. (1980a). On multiple covering of a circle with random arcs. *J. Appl. Prob.* **17**, 284-290.

Holst, L. (1980b). On the lengths of the pieces of a stick broken at random. *J. Appl. Prob.* **17**, 623-634.

Holst, L. (1981). On convergence of the coverage by random arcs on a circle and the largest spacing. *Ann. Prob.* **9**, 648-655.

Holst, L. (1983). A note on random arcs on the circle. In: *Probability and Mathematical Statistics*, Eds. A. Gut and L. Holst, pp. 40-46. Uppsala: Department of Mathematics, Uppsala University.

Holst, L., and Hüsler, J. (1984). On the random coverage of the circle. *J. Appl. Prob.* **21**, 558-566.

Hope, A. C. A. (1968). A simplified Monte Carlo significance test procedure. *J. Roy. Statist. Soc. Ser. B* **30**, 582-598.

Hua, L. K. (1942). The lattice-points on a circle. *Quart. J. Math.* **13**, 18-29.

Hüsler, J. (1982). Random coverage of the circle and asymptotic distributions. *J. Appl. Prob.* **19**, 578-587.

Indritz, J. (1963). *Methods in Analysis.* New York: Macmillan.

Irwin, J. O., Armitage, P., and Davies, C. N. (1949). The overlapping of dust particles on a sampling plate. *Nature (London)* **163**, 809.

Itoh, Y. (1980). On the minimum of gaps generated by one-dimensional random packing. *J. Appl. Prob.* **17**, 134-144.

Itoh, Y., and Ueda, S. (1979). A random packing model for elections. *Ann. Inst. Statist. Math.* **31**, 157-167.

Jackson, J. L., and Montroll, E. W. (1958). Free radical statistics. *J. Chem. Phys.* **28**, 1101-1109.

Janson, S. (1983a). Random coverings of the circle with arcs of random lengths. In: *Probability and Mathematical Statistics*, Eds. A. Gut and L. Holst, pp. 62-73. Uppsala: Department of Mathematics, Uppsala University.

Janson, S. (1983b). Random coverings in several dimensions. Uppsala University Department of Mathematics Report 1983:14.

Janson, S. (1984). Maximal spacings in several dimensions. Uppsala University Department of Mathematics Report 1984:3.

Jarnagin, M. P. (1965). Expected coverage of a circular target by bombs all aimed at the center. N. W. L. Report No. 1941, U.S. Naval Weapons Laboratory, Dahlgren, Virginia.

Jarnagin, M. P. (1966). Expected coverage of a circular target by bombs all aimed at the center. *Oper. Res.* **14**, 1139-1143.

Jensen, E. B. (1984). A design-based proof of Wicksell's integral equation. *J. Microsc.* **136**, 345.

Jensen, E. B., Baddeley, A. J., Gundersen, H. J. G., and Sundberg, R. (1985). Recent trends in stereology. *Int. Statist. Rev.* **53**, 99-108.

Jensen, E. B., and Sundberg, R. (1986). Generalized associated point methods for sampling planar objects. *J. Microsc.*, **144**, 55-70.

Jewell, N. P., and Romano, J. P. (1982). Coverage problems and random convex hulls. *J. Appl. Prob.* **19**, 546-561.

Jodrey, W. S., and Tory, G. M. (1980). Random sequential packing in $\mathbb{R}^n$. *J. Statist. Comp. Simul.* **10**, 87-93.

Kahane, J.-P. (1968). *Some Random Series of Functions.* Lexington, Massachusetts: D. C. Heath.

Kallenberg, O. (1983). *Random Measures*, 3rd Ed. Berlin: Akademie-Verlag/New York: Academic Press.

Kaplan, N. (1978). A limit theorem for random coverings of a circle which do not quite cover. *J. Appl. Prob.* **15**, 443–446.

Karlin, S., and Taylor, H. M. (1975). *A First Course in Stochastic Processes.* New York: Academic Press.

Karr, A. F. (1986). *Point Processes and Their Statistical Inference.* New York: Dekker.

Kellerer, A. M. (1983). On the number of clumps resulting from the overlap of randomly placed figures in a plane. *J. Appl. Prob.* **20**, 126–135.

Kellerer, A. M. (1985). Counting figures in planar random configurations. *J. Appl. Prob.* **22**, 68–81.

Kellerer, A. M. (1986). The variance of a Poisson process of domains. *J. Appl. Prob.* **23**, 307–321.

Kelly, F. P., and Ripley, B. D. (1976). A note on Strauss's model for clustering. *Biometrika* **63**, 357–360.

Kendall, D. G. (1948). On the number of lattice points inside a random oval. *Quart. J. Math.* **19**, 1–26.

Kendall, D. G. (1974). Foundations of a theory of random sets. In: *Stochastic Geometry*, Eds. E. F. Harding and D. G. Kendall, pp. 322–376. New York: Wiley.

Kendall, D. G., and Rankin, R. A. (1953). On the number of points of a given lattice in a random hypersphere. *Quart. J. Math.* **4**, 178–189.

Kendall, M. G., and Moran, P. A. P. (1963). *Geometrical Probability.* London: Griffin.

Kendall, M. G., and Stuart, A. (1977). *The Advanced Theory of Statistics* **1**, 4th Ed. London: Griffin.

Kendall, M. G., and Stuart, A. (1979). *The Advanced Theory of Statistics* **2**, 4th Ed. London: Griffin.

Kershaw, K. A. (1957). The use of cover and frequency in the detection of pattern in plant communities. *Ecology* **38**, 291–299.

Kertész, J. (1981). Percolation of holes between overlapping spheres—Monte Carlo calculation of the critical volume fraction. *J. Phys. Lett.* **42**, L393–L395.

Kertész, J., and Vicsek, T. (1982). Monte Carlo renormalization group study of the percolation problem of disks with a distribution of radii. *Z. Phys. B* **45**, 345–350.

Kesten, H. (1982). *Percolation Theory for Mathematicians.* Progress in Probability and Statistics **2**. Boston: Birkhäuser.

Kester, A. (1975). Asymptotic normality of the number of small distances between random points in a cube. *Stoch. Proc. Appl.* **3**, 45–54.

Kirkpatrick, S. (1971). Classical transport in disordered media—scaling and effective-medium theories. *Phys. Rev. Lett.* **27**, 1722–1725.

Kirkwood, J. R., and Wayne, C. E. (1983). Percolation in continuous systems. IMA Preprint No. 34, Institute for Mathematics and Its Applications, University of Minnesota.

Klamkin, M. S. (1966). On a probability of overlap. *SIAM Rev.* **8**, 112-115.

Krishna Iyer, P. V. (1949). The first and second moments of some probability distributions arising from some points on a lattice and their application. *Biometrika* **36**, 135-141.

Kurkijärvi, J. (1974). Conductivity in random systems, II. Finite-size-system percolation. *Phys. Rev. B* **9**, 770-774.

Lancaster, H. O. (1950). Statistical control in haematology. *J. Hyg.* **48**, 402-417.

Laslett, G. M., and Liow, S. (1984). A simple transformation of the Boolean model. Manuscript.

Le Cam, L. (1958). Un théorème sur la division d'une intervalle par des points pris au hasard. *Publ. Inst. Statist. Univ. Paris* **7**, 7-16.

Le Cam, L. (1970). On the assumptions used to prove asymptotic normality of maximum likelihood estimates. *Ann. Math. Statist.* **41**, 802-828.

Lévy, P. (1939). Sur la division d'un segment par les points choisis au hasard. *Compt. Rend.* **208**, 147.

Ling, K. D. (1971). Small sample distribution of the measure of a random linear set. *Nanta Math.* **5**, 41-48.

Little, D. V. (1974). A third note on recent research in geometrical probability. *Adv. Appl. Prob.* **6**, 103-130.

Lomnicki, Z. A and Zaremba, S. K. (1957). A further instance of the central limit theorem for dependent random variables. *Math. Z.* **66**, 490-494.

Loomis, L. H. (1953). *An Introduction to Abstract Harmonic Analysis.* London: Van Nostrand.

Lotwick, H. W. (1982). Simulations of some spatial hard core models, and the complete packing problem. *J. Statist. Comp. Simul.* **15**, 295-314.

Lucas, H. A and Seber, G. A. F. (1977). Estimating coverage and particle density using the line intercept method. *Biometrika* **64**, 618-622.

McDonald, L. L. (1980). Line-intercept sampling for attributes other than coverage and density. *J. Wild. Management* **44**, 530-533.

McIntyre, G. A (1953). Estimation of plant density using line transects. *J. Ecol.* **41**, 319-330.

Mack, C. (1948). An exact formula for $Q_k(n)$, the probability of k-aggregates in a random distribution of n points. *Phil. Mag.* **39**, 778-790.

Mack, C. (1949). The expected number of aggregates in a random distribution of points. *Proc. Cambridge Phil. Soc.* **46**, 285-292.

Mack, C. (1953). The effect of overlapping in bacterial counts of incubated colonies. *Biometrika* **40**, 220–222.

Mack, C. (1954). The expected number of clumps when convex laminae are placed at random and with random orientation on a plane area. *Proc. Cambridge Phil. Soc.* **50**, 581–585.

Mack, C. (1956). On clumps formed when convex laminae or bodies are placed at random in two or three dimensions. *Proc. Cambridge Phil. Soc.* **52**, 246–250.

Mackay, A. L. (1980). The packing of three-dimensional spheres on the surface of a four-dimensional hypersphere. *J. Phys. A* **13**, 3373–3379.

Mackenzie, J. K. (1962). Sequential filling of a line by intervals placed at random and its application to linear adsorption. *J. Chem. Phys.* **37**, 723–728.

Mandelbrot, B. B. (1972a). Renewal sets and random cutouts. *Z. Wahrscheinlichkeitstheor. verw. Geb.* **22**, 145–157.

Mandelbrot, B. B. (1972b). On Dvoretzky coverings for the circle. *Z. Wahrscheinlichkeitstheor. verw. Geb.* **22**, 158–160.

Mannion, D. (1964). Random space-filling in one dimension. *Publ. Math. Inst. Hungar. Acad. Sci. Ser. A* **9**, 143–154.

Mannion, D. (1976). Random packing of an interval. *Adv. Appl. Prob.* **8**, 477–501.

Mannion, D. (1979). Random packing of an interval, II. *Adv. Appl. Prob.* **11**, 591–602.

Mannion, D. (1983). Sequential random displacements of points in an interval. *J. Appl. Prob.* **20**, 251–263.

Marriott, F. H. C. (1979). Barnard's Monte Carlo tests: How many simulations? *Appl. Statist.* **28**, 75–77.

Mase, S. (1982). Asymptotic properties of stereological estimators of volume fraction for stationary random sets. *J. Appl. Prob.* **19**, 111–126.

Matérn, B. (1960). *Spatial Variation,* **49**, No. 5. Stockholm: Meddelanden fran Statens Skogsforskningsinstitut.

Matheron, G. (1967). *Eléments pour une Théorie des Mileux Poreux.* Paris: Masson.

Matheron, G. (1975). *Random Sets and Integral Geometry.* New York: Wiley.

Matheson, A. J. (1974). Computation of a random packing of hard spheres. *J. Phys. Chem.* **7**, 2569–2576.

Matthes, K. (1963). Stationäre zufällige Punktfolgen, I. *Jahresbericht DMV* **66**, 66–79.

Matthes, K., Kerstan, J., and Mecke, J. (1978). *Infinitely Divisible Point Processes.* New York: Wiley-Interscience.

Mead, R. (1974). A test for spatial pattern at several scales using data from a grid of contiguous quadrats. *Biometrics* **30**, 295-307.

Mecke, J. (1981). Formulas for stationary planar fibre processes, III—Intersection with fibre systems. *Math. Oper. Statist. Ser. Statist.* **12**, 201-210.

Mecke, J. (1984). Parametric representation of mean values for stationary random mosaics. *Math. Oper. Statist. Ser. Statist.* **15**, 437-442.

Mecke, H., and Stoyan, D. (1980a). Formulas for stationary planar fibre processes, I—General theory. *Math. Oper. Statist. Ser. Statist.* **11**, 267-281.

Mecke, H., and Stoyan, D. (1980b). Stereological problems for spherical particles. *Math. Nachr.* **96**, 311-317.

Melnyk, T. W., and Rowlinson, J. S. (1971). The statistics of the volumes covered by systems of penetrating spheres. *J. Comp. Phys.* **7**, 385-393.

Miles, R. E. (1964a). Random polygons determined by random lines in a plane, I. *Proc. Natl. Acad. Sci. U.S.A.* **52**, 901-907.

Miles, R. E. (1964b). Random polygons determined by random lines in a plane, II. *Proc. Natl. Acad. Sci. U.S.A.* **52**, 1157-1160.

Miles, R. E. (1969). The asymptotic values of certain coverage probabilities. *Biometrika* **56**, 661-680.

Miles, R. E. (1971). Poisson flats in Euclidean spaces. Part II: Homogeneous Poisson flats and the complementary theorem. *Adv. Appl. Prob.* **3**, 1-43.

Miles, R. E. (1972). The random division of space. *Suppl. Adv. Appl. Prob.* **4**, 243-266.

Miles, R. E. (1974a). The fundamental formula of Blaschke in integral geometry and geometrical probability, and its iteration, for domains with fixed orientations. *Aust. J. Statist.* **16**, 111-118.

Miles, R. E. (1974b). A synopsis of "Poisson flats in Euclidean spaces." In: *Stochastic Geometry*, Eds. E. F. Harding and D. G. Kendall, pp. 202-227. Chichester: Wiley.

Miles, R. E. (1974c). On the elimination of edge effects in planar sampling. In: *Stochastic Geometry*, Eds. E. F. Harding and D. G. Kendall, pp. 228-247. Chichester: Wiley.

Miles, R. E. (1976). Estimating aggregate and overall characteristics from thick sections by transmission microscopy. *J. Microsc.* **107**, 227-233.

Miles, R. E. (1978). The sampling, by quadrats, of planar aggregates. *J. Microsc.* **113**, 257-267.

Miles, R. E. (1980). A survey of geometrical probability in the plane, with emphasis on stochastic image modeling. *Comp. Vision, Graph., Image Process.* **12**, 1-24.

Miles, R. E., and Davy, P. J. (1976). Precise and general conditions for the validity of a comprehensive set of stereological fundamental formulas. *J. Microsc.* **107**, 211-226.

Miles, R. E., and Serra, J. (Eds.) (1978). *Geometrical Probability and Biological Structures: Buffon's 200th Anniversary.* Berlin: Springer. (Lecture Notes in Biomathematics **23**.)

Miser, H. J. (1951). Bombing accuracy and damage from atomic bombs: an introduction. *Operations Analysis Technical Memorandum No. 25*, USAF, Washington, DC.

Modestino, J. W., Fries, R. W., and Vickers, A. L. (1980). Stochastic image models generated by random tesselations of the plane. *Comp. Vision, Graph., Image Process.* **12**, 74-98.

Moran, P. A. P. (1947). Random associations on a lattice. *Math. Proc. Cambridge Phil. Soc.* **43**, 321-328.

Moran, P. A. P. (1948). The interpretation of statistical maps. *J. Roy. Statist. Soc. Ser. B* **10**, 243-251.

Moran, P. A. P. (1950a). Notes on continuous stochastic phenomena. *Biometrika* **37**, 17-23.

Moran, P. A. P. (1950b). Numerical integration by systematic sampling. *Proc. Cambridge Phil. Soc.* **46**, 111-115.

Moran, P. A. P. (1966a). A note on recent research in geometrical probability. *J. Appl. Prob.* **3**, 453-463.

Moran, P. A. P. (1966b). Measuring the length of a curve. *Biometrika* **53**, 359-364.

Moran, P. A. P. (1968a). Statistical theory of a high-speed photoelectric planimeter. *Biometrika* **55**, 419-422.

Moran, P. A. P. (1968b). *An Introduction to Probability Theory.* Oxford: Oxford University Press.

Moran, P. A. P. (1969). A second note on recent research in geometrical probability. *Adv. Appl. Prob.* **1**, 73-79.

Moran, P. A. P. (1972). The probabilistic basis of stereology. *Suppl. Adv. Appl. Prob.* **4**, 69-91.

Moran, P. A. P. (1973a). The random volume of interpenetrating spheres in space. *J. Appl. Prob.* **10**, 483-490.

Moran, P. A. P. (1973b). A central limit theorem for exchangeable variates with geometric applications. *J. Appl. Prob.* **10**, 837-846.

Moran, P. A. P. (1974). The volume occupied by normally distributed spheres. *Acta Math.* **133**, 273-286.

Moran, P. A. P., and Fazekas de St. Groth, S. (1962). Random circles on a sphere. *Biometrika* **49**, 389-396.

Morgenthaler, G. W. (1961). Some circular coverage problems. *Biometrika* **48**, 313-324.

Morton, R. R. A. (1966). The expected number and angle of intersections between random curves in a plane. *J. Appl. Prob.* **3**, 559-562.

Mullins, W. W. (1976). An estimator of the underlying size distribution of overlapping impact-craters. *Icarus* **29**, 113-123.

Mullins, W. W. (1978). Use of equivalency in estimating the historical size distribution of impact craters. *Icarus* **33**, 624-629.

Mullooly, J. P. (1968). A one-dimensional random space filling problem. *J. Appl. Prob.* **5**, 427-435.

Naus, J. I. (1979). An indexed bibliography of clusters, clumps and coincidences. *Int. Statist. Rev.* **47**, 47-78.

Ney, P. E. (1962). A random interval filling problem. *Ann. Math. Statist.* **33**, 702-718.

Nicholson, W. L. (1970). Estimation of linear properties of particle size distributions. *Biometrika* **57**, 273-297.

Norburg, T. (1984). Convergence and existence of random set distributions. *Ann. Prob.* **12**, 726-732.

Ohser, J. (1980). On statistical analysis of the Boolean model. *J. Inform. Process. Cybernet.* **16**, 651-653.

O'Neill, B. (1966). *Elementary Differential Geometry.* New York: Academic Press.

Ottavi, H., and Gayda, J. P. (1973). Site percolation: Frontier curvature of clusters. *J. Physique* **34**, 341-344.

Ottavi, H., and Gayda, J. P. (1974). Percolation in a continuous 2-dimensional medium. *J. Phys.* **35**, 631-633.

Page, E. S. (1959). The distribution of vacancies on a line. *J. Roy. Statist. Soc. Ser. A* **21**, 364-374.

Palásti, I. (1960). On some random space filling problems. *Publ. Math. Inst. Hungar. Acad. Sci.* **5**, 353-360.

Palásti, I. (1976). On a two-dimensional random space filling problem. *Studia Sci. Math. Hungar.* **11**, 247-252.

Palmer, R. C. (1948). The dye sampling method of measuring fibre length distribution. *J. Textile Inst.* **39**, T8-T22.

Parker, P., and Cowan, R. (1976). Some properties of line segment processes. *J. Appl. Prob.* **13**, 96-107.

Percus, J. K., and Yevick, G. J. (1958). Analysis of classical statistical mechanics by means of collective coordinates. *Phys. Rev.* **110**, 1-13.

Petrov, V. V. (1975). *Sums of Independent Random Variables.* Berlin: Springer.

Phani, M. K., and Dhar, D. (1984). Continuum percolation with disks having a distribution of radii. *J. Phys. A* **17**, L645-L649.

Pielou, E. C. (1961). Segregation and symmetry in two-species populations as studied by nearest neighbor relationships. *J. Ecol.* **49**, 255-269.

Pielou, E. C. (1964). The spatial pattern of two-phase patchworks of vegetation. *Biometrics* **20**, 156-167.

Pielou, E. C. (1965). The concept of randomness in the patterns of mosaics. *Biometrics* **21**, 908-920.

Pielou, E. C. (1967). A test for random mingling of the phases of a mosaic. *Biometrics* **23**, 657-670.

Pike, G. E., and Seager, C. H. (1974). Percolation and conductivity: A computer study, I. *Phys. Rev.* **10**, 1421-1434.

Pitts, E. (1981). The overlap of random particles and similar problems: Expressions for variance of coverage and its analogue. *SIAM J. Appl. Math.* **41**, 493-498.

Pyke, R. (1965). Spacings. (With discussion.) *J. Roy. Statist. Soc. Ser. B* **27**, 395-449.

Pyke, R. (1972). Spacings revisited. In: *Proceedings of the Sixth Berkeley Symposium on Mathematical Statistics and Probability*, Eds. L. M. LeCam, J. Neyman, and E. L. Scott, Vol. I, pp. 417-427. Berkeley, California: University of California.

Quickenden, T. I., and Tan, G. K. (1974). Random packing in two dimensions and structure of monolayers. *J. Colloid Interface Sci.* **48**, 382-393.

Ramalhoto, M. F. (1984). Bounds for the variance of the busy period of the M/G/∞ queue. *Adv. Appl. Prob.* **16**, 929-932.

Rand Corporation (1952). *Offset Circle Probabilities.* Publication No. R-234. Santa Monica, California: Rand.

Rao, J. S. (1976). Some tests based on arc-lengths for the circle. *Sankhyā Ser. B* **38**, 329-338.

Renshaw, E., and Ford, E. D. (1983). The interpretation of process from pattern using two-dimensional spectral analysis: Methods and problems of interpretation. *Appl. Statist.* **32**, 51-63.

Rényi, A. (1958). On a one-dimensional problem concerned with random space-filling. (In Hungarian.) *Mag. Tud. Akad. Kut. Mat. Intézet Kozlemenyei*, 109-127.

Rényi, A. (1962). Three new proofs and a generalisation of a theorem of Irving Weiss. *Publ. Math. Inst. Hungar. Acad. Sci. Ser. A* **7**, 203-214.

Rényi, A. (1976). *Selected Papers of Alfréd Rényi*, Vol. 2. Ed. P. Turán. Budapest: Akadémiai Kiadó.

Rényi, A., and Sulanke, R. (1963). Uber die konvexe Hülle von n zufällig gewählten Pankten. *Z. Wahrscheinlichkeitstheor. verw. Geb.* **2**, 75-84.

Rényi, A., and Sulanke, R. (1964). Uber die konvexe Hülle von n zufällig gewählten Pankten, II. *Z. Wahrscheinlichkeitstheor. verw. Geb.* **3**, 138-147.

Rice, J. (1977). On generalized shot noise. *J. Appl. Prob.* **9**, 178-184.

Ripley, B. D. (1976). The foundations of stochastic geometry. *Ann. Prob.* **4**, 995-998.

Ripley, B. D. (1977). Modelling spatial patterns. (With discussion.) *J. Roy. Statist. Soc. Ser. B* **39**, 172-212.

Ripley, B. D. (1981). *Spatial Statistics.* New York: Wiley.

Ripley, B. D. (1984). Spatial statistics: Developments 1980-3. *Int. Statist. Rev.* **52**, 141-150.

Ripley, B. D. (1986). Statistics, images and pattern recognition. *Can. J. Statist.* **14**, 83-111.

Roach, S. A. (1968). *The Theory of Random Clumping.* London: Methuen.

Robbins, H. E. (1944). On the measure of a random set, I. *Ann. Math. Statist.* **15**, 70-74.

Robbins, H. E. (1945). On the measure of a random set, II. *Ann. Math. Statist.* **16**, 342-347.

Roberts, F. D. K. (1967). A Monte Carlo study of a two-dimensional unstructured cluster problem. *Biometrika* **54**, 625-628.

Roberts, F. D. K., and Storey, S. H. (1968). A three-dimensional cluster problem. *Biometrika* **55**, 258-260.

Rogers, C. A. (1964). *Packing and Covering.* London: Cambridge University Press.

Rosenblatt, M. (1971). Curve estimation. *Ann. Math. Statist.* **42**, 1815-1842.

Runneburg, J. Th. (1982). Asymptotic normality in vacancies on a line. *Statist. Neerland.* **36**, 135-148.

Russell, A. M., and Josephson, N. S. (1965). Measurement of area by counting. *J. Appl. Prob.* **2**, 339-351.

Ryle, M., and Scheuer, P. A. G. (1960). The statistics of overlapping images and its applications in radio astronomy. *Bull. Int. Statist. Inst.* **37**, 55-67.

Santaló, L. A. (1947). On the first two moments of the measure of a random set. *Ann. Math. Statist.* **18**, 37-49.

Santaló, L. A. (1953). *Introduction to Integral Geometry.* Paris: Hermann.

Santaló, L. A. (1976). *Integral Geometry and Geometric Probability.* Reading, Massachusetts: Addison-Wesley.

Sathe, Y. S. (1985). Improved bounds for the variance of the busy period of the M/G/∞ queue. *Adv. Appl. Prob.* **17**, 913-914.

Saxl, I. (1981). Overview of contemporary theoretical approaches to stereology. *Proc. Third Eur. Symp. Stereol.*, Ljubljana, pp. 19-41.

Saxl, I. (1986). *Stereology of Objects with Internal Structure.* Amsterdam: Elsevier.

Schacter, B., and Ahuja, N. (1979). Random pattern generation processes. *Comp. Graphics Image Process.* **10**, 95-114.

Schmidt, V. (1984). On shot noise processes induced by stationary marked point processes. *Elektron Inform. Kybernet.* **20**, 397-406.

Schmidt, V. (1985a). On finiteness and continuity of shot noise processes. *Optimization* **16**, 921-933.

Schmidt, V. (1985b). Poisson bounds for moments of shot noise processes. *Statistics* **16**, 253-262.

Schneider, R. (1979a). Boundary structure and curvature of convex bodies. In: *Contributions to Geometry*, Eds. J. Tölke and J. M. Wills, pp. 13-59. Basel: Birkhäuser.

Schneider, R. (1979b). Bestimmung konvexer Körper durch Krümmungsmasse. *Comment. Math. Helvet.* **54**, 42-60.

Schneider, R. (1980). Parallelmengen mit Vielfachheit und Steiner-Formeln. *Geom. Ded.* **9**, 111-127.

Schneider, T. (1979). The influence of counting rules on the number and on the size distribution of fibers. *Ann. Occup. Hyg.* **21**, 341-350.

Schroeter, G. (1982). The variance of the coverage of a randomly located target by a salvo of weapons. *Naval Res. Log. Quart.* **29**, 97-111.

Schroeter, G. (1984). Distribution of number of point targets killed and higher moments of coverage of area targets. *Naval Res. Log. Quart.* **31**, 373-385.

Schwandtke, A., Ohser, J., and Stoyan, J. (1987). Improved estimation in planar sampling. *Acta Stereol.*, in press.

Scott, G. D. (1960). Packing of spheres. *Nature (London)* **188**, 908-909.

Scott, G. D., and Kilgour, D. M. (1969). Density of random close packing of spheres. *Br. J. Appl. Phys. (J. Phys. Ser. D)* **2**, 863-866.

Seager, C. H., and Pike, G. E. (1974). Percolation and conductivity: A computer study, II. *Phys. Rev. B* **10**, 1435-1446.

Serra, J. (1978). One, two, three...infinity. *Lect. Notes Biomath.* **23**, 137-152.

Serra, J. (1980). The Boolean model and random sets. *Comp. Vision, Graph., Image Process.* **12**, 99-126.

Serra, J. (1982). *Image Analysis and Mathematical Morphology.* New York: Academic Press.

Shanbhag, D. N. (1966). On infinite server queues with batch arrivals. *J. Appl. Prob.* **3**, 274-279.

Shanks, D. (1962). *Solved and Unsolved Problems in Number Theory.* Washington: Spartan.

Shante, V. K. S., and Kirkpatrick, S. (1971). An introduction to percolation theory. *Adv. Phys.* **20**, 325-357.

Shepp, L. A. (1972a). Covering the circle with random arcs. *Israel J. Math.* **11**, 328-345.

Shepp, L. A. (1972b). Covering the line with random intervals. *Z. Wahrscheinlichkeitstheor. verw. Geb.* **23**, 163-170.

Sibson, R. (1980a). The Dirichlet tessellation as an aid in data analysis. *Scand. J. Statist.* **7**, 14-20.

Sibson, R. (1980b). A vector identity for the Dirichlet tessellation. *Math. Proc. Cambridge Phil. Soc.* **87**, 151-155.

Siegel, A. F. (1978a). Random space filling and moments of coverage in geometrical probability. *J. Appl. Prob.* **15**, 340-355.

Siegel, A. F. (1978b). Random arcs on the circle. *J. Appl. Prob.* **15**, 774-789.

Siegel, A. F. (1979a). Asymptotic coverage distributions on the circle. *Ann. Prob.* **7**, 651-661.

Siegel, A. F. (1979b). The noncentral chi-squared distribution with zero degrees of freedom and testing for uniformity. *Biometrika* **66**, 381-386.

Siegel, A. F., and Holst, L. (1982). Covering the circle with random arcs of random sizes. *J. Appl. Prob.* **19**, 373-381.

Siegel, S. (1956). *Nonparametric Statistics for the Behavioural Sciences.* New York: McGraw-Hill.

Silberstein, L. (1945). The probable number of aggregates in random distributions of points. *London, Edinburgh, Dublin Phil. Mag. J. Sci.* **36**, 319-336.

Silverman, D. L. (1964). Problem E1658. *Am. Math. Monthly* **71**, 1135-1136.

Skal, A. S., and Shklovskii, B. I. (1973). Influence of impurity concentration on hopping conduction in semiconductors. *Soviet Phys.-Semicond.* **7**, 1058-1059.

Smythe, R. T. (1973). Strong laws of large numbers for r-dimensional arrays of random variables. *Ann. Prob.* **1**, 164-170.

Smythe, R. T., and Wierman, J. C. (1978). *First-Passage Percolation on the Square Lattice.* Lecture Notes in Mathematics **671**. Berlin: Springer.

Snyder, D. L. (1975). *Random Point Processes.* New York: Wiley-Interscience.

Solomon, H. (1950). A coverage distribution. *Ann. Math. Statist.* **21**, 139-140.

Solomon, H. (1953). Distribution of the measure of a random two-dimensional set. *Ann. Math. Statist.* **24**, 650-656.

Solomon, H. (1967). Random packing density. *Proc. Fifth Berkeley Symp. Math. Statist. Prob.* III, 119-134.

Solomon, H. (1978). *Geometric Probability.* Philadelphia: SIAM.

Solomon, H., and Sutton, C. (1986). A simulation study of random caps on a sphere. *J. Appl. Prob.* **23**, 951-960.

Solomon, H., and Weiner, H. (1986). A review of the packing problem. *Commun. Statist.-Theor. Meth.* **15**, 2571-2607.

Stadje, W. (1985). The busy period of the queueing systems $M/G/\infty$. *J. Appl. Prob.* **22**, 697-704.

Stam, A. J. (1981). A nonrandom limit for the volume covered k times. *J. Appl. Prob.* **18**, 148-156.

Stam, A. J. (1984). Expectation and variance of the volume covered by a large number of independent random sets. *Comp. Math.* **52**, 57-83.

Steutel, F. W. (1967). Random division of an interval. *Statist. Neerland.* **21**, 231-244.

Stevens, W. L. (1939). Solution to a geometrical problem in probability. *Ann. Eugen.* **9**, 315-320.

Stillinger, F. H., Jr., Di Marzio, E. A., and Kornegay, R. L. (1964). Systematic approach to explanation of the rigid disk phase transition. *J. Chem. Phys.* **40**, 1564-1576.

Stoyan, D. (1979a). On the accuracy of lineal analysis. *Biometric. J.* **21**, 439-449.

Stoyan, D. (1979b). Applied stochastic geometry: A survey. *Biometric. J.* **21**, 693-715.

Stoyan, D. (1979c). Proofs of some fundamental formulas of stereology for non-Poisson grain models. *Math. Oper. Statist. Ser. Optim.* **10**, 575-583.

Stoyan, D. (1986). On generalized planar random tessellations. *Math. Nachr.* **128**, 215-219.

Stoyan, D., and Mecke, J. (1983). *Stochastische Geometrie: Eine Einführung.* Berlin: Akademie-Verlag.

Stoyan, D., Kendall, W. S., and Mecke, J. (1987). *Stochastic Geometry and Its Applications.* Berlin: Akademie-Verlag.

Stoyan, H., and Stoyan, D. (1986). Simple stochastic models for the analysis of dislocation distributions. *Phys. Stat. Sol. A* **97**, 163-172.

Switzer, P. (1965). A random set process in the plane with a Markovian property. *Ann. Math. Statist.* **36**, 1859-1863.

Takács, L. (1958). On the probability distribution of the measure of the union of random sets placed in a Euclidean space. *Ann. Univ. Sci. Budapest, Eötvös Sect. Math.* **1**, 89-95.

Takács, L. (1962). *An Introduction to Queueing Theory.* New York: Oxford University Press.

Tallis, G. M. (1970). Estimating the distribution of spherical and elliptical bodies in conglomerates from plane sections. *Biometrics* **26**, 87-103.

Tanemura, M. (1979). On random complete packing by disks. *Ann. Inst. Statist. Math.* **31**, 351-365.

Tate, C. (1969). Aggregates in a random distribution of points in a cylinder. *SIAM J. Appl. Math.* **17**, 1177-1189.

Thompson, T. M. (1983). *From Error-Correcting Codes Through Sphere Packings to Simple Groups.* Carus Mathematical Monographs, Mathematical Association of America.

Underwood, E. E. (1970). *Quantitative Stereology.* Reading, Massachusetts: Addison-Wesley.

Vicsek, T., and Kertész, J. (1981). Monte Carlo renormalisation-group approach to percolation on a continuum: Test of universality. *J. Phys. A* **14**, L31-L37.

Visscher, W. M., and Bolsterli, M. (1972). Random packing of equal and unequal spheres in two and three dimensions. *Nature (London)* **239**, 504-507.

Votaw, D. F., Jr. (1946). The probability distribution of the measure of a random linear set. *Ann. Math. Statist.* **17**, 240-244.

Wald, A. (1943). Tests of statistical hypotheses concerning several parameters when the number of observations is large. *Trans. Am. Math. Soc.* **54**, 426-482.

Watson, G. S. (1971). Estimating functionals of particle size distribution. *Biometrika* **58**, 483-490.

Watson, G. S. (1975). Texture analysis. *Geol. Soc. Am. Mem.* **142**, 367-391.

Watt, A. S. (1947). Pattern and process in the plant community. *J. Ecol.* **35**, 1-22.

Weibel, E. R. (1973). Stereological techniques for electron microscopy morphometry. In: *Principles and Techniques of Electron Microscopy* **3**, Ed. M. A. Heyat, pp. 237-296. New York: Van Nostrand.

Weibel, E. R. (1979, 1980). *Stereological Methods.* Vols. I and II. London: Academic Press.

Weil, W. (1983). Stereology: A survey for geometers. In: *Convexity and Its Applications*, Eds. P. Gruber and J. M. Wills, pp. 360-412. Basel: Birkhäuser.

Weil, W. (1984). Densities of quermassintegrals for stationary random sets. In: *Stochastic Geometry, Geometric Statistics, Stereology*, Eds. R. V. Ambartzumian and W. Weil, pp. 233-247. Leipzig: Teubner-Verlag.

Weil, W., and Wieacker, J. A. (1983). Densities for stationary random sets and point processes. *Adv. Appl. Prob.* **16**, 324-346.

Weiss, L. (1959). The limiting joint distribution of the largest and smallest sample spacings. *Ann. Math. Statist.* **30**, 590-593.

Wendel, J. G. (1962). A problem in geometric probability. *Math. Scand.* **11**, 109-111.

Westcott, M. (1976). On the existence of a generalized shot noise process. In: *Studies in Probability and Statistics*, Ed. E. J. Williams, pp. 73-88. Amsterdam: North-Holland.

Westcott, M. (1982). Approximations to hard-core models and their application to statistical analysis. In: *Essays in Statistical Science*, Eds. J. Gani and E. J. Hannan, pp. 281-292. Sheffield: Applied Probability Trust.

White, J. A., Schmidt, J. W., and Bennett, G. K. (1975). *Analysis of Queueing Systems.* New York: Academic Press.

Whitt, W. (1983). The queueing network analyzer. *Bell Syst. Tech. J.* **62**, 2779-2815.

Whittaker, E. T., and Watson, G. N. (1927). *A Course in Modern Analysis.* London: Cambridge University Press.

Whittle, P. (1965). Statistical processes of aggregation and polymerization. *Proc. Cambridge Phil. Soc.* **61**, 475-495.

Whitworth, W. A. (1897). *DCC Exercises in Choice and Chance.* (Reprinted 1965). New York: Hafner.

Whitworth, W. A. (1901). *Choice and Chance.* (Reprinted 1959). New York: Hafner.

Wicksell, S. D. (1925). The corpuscle problem, I. *Biometrische Z.* **17**, 84-99.

Wicksell, S. D. (1926). The corpuscle problem, II. *Biometrische Z.* **18**, 151-172.

Widom, B. (1966). Random sequential addition of hard spheres to a volume. *J. Chem. Phys.* **44**, 3888-3894.

Widom, B., and Rowlinson, J. S. (1970). New model for the study of liquid-vapour phase transitions. *J. Chem. Phys.* **52**, 1670-1684.

Wieacker, J. A. (1986). Intersections of random hypersurfaces and visibility. *Z. Wahrscheinlichkeitstheor. verw. Geb.* **71**, 405-433.

Wierman, J. C. (1978). On critical probabilities in percolation theory. *J. Math. Phys.* **19**, 1979-1982.

Woodcock, L. V. (1976). Glass transition in the hard-sphere model. *J. Chem. Soc. Faraday Trans. II* **74**, 11-16.

Wschebor, M. (1973a). Sur le recouvrement du cercle par des ensembles placées au hasard. *Israel J. Math.* **15**, 1-11.

Wschebor, M. (1973b). Sur un théorème de Léonard Shepp. *Z. Wahrscheinlichkeitstheor. verw. Geb.* **27**, 179-184.

Yadin, M., and Zacks, S. (1982). Random coverage of a circle with applications to a shadowing problem. *J. Appl. Prob.* **19**, 562–577.

Yadin, M., and Zacks, S. (1985). The visibility of stationary and moving targets in the plane subject to a Poisson field of shadowing elements. *J. Appl. Prob.* **22**, 776–786.

Zacks, S., and Yadin, M. (1984). The distribution of the random lighted portion of a curve in a plane shadowed by a Poisson random field of obstacles. In: *Statistical Signal Processing*, Eds. E. J. Wegman and J. G. Smith, pp. 273–286. New York: Dekker.

Zähle, M. (1984a). Curvature measures and random sets, I. *Math. Nachr.* **119**, 327–339.

Zähle, M. (1984b). Properties of signed curvature measures. In: *Stochastic Geometry, Geometric Statistics, Stereology*, Eds. R. V. Ambartzumian and W. Weil, pp. 256–266. Leipzig: Teubner-Verlag.

Zähle, M. (1986). Curvature measures and random sets, II. *Z. Wahrscheinlichkeitstheor. verw. Geb.* **71**, 37–58.

Zallen, R., and Scher, H. (1971). Percolation on a continuum and localization-delocalization transition in amorphous semiconductors. *Phys. Rev.* B **4**, 4471–4479.

Zygmund, A. (1951). An individual ergodic theorem for non-commutative transformations. *Acta Sci. Math. Szeged* **14**, 103–110.

Author Index

Subject Index

Abel integral equation, 49, 300
Aiming error, 35, 36
Alignment, preferred, 292
Ammunition, round of, 32
Analyser:
  image, 287, 290, 321
  texture, 290
Angle, turning, 220, 223, 224, 226
Anistrophy, 292, 293, 307
"Arbitrary", 126, 200, 201, 202
Arcs:
  on circle, 22, 352
  random, 63, 70, 71, 76
Associated point method, 218, 287
Asbestos fibres, 217, 287
Atom, of set distribution, 198
Autocorrelation, 307, 338

Ballistics, 32, 71, 72
Bernstein's inequality, 335
Berry-Esseen theorem, 29, 347
Bessel function, 7
Binary pattern, 43
Blaschke's fundamental formula, 208, 286
Blood cells, 53
Blur, 17, 66
Bomb damage, 32

Boolean grain, 44
Boolean model, 4, 46, 68, 183, 195, 198, 206, 209, 217, 252, 267, 268, 269, 271, 278, 286, 289, 291, 296, 303, 307, 312, 318, 321
definition, 17, 43, 72, 125, 192
distribution, 126
high-intensity, 160, 190, 235
  critical, 171
invariance under translations of centers, 125
linear, 126
lower dimensional section, 285
low-intensity, 191, 235, 242, 252, 253, 265
vs. Markovian, 44, 45
moderate-intensity, 160, 178, 235
scale-changed, 141, 142, 160, 178, 233, 235
simple linear, 80, 102, 107, 115, 116, 117, 126
  inference for, 102
sparse, 178, 242, 252, 253, 265
sub-critical, 178
test, 293, 294
testing for isotropy, 292, 307
Boolean set, 44
Bootstrap, 292
Borel measurable set, 3